DATA PROCESSING AND RECONCILIATION FOR CHEMICAL PROCESS OPERATIONS

This is Volume 2 of
PROCESS SYSTEMS ENGINEERING
A Series edited by George Stephanopoulos and John Perkins

DATA PROCESSING AND RECONCILIATION FOR CHEMICAL PROCESS OPERATIONS

José A. Romagnoli
Department of Chemical Engineering
University of Sydney
Sydney, Australia

Mabel Cristina Sánchez
Planta Piloto de Ingeniería Química
Bahia Blanca, Argentina

ACADEMIC PRESS
San Diego London Boston New York Sydney Tokyo Toronto

This book is printed on acid-free paper. ∞

Academic Press
A Harcourt Science and Technology Company
525 B Street, Suite 1900, San Diego, California 92101-4495, USA
http://www.apnet.com

Academic Press
24-28 Oval Road, London NW1 7DX, UK
http://www.hbuk.co.uk/ap/

Library of Congress Catalog Card Number: 99-60407

International Standard Book Number: 0-12-594460-8

PRINTED IN THE UNITED STATES OF AMERICA
99 00 01 02 03 04 MM 9 8 7 6 5 4 3 2 1

To Liliana, Juliana, José Ignacio, my mother,
and the memory of my father
José A. Romagnoli
July 1998

To Brenda, Guillermo, Fernanda, Gabriel,
and the memory of my father
Mabel C. Sánchez
July 1998

It is good to have an end
to journey towards
but it is the journey that matters,
in the end.

Ursula Le Guin
The Left Hand of Darkness

CONTENTS

3 Classification of the Process Variables for Chemical Plants

4 Decomposition Using Orthogonal Transformations

5 Steady-State Data Reconciliation

6 Sequential Processing of Information

7 Treatment of Gross Errors

8 Rectification of Process Measurement Data in Dynamic Situations

9 Joint Parameter Estimation–Data Reconciliation

10 Estimation of Measurement Error Variances from Process Data

11 New Trends

12 Case Studies

APPENDIX: STATISTICAL CONCEPTS

PREFACE

Data processing and reconciliation deal with the problem of improving process knowledge to enhance plant operations and general management planning. Today, data storage facilities, advances in computational tools, and implementation of distributed control systems to chemical processes allow processing of large volumes of data, transforming them into reliable process information.

Chemical process data inherently contain some degree of error, and this error may be random or systematic. Thus, the application of data reconciliation techniques allows optimal adjustment of measurement values to satisfy material and energy constraints. It also makes possible the estimation of unmeasured variables. It should be emphasized that, in today's highly competitive world market, resolving even small errors can lead to significant improvements in plant performance and economy. This book attempts to provide a comprehensive statement, analysis, and discussion of the main issues that emerge in the treatment and reconciliation of plant data.

The concepts of estimability and redundancy in steady-state processes are introduced in Chapter 2 because of their importance in decomposing the general estimation problem. The two subsequent chapters deal with process variable classification for linear and nonlinear plant models. This topic is of great importance in the design of monitoring systems and also in decomposing the data reconciliation problem. The adjustment of measurements for different kinds of plant models is addressed in Chapter 5, and the next chapter introduces some of the advantages of sequential processing of measurements and constraints. Chapter 7 starts by defining the data reconciliation problem in the presence of gross errors. It contains an analysis of the steps to follow for the treatment of biased data.

All of the previous ideas are developed further in Chapter 8, where the analysis of dynamic and quasi-steady-state processes is considered. Chapter 9 is devoted to the general problem of joint parameter estimation–data reconciliation, an important issue in assessing plant performance. In addition, some techniques for estimating the covariance matrix from the measurements are discussed in Chapter 10. New trends in this field are summarized in Chapter 11, and the last chapter is devoted to illustrations of the application of the previously presented techniques to various practical cases.

This book can be used as a text covering plant operation, control, and optimization in courses for graduate students and advanced undergraduates in chemical engineering. We hope that this book will also be of special benefit to those engaged in the industrial application of reconciliation techniques.

In preparing the book, a special effort has been made to create self-contained chapters. Within each one, numerical examples and graphics have been provided to aid the reader in understanding the concepts and techniques presented. Notation, references, and material related to that covered in the text are included at the end of each chapter. It is assumed that the reader has a basic knowledge of matrix algebra and statistics; however, an appendix covering pertinent statistical concepts is included at the end of the book.

José A. Romagnoli

Mabel Cristina Sánchez

ACKNOWLEDGMENTS

We are especially grateful to Professor George Stephanopoulos for his continuing encouragement and constructive comments throughout all the stages of manuscript preparation. We also thank Professor Ahmet Palazoglu for reading the initial draft and for providing helpful suggestions. Special thanks are due to Dr. James Chen, who, during his Ph.D. work, contributed to many of the subjects developed in this book. The contributions of Dr. Gordon Weiss from Orica Ltd., Australia, especially to the industrial applications, and of Dr. Ali Nooraii to the application of the on-line exercise are gratefully acknowledged.

During the years this book was being developed, a number of colleagues were of considerable assistance in many ways. Professor Esteban Brignole, Professor Miguel Bagajewicz, Dr. José Figueroa, Dr. Alberto Bandoni, Professor Rolf Prince, and Professor Mike Brisk should be mentioned particularly.

Acknowledgments are also due to the Department of Chemical Engineering at the University of Sydney (Australia), Orica Ltd., and the Planta Piloto de Ingeniería Química (UNS—CONICET, Argentina) for financial support and for providing the appropriate environment for many productive years of our academic lives.

Last, but certainly not least, we thank our families for their understanding and support throughout the whole endeavor.

1 GENERAL INTRODUCTION

The highly competitive nature of the world market, the increasing importance of product quality, and the growing number of environmental and safety regulations have increased the need to introduce fast and low-cost changes in chemical processes to improve their performance.

The decision-making process about possible modifications in a system requires *knowledge about its actual state*. This information is obtained by collecting a data set, improving its accuracy, and storing it. This procedure, called *monitoring*, constitutes a critical tool for further control action, optimization, and general management planning, according to the hierarchical order shown in the typical activities pyramid of a chemical plant of Fig. 1 (Brisk, 1993).

Since the main activities involved in attaining the final goals of the company are based on this monitoring, it is essential that it provide reliable and complete information about the process.

Because measurements always contain some type of error, it is necessary to correct their values to know objectively the operating state of the process. Two types of errors can be identified in plant data: random and systematic errors. Random errors are small errors due to the normal fluctuation of the process or to the random variation inherent in instrument operation. Systematic errors are larger errors due to incorrect calibration or malfunction of the instruments, process leaks, etc. The systematic errors, also called gross errors, occur occasionally, that is, their number is small when compared to the total number of instruments in a chemical plant.

Furthermore, the information provided by the monitoring system must allow the complete estimation of all of the required process variables. This can only be achieved through the selection of an adequate set of instruments. The selection of

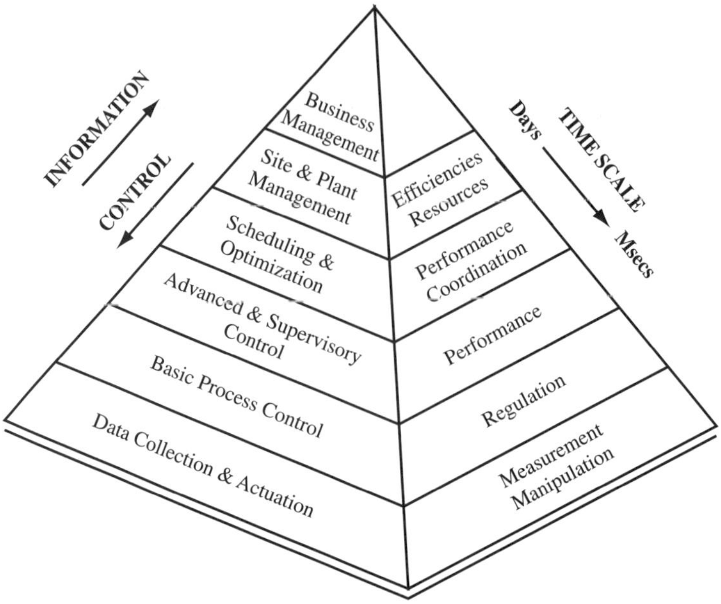

FIGURE 1 Activity pyramid in a processing plant (from Busk, 1993).

the necessary instrumentation thus has a fundamental role during the design of the monitoring system of a new plant, as well as in the case of adapting existing plants to satisfy new environmental and production requirements.

With the advance of computer techniques, especially implementation of distributed control systems (DCS) to chemical processes, a large set of on-line measurements are available at every sampling period. The rational use of this large volume of data requires the application of suitable techniques to improve their accuracy. This goal has triggered the focus on research and development, during the last ten years, in the area of plant data reconciliation. Complete reviews on the subject can be found in the works of Mah (1990), Madron (1992), and Crowe (1996).

The treatment of plant data involves a set of tasks that allows the processing of data arising from different sources (on-line acquisition system, laboratory, direct reading from the operators, etc.), transforming them into reliable process information. This information can be used by the company for different purposes: management planning, modeling, optimization, design of monitoring systems, instrument maintenance, analysis of equipment performance, etc., as shown in Fig. 2 (Simulation Sciences Inc., 1989).

1.1. RELIABLE AND COMPLETE PROCESS KNOWLEDGE

To control, optimize, or evaluate the behavior of a chemical plant, it is important to know its current status. This is determined by the values of the process variables contained in the model chosen to represent the operation of the plant. This model is constituted, in general, by the equations of conservation of mass and energy.

Whenever measurements are planned, some functional model is usually chosen to represent either a physical or ideal system with which the measurements are

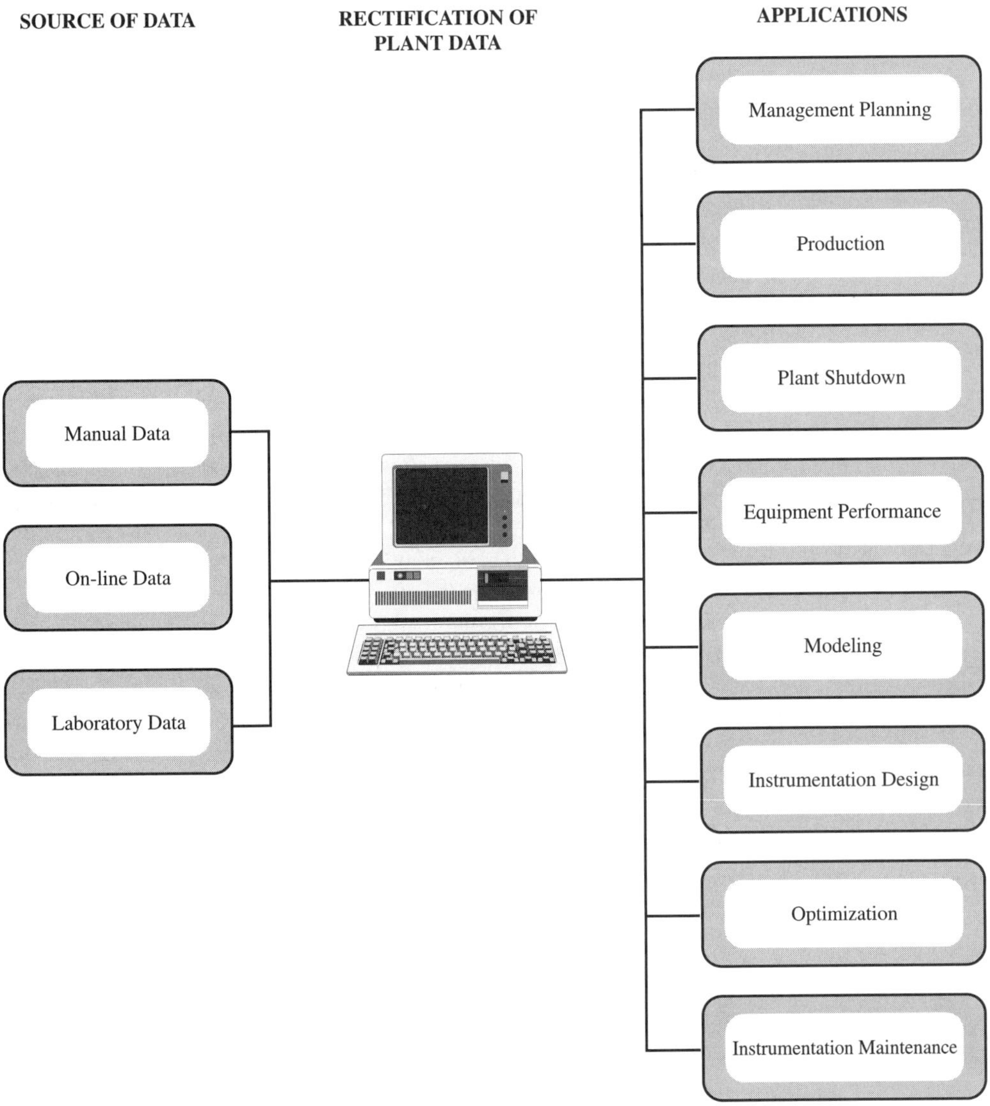

FIGURE 2 Applications of the plant data rectification procedure (from Simulation Sciences, Inc., 1989).

associated. In fact, the measurements are usually made in order to assess the values of some or all of the parameters of the functional model.

The mathematical model is often thought of as being composed of two parts (Mikhail, 1976): the functional model, and the stochastic model.

Functional Model: The functional model will in general describe the deterministic properties of the physical situation or event under consideration.

Stochastic Model: The stochastic model, on the other hand, designates and describes the nondeterministic or stochastic (probabilistic) properties of the variables involved, particularly those representing the measurements.

As pointed out by Mikhail, both functional and stochastic models must be considered together at all times, as there may be several possible combinations, each representing a possible mathematical model. The functional model describes the physical events using an intelligible system, suitable for analysis. It is linked to physical realities by measurements that are themselves physical operations. In simpler situations, measurements refer directly to at least some elements of the functional model. However, it is not necessary, and often not practical, that all the elements of the model be observable. That is, from practical considerations, direct access to the system may not be possible or in some cases may be very poor, making the selection of the measurements of capital importance.

During normal operation of a chemical plant it is common practice to obtain data from the process, such as flowrates, compositions, pressures, and temperatures. The numerical values resulting from the observations do not provide consistent information, since they contain some type of error, either random measurement errors or gross biased errors. This means that the conservation equations (mass and energy), the common functional model chosen to represent operation at steady state, are not satisfied exactly.

It is becoming common practice, in today's chemical plants, to incorporate some kind of technique to rectify or reconcile the plant data. These techniques allow adjustment of the measurement values so that the corrected measurements are consistent with the corresponding balance equations. In this way, the simulation, optimization, and control tasks are based on reliable information. Figure 3 shows schematically a typical interconnection between the previous mentioned activities (Simulation Sciences Inc., 1989).

The integrated approach for data treatment or reconciliation involves a set of mathematical procedures applied on the process instrumentation and the measurements (observations). This provides complete and consistent estimation of the process variables. The general methodology can be divided into three main steps:

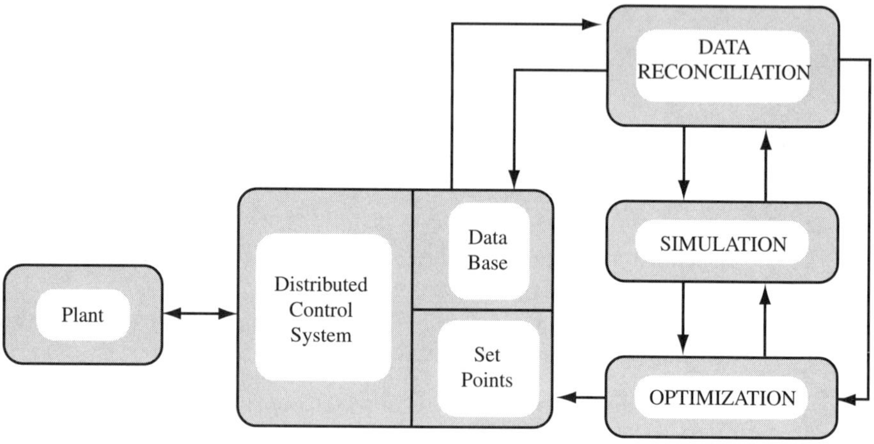

FIGURE 3 Typical arrangement between the DCS and the Data Reconciliation, Simulation, and Optimization procedures (from Simulation Sciences, Inc., 1989).

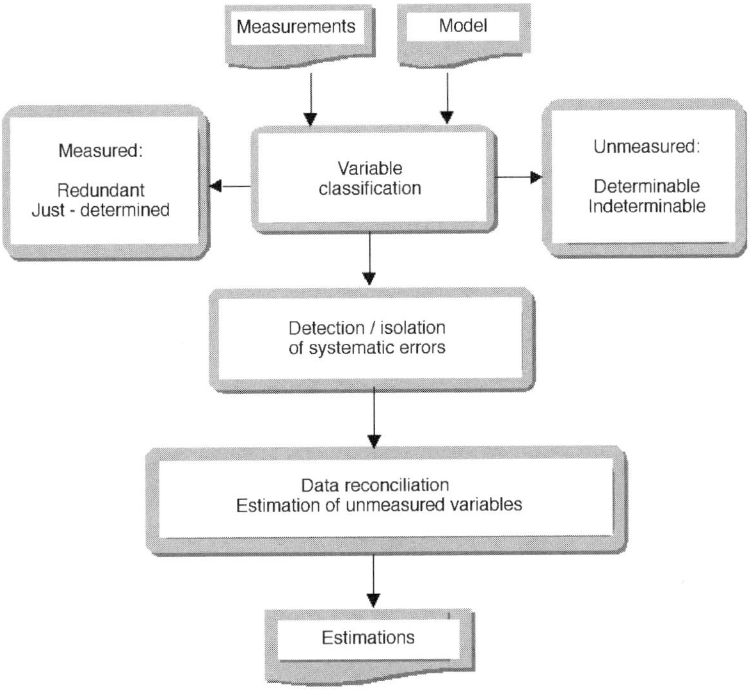

FIGURE 4 Scheme for the integral treatment of plant data.

- Classification of process variables and problem decomposition
- Detection, identification, and estimation of gross errors
- Measurement adjustment and estimation of the unmeasured process variables

A simplified diagram of the general procedure for data reconciliation in chemical plants is given in Fig. 4.

1.2. SOME ISSUES ASSOCIATED WITH A GENERAL DATA RECONCILIATION PROBLEM

A processing plant is a physical system containing a large number of units and streams. For example, counting the equipment in the processing section as well as in the service section (considering mixers and stream dividers) of a petrochemical plant reveals the existence of approximately 1000 interconnected units and about 2500 streams. If we also take into account that in each stream the variables of interest can be flowrate, composition, temperature, pressure, and enthalpy, it is evident that data treatment in a typical plant involves the solution of a large-scale problem.

The original idea of reducing the systems of equations used in the reconciliation problem is due to Vaclavek (1969), who proposed a correction procedure based only on a reduced subset of equations and measurements. The idea basically consists of exploiting the process topology to classify the process variables and eliminate from the general problem the unmeasured ones, resulting in a subset of equations involving only measured variables. Several strategies have been developed since then to achieve the same goal, that is, process decomposition to reduce the dimensionality of the problem. Some of these strategies are based on graph theory (Mah *et al.*, 1976; Stanley and Mah, 1981a,b; Kretsovalis and Mah, 1988a,b; Meyer *et al.*, 1993), and others are equation-oriented approaches (Crowe *et al.*, 1983; Crowe, 1986, 1989); Romagnoli and Stephanopoulos, 1980; Joris and Kaliventzeff, 1987).

From the foregoing it is clear that the application of data reconciliation techniques to large plants, represented by nonlinear complex models, is a challenging problem. The decomposition of the problem through the classification of the process variables appears to be an important tool in dealing with the dimensionality of the problem. What is more important is that understanding the topological structure of the plant not only allows us to decompose it, but can be very important in the process of designing or redesigning the complete monitoring system.

The adjustment of measurements to compensate for random errors involves the resolution of a constrained minimization problem, usually one of constrained least squares. Balance equations are included in the constraints; these may be linear but are generally nonlinear. The objective function is usually quadratic with respect to the adjustment of measurements, and it has the covariance matrix of measurements errors as weights. Thus, this matrix is essential in the obtaining of reliable process knowledge. Some efforts have been made to estimate it from measurements (Almasy and Mah, 1984; Darouach *et al.*, 1989; Keller *et al.*, 1992; Chen *et al.*, 1997). The difficulty in the estimation of this matrix is associated with the analysis of the serial and cross correlation of the data.

The presence of gross errors invalidates the statistical basis of the common data reconciliation procedures, so they must be identified and removed. Gross error detection has received considerable attention in the past 20 years. Statistical tests in combination with an identification strategy have been used for this purpose. A good survey of the available methodologies can be found in Mah (1990) and Crowe (1996).

Parameter estimation is also an important activity in process design, evaluation, and control. Because data taken from chemical processes do not satisfy process constraints, error-in-variable methods provide both parameter estimates and reconciled data estimates that are consistent with respect to the model. These problems represent a special class of optimization problem because the structure of least squares can be exploited in the development of optimization methods. A review of this subject can be found in the work of Biegler *et al.* (1986).

Finally, approaches are emerging within the data reconciliation problem, such as Bayesian approaches and robust estimation techniques, as well as strategies that use Principal Component Analysis. They offer viable alternatives to traditional methods and provide new grounds for further improvement.

1.3. ABOUT THIS BOOK

It is our goal in this book to address the problems, introduced earlier, that arise in a general data reconciliation problem. It is the culmination of several years of research and implementation of data reconciliation aspects in Argentina, the United States, and Australia. It is designed to provide a simple, smooth, and readable account of all aspects involved in data classification and reconciliation, while providing the interested reader with material, problems, and directions for further study.

Chapter 2 offers a unifying exposure to the relevant concepts of estimability and redundancy, in particular their importance in the decomposition of the general data processing problem.

In **Chapter 3** these concepts are extended, and by exploiting the structural topology of a chemical process, the operational parameters or process variables are classified. These ideas, making use of classification strategy, allow the general reduction of a large-scale problem. The decomposition problem is further investigated in **Chapter 4** using orthogonal transformations for both linear and bilinear balances.

Chapter 5 deals with steady-state data reconciliation problem, from both a linear and a nonlinear point of view. Special consideration is given, in **Chapter 6**, to the problem of sequential processing of information. This has several advantages when compared with classical batch processing.

In **Chapter 7** the problem of dealing with systematic gross biased errors is addressed. Systematic techniques are described for the identification of the source of gross errors and for their estimation. These techniques are computationally simple, they are well suited for on-line implementation, and they conform to the general process of variable monitoring in a chemical plant.

All the previous ideas are developed further in **Chapter 8**, where the analysis of dynamic and quasi-steady-state processes is considered.

Chapter 9 deals with the general problem of joint parameter estimation data reconciliation. Starting from the typical parameter estimation problem, the more general formulation in terms of the error-in-variable methods is described, where measurement errors in all variables are considered. Some solution techniques are also described here.

Most techniques for process data reconciliation start with the assumption that the measurement errors are random variables obeying a known statistical distribution, and that the covariance matrix of measurement errors is given. In **Chapter 10** direct and indirect approaches for estimating the variances of measurement errors are discussed, as well as a robust strategy for dealing with the presence of outliers in the data set.

In **Chapter 11** some recent approaches for dealing with different aspects of the data reconciliation problem are discussed. A more general formulation in terms of a probabilistic framework is first introduced and its application in dealing with gross error is discussed in particular. In addition, robust estimation approaches are considered, in which the estimators are designed so they that are insensitive to outliers. Finally, an alternative strategy that uses Principal Component Analysis is reviewed.

At last, in **Chapter 12** several application case studies are given that highlight implementation aspects as well as the relative improvements of the different techniques used. Emphasis is given to industrial applications and on-line exercises.

REFERENCES

Almasy, G. A., and Mah, R. S. H. (1984). Estimation of measurement error variances from process data. *Ind. Eng. Chem. Process Des. Dev.* **23**, 779–784.

Biegler, L. T., Damiano, J. J., and Blau, G. E. (1986). Non-linear parameter estimation: A case study comparison. *AIChE J.* **32**, 29–43.

Busk, M. (1993). Process control: Theories and profits. *IFAC World Congress, Sydney, Australia, July*, **7**, 241–250.

Chen, J., Bandoni, A., and Romagnoli, J. A. (1997). Robust estimation of measurement error variance/covariance from process sampling data. *Comput. Chem. Eng.* **21**, 593–600.

Crowe, C. M. (1986). Reconciliation of process flow rates by matrix projection. Part II: The non-linear case. *AIChE J.* **32**, 616–623.

Crowe, C. M. (1989). Observability and redundancy of process data for steady state reconciliation. *Chem. Eng. Sci.* **44**, 2909–2917.

Crowe, C. M. (1996). Data reconciliation—Progress and challenges. *J. Proc. Control* **6**, 89–98.

Crowe, C. M., García Campos, Y. A., and Hrymak, A. (1983). Reconciliation of process flow rates by matrix projection. Part I: Linear case. *AIChE J.* **29**, 881–888.

Darouach, M., Ragot, R., Zasadzinski, M., and Krzakala, G. (1989). Maximum likelihood estimator of measurement error variances in data reconciliation. *IFAC, AIPAC Symp.* **2**, 135–139.

Joris, P., and Kalitventzeff, B. (1987). Process measurements analysis and validation. *Proc. CEF'87: Use Comput. Chem. Eng.*, Italy, pp. 41–46.

Keller, J. Y., Zasadzinski, M., and Darouach, M. (1992). Analytical estimator of measurement error variances in data reconciliation. *Comput. Chem. Eng.* **16**, 185–188.

Kretsovalis, A., and Mah, R. S. H. (1988a). Observability and redundancy classification in generalised process networks. I: Theorems. *Comput. Chem. Eng.* **12**, 671–687.

Kretsovalis, A., and Mah, R. S. H. (1988b). Observability and redundancy classification in generalised process networks. II. Algorithms. *Comput. Chem. Eng.* **12**, 689–703.

Madron, F. (1992). "Process Plant Performance. Measurement and Data Processing for Optimisation and Retrofits." Ellis Horwood, Chichester, England.

Mah, R. S. H. (1990). "Chemical Process Structures and Information Flows," Chem. Eng. Ser. Butterworth, Boston.

Mah, R. S. H., Stanley, G., and Downing, D. (1976). Reconciliation and rectification of process flow and inventory data. *Ind. Eng. Chem. Process Des. Dev.* **15**, 175–183.

Meyer, M., Koehret, B., and Enjalbert, M. (1993). Data reconciliation on multicomponent network process. *Comput. Chem. Eng.* **17**, 807–817.

Mikhail, E. (1976). "Observation and Least Squares." *IEP Series.* Harper & Row, New York.

Romagnoli, J., and Stephanopoulos, G. (1980). On the rectification of measurement errors for complex chemical plants. *Chem. Eng. Sci.* **35**, 1067–1081.

Simulation Sciences Inc. (1989). "DATACON: A Critical Link to Better Process Monitoring, Control and Analysis." Simulation Sciences Inc., Houston, TX.

Stanley, G., and Mah, R. S. H. (1981a). Observability and redundancy in process data estimation. *Chem. Eng. Sci.* **36**, 259–272.

Stanley, G., and Mah, R. S. H. (1981b). Observability and redundancy classification in process networks—Theorems and algorithms. *Chem. Eng. Sci.* **36**, 1941–1954.

Václavek, V. (1969). Studies on system engineering. III. Optimal choice of the balance measurements in complicated chemical engineering systems. *Chem. Eng. Sci.* **24**, 947–955.

2 ESTIMABILITY AND REDUNDANCY WITHIN THE FRAMEWORK OF THE GENERAL ESTIMATION THEORY

In this chapter, the mathematical tools and fundamental concepts utilized in the development and application of modern estimation theory are considered. This includes the mathematical formulation of the problem and the important concepts of redundancy and estimability: in particular, their usefulness in the decomposition of the general optimal estimation problem. A brief discussion of the structural aspects of these concepts is included.

2.1. INTRODUCTION

Throughout this chapter, we will refer to estimation in a very general sense. We will see later that data reconciliation is only a particular case within the framework of the optimal estimation theory.

The estimation problem may be posed in terms of a single sensor making measurements on a single process or, more generally, in terms of multiple sensors and multiple processes. When relating the observations to an estimator, several questions arise. First, how does one determine whether a measurement is redundant? Second, what is the effect of measurement placement on the estimator's performance? Third, what if there are measurements which are grossly faulty? These questions are of paramount importance in any general estimation problem, and in selecting the measurements' structure for monitoring or controlling a given process. It is clear that the concepts of redundancy and the allocation of the measurements play an important role in the estimation problem. Also, redundancy is useful as safety when there are biases in the measurements or imperfections in the model of the physical situation under consideration.

Since the concept of observability was primarily defined for dynamic systems, observability as a property of steady-state systems will be defined in this chapter. Instead of a measurement trajectory, only a measurement vector is available for steady-state systems. Estimability of the state process variables is the concept associated with the analysis of a steady-state situation.

In this chapter we will present a discussion of those points, leading us directly to the decomposition of the general problem into estimable, nonestimable, redundant, and nonredundant subsystems. This allows us to reduce the size of the commonly used least squares estimation technique and allows easy classification of the process variables: the topic of the next chapter.

2.2. BASIC CONCEPTS AND DEFINITIONS

In order to have any sort of estimation problem in the first place, there must be a system, various measurements of which are available. Rather than develop the notion of a system with a large amount of mathematical formalism, we prefer here to appeal to intuition and common sense in pointing out what we mean.

The system is some physical object, and its behavior can normally be described by equations. The system can be dynamic (discrete or continuous) or static. Here, we will refer to a process under steady-state behavior. Later in this book we will extend our attention to considering dynamic or quasi-steady-state situations.

Now let us consider exactly what we mean by estimation. Suppose that there is some quantity (possibly a vector quantity), associated with a system operation, whose value we would like to know at each instant of time. It may be that this quantity is not directly measurable, or that it can only be measured with error. In any case, we shall assume that noisy measurements, $\mathbf{y}$, are available. Suppose, furthermore, that an experiment is designed to measure, or estimate, a set of system variables $x_1, x_2, \ldots, x_g$. The set of variables can be written as the vector

$$\mathbf{x}^{\mathrm{T}} = [x_1, x_2, \ldots, x_g]. \tag{2.1}$$

The most general situation is that in which the desired variables cannot be observed (measured) directly and must therefore be indirectly measured as functions of the direct observations. Thus, let us assume that the set of l measurements $\mathbf{y}$ can be expressed as a function of the g elements of a constant vector $\mathbf{x}$ plus a random, additive measurement error ε. Then the process measurements are modeled as

$$\mathbf{y} = \phi(\mathbf{x}) + \varepsilon, \quad \mathbf{y} \in \Re^{l}, \quad \mathbf{x} \in \Re^{g}, \tag{2.2}$$

where ϕ represents the measurement functional model.

If $\varepsilon = \mathbf{0}$, then $\mathbf{y} = \phi(\mathbf{x})$ and we say that the measurements are perfect. If $\varepsilon \neq \mathbf{0}$, then they are noisy. In cases where ϕ is assumed to be differentiable at a point $\mathbf{x}^0$, we can define the matrix $\mathbf{C}$:

$$\mathbf{C}(\mathbf{x}^0) = \left.\frac{\partial \phi}{\partial \mathbf{x}}\right|_{\mathbf{x}=\mathbf{x}^0}, \tag{2.3}$$

where $\mathbf{C}$ is used for the linearized version of the nonlinear measurement equations. For linear systems the matrix $\mathbf{C}$ is constant and independent of $\mathbf{x}$. In general, though, we will refer to the linear or linearized system described by

$$\mathbf{y} = \mathbf{C}\mathbf{x} + \varepsilon, \quad \mathbf{C} \in \Re^{l \times g}, \tag{2.4}$$

where $\mathbf{C}$ is the $(l \times g)$ matrix of the Jacobian of ϕ. Thus, when planning the observations, a general functional model about the system to be assessed (matrix $\mathbf{C}$) must be specified. Such a functional model, which refers to a finite closed system, is determined by a certain number of variables and the relationships between them. There is always a minimum number of independent variables that uniquely determine a chosen model. In our case it will always be denoted by g. Unless the observations are sufficient for determining the g variables, the situation will be obviously deficient. These observations must be functionally independent, that is, not one of the l observations can be derived from any or all of the remaining $(l - 1)$ measurements. Let us now introduce the concept of redundancy.

DEFINITION 2.1
We define a system as redundant when the amount of available data (information) exceeds the minimum amount necessary for a unique determination of the independent variables that determine a chosen model.

For the system in Eq. (2.2), when l is larger than g, redundancy is said to exist. This redundancy, which is denoted by r, is given by

$$r = l - g \tag{2.5}$$

and is equal to the (statistical) degrees of freedom.

Since the data are usually obtained from observations (measurements) that are subject to probabilistic fluctuations, redundant data are usually inconsistent in the sense that each sufficient subset yields different results from other subsets. To obtain a unique solution, an additional criterion is necessary. If the least square principle is applied, among all the solutions that are consistent with the measurement model, the estimations that are as close as possible to the measurements are considered to be the solution of the estimation problem. We define a least squares estimation problem as follows:

$$\text{Min } J = (\mathbf{y} - \mathbf{C}\mathbf{x})^{\mathrm{T}}(\mathbf{y} - \mathbf{C}\mathbf{x}). \tag{2.6}$$

Then the least squares solution is that which minimizes the sum of the squares of the residual $J = \varepsilon^{\mathrm{T}}\varepsilon$. The equation in $\mathbf{x}$,

$$\mathbf{C}^{\mathrm{T}}\mathbf{C}\mathbf{x} = \mathbf{C}^{\mathrm{T}}\mathbf{y}, \tag{2.7}$$

obtained by differentiating (to minimize) J, is called the normal equation. We can now define the estimability property as follows.

DEFINITION 2.2
We say a system is estimable if the normal equation admits a unique solution and, naturally, $\mathbf{x}$ is unique.

Accordingly, the necessary conditions for estimability can be stated. In order for the process variables $\mathbf{x}$ to be estimable, the following must be true (Rao, 1973).

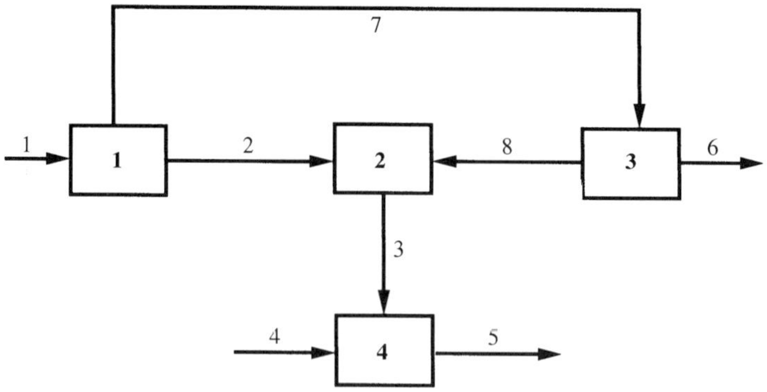

FIGURE 1 A simplified process flowsheet for Example 2.1 (from Madron, 1985).

THEOREM 2.1
The system described by Eq. (2.7) is globally estimable if and only if

$$\text{rank } \mathbf{C} = g, \tag{2.8}$$

where g is the dimension of the system.
Conversely, if

$$\text{rank } \mathbf{C} < g, \tag{2.9}$$

the system is globally nonestimable or confounded.

EXAMPLE 2.1
Let us introduce a simple example to illustrate the previous concepts. The simplified process flowsheet presented in Fig. 1 (Madron, 1985) consists of four units interconnected by eight streams. We are interested in the estimation of the total flowrates of the system. If these variables are measured for streams 1, 7, and 8, then the measurement matrix $\mathbf{C}$ is of dimension $(l \times g)$, where $l = 3$ and $g = 8$.

$$\mathbf{C} = \begin{bmatrix} 1 & 0 & 0 & 0 & 0 & 0 & 0 & 0 \\ 0 & 0 & 0 & 0 & 0 & 0 & 1 & 0 \\ 0 & 0 & 0 & 0 & 0 & 0 & 0 & 1 \end{bmatrix}$$

The rank of $\mathbf{C}$ is 3, so the system is globally nonestimable. Obviously, a globally estimable system can be obtained by increasing the instrumentation cost.

When the system is nonestimable, the estimated value of $\mathbf{x}$ ($\hat{\mathbf{x}}$) is not a unique solution to the least squares problem. In this case a solution is only possible if additional information is incorporated. This must be introduced via the process model equations (constraint equations). They occur in practice when some or all of the system variables must conform to some relationships arising from the physical constraints of the process.

Note: In some cases, the introduction of the additional process model equations will augment the number of variables to be estimated and thus may not alleviate the estimability deficiency. ♣

Thus, by introducing the additional constraints we have

$$\begin{aligned} \mathbf{0} &= \varphi(\mathbf{x}), & \mathbf{x} &\in \Re^g \\ \mathbf{y} &= \phi(\mathbf{x}) + \varepsilon, & \mathbf{y} &\in \Re^l, \end{aligned} \tag{2.10}$$

where $\varphi \in \Re^m$, m being the number of additional constraint equations.

The functional relationships that characterize the real process behavior are never known exactly. A conventional way to account for the inaccuracies generated by approximations is to introduce additive noise, which in some sense reflects the expected degree of modeling errors, that is,

$$\begin{aligned} \mathbf{0} &= \varphi(\mathbf{x}) + \mathbf{w}, & \mathbf{x} &\in \Re^g \\ \mathbf{y} &= \phi(\mathbf{x}) + \varepsilon, & \mathbf{y} &\in \Re^l. \end{aligned} \tag{2.11}$$

Assuming $\varphi(\mathbf{x})$ and $\phi(\mathbf{x})$ are differentiable at $\mathbf{x}_0$, and applying a Taylor series expansion using only zero- and first-order terms (dropping second and higher order terms), we arrive at the linear or linearized system described by

$$\begin{aligned} \mathbf{0} &= \mathbf{A}\mathbf{x} + \mathbf{w} \\ \mathbf{y} &= \mathbf{C}\mathbf{x} + \varepsilon, \end{aligned} \tag{2.12}$$

where $\mathbf{A}$ and $\mathbf{C}$ are the $(m \times g)$ and $(l \times g)$ matrices of the Jacobian of φ and ϕ. In this case the redundancy condition will be satisfied when $(m + l) > g$. We can now define the least squares problem as follows:

$$\text{Min } J = (\mathbf{z} - \mathbf{M}\mathbf{x})^{\mathrm{T}}(\mathbf{z} - \mathbf{M}\mathbf{x}), \tag{2.13}$$

where

$$\mathbf{M} = \begin{bmatrix} \mathbf{A} \\ \mathbf{C} \end{bmatrix}, \quad \mathbf{z} = \begin{bmatrix} \mathbf{0} \\ \mathbf{y} \end{bmatrix}. \tag{2.14}$$

The normal equation is given now by

$$\mathbf{M}^{\mathrm{T}}\mathbf{M}\mathbf{x} = \mathbf{M}^{\mathrm{T}}\mathbf{z}. \tag{2.15}$$

In a similar fashion to the previous case, the general conditions for estimability can be stated as follows.

THEOREM 2.2

The system described by Eqs. (2.14) and (2.15) is globally estimable if, and only if,

$$\text{rank } \mathbf{M} = \text{rank} \begin{bmatrix} \mathbf{A} \\ \mathbf{C} \end{bmatrix} = g. \tag{2.16}$$

Conversely, if

$$\text{rank } \mathbf{M} = \text{rank} \begin{bmatrix} \mathbf{A} \\ \mathbf{C} \end{bmatrix} < g, \tag{2.17}$$

the system is globally nonestimable or confounded.

EXAMPLE 2.2

If the total flowrates of streams 1, 4, 7, and 8 are measured for the process flowsheet presented in Example 2.1, the matrix **M** is defined as follows:

$$\mathbf{M} = \begin{bmatrix} 1 & -1 & 0 & 0 & 0 & 0 & -1 & 0 \\ 0 & 1 & -1 & 0 & 0 & 0 & 0 & 1 \\ 0 & 0 & 0 & 0 & 0 & -1 & 1 & -1 \\ 0 & 0 & 1 & 1 & -1 & 0 & 0 & 0 \\ 1 & 0 & 0 & 0 & 0 & 0 & 0 & 0 \\ 0 & 0 & 0 & 1 & 0 & 0 & 0 & 0 \\ 0 & 0 & 0 & 0 & 0 & 0 & 1 & 0 \\ 0 & 0 & 0 & 0 & 0 & 0 & 0 & 1 \end{bmatrix},$$

where the first four rows of **M** relate to the total mass balance relationships around each unit (matrix **A**) and the last ones correspond to the measurement matrix **C**. For this case, the rank of matrix **M** is 8 and the system is globally estimable.

Now let us define a more general form of the quadratic objective function, which permits us to assign predetermined weights to the components. Consider the general quadratic objective

$$J = \begin{bmatrix} \mathbf{w} \\ \varepsilon \end{bmatrix} \mathbf{W} \, [\mathbf{w} \quad \varepsilon], \tag{2.18}$$

where **W** is a weighting matrix that is restricted to being both symmetric and positive definite, that is, $\mathbf{W} = \mathbf{W}^{\mathrm{T}} > \mathbf{0}$. The introduction of the weighting matrix defines the weighted least squares problem and the same conditions established by Theorems 2.1 and 2.2 hold also in this situation. It can be demonstrated (Deutsch, 1973) that if the quadratic objective is weighted by the covariance matrix of the noises, the result is also a minimum variance estimate or Markov estimate.

2.3. DECOMPOSITION OF THE GENERAL ESTIMATION PROBLEM

The preceding section discusses the mathematical formulation of the problem under consideration and the general conditions for redundancy and estimability. Now, we are ready to analyze the decomposition of the general estimation problem. The division of linear dynamic systems into their observable and unobservable parts was first suggested by Kalman (1960). The same type of arguments can be extended here to decompose a system considered to be at steady-state conditions.

When the results of matrix theory are applied to the general estimation problem (see Appendix A), the following can be stated.

THEOREM 2.3

For the system given by Eq. (2.14) and (2.15), if

$$\operatorname{rank} \mathbf{M} = \operatorname{rank} \begin{bmatrix} \mathbf{A} \\ \mathbf{C} \end{bmatrix} = j < g, \tag{2.19}$$

then there exists a nonsingular matrix **T** *such that*

$$\mathbf{MT} = \begin{bmatrix} \mathbf{A}_\mathrm{U} & \mathbf{0} \\ \mathbf{C}_\mathrm{U} & \mathbf{0} \end{bmatrix}, \tag{2.20}$$

where $\mathbf{A}_\mathrm{U}$ *and* $\mathbf{C}_\mathrm{U}$ *each have* j *columns and*

$$\operatorname{rank} \begin{bmatrix} \mathbf{A}_\mathrm{U} \\ \mathbf{C}_\mathrm{U} \end{bmatrix} = j. \tag{2.21}$$

Proof. See Appendix. ■

The system of equations (2.12) can be written using the column echelon form of matrix **M** as follows:

$$\begin{bmatrix} \mathbf{0} \\ \mathbf{y} \end{bmatrix} = \mathbf{MTT}^{-1}\mathbf{x} + \begin{bmatrix} \mathbf{w} \\ \varepsilon \end{bmatrix} = \begin{bmatrix} \mathbf{A}_\mathrm{U} & \mathbf{0} \\ \mathbf{C}_\mathrm{U} & \mathbf{0} \end{bmatrix} \mathbf{T}^{-1}\mathbf{x} + \begin{bmatrix} \mathbf{w} \\ \varepsilon \end{bmatrix}. \tag{2.22}$$

Depending on the structure of $\mathbf{T}^{-1}$, two situations arise:

1. If each row of $\mathbf{T}^{-1}$ has only one nonzero element, then physically this means that in the new coordinates $\mathbf{x}_\mathrm{c} = [\mathbf{x}_\mathrm{r} \;\; \mathbf{x}_{\mathrm{g-r}}]$, where $\mathbf{x}_\mathrm{r}$ is a j-dimensional vector, the subsystem

$$\begin{aligned} \mathbf{0} &= \mathbf{A}_\mathrm{U}\mathbf{x}_\mathrm{r} + \mathbf{w} \\ \mathbf{y} &= \mathbf{C}_\mathrm{U}\mathbf{x}_\mathrm{r} + \varepsilon \end{aligned} \tag{2.23}$$

is estimable. The whole system admits a decomposition into two smaller subsystems: one estimable, of dimension j, and the other nonestimable, of dimension $(g - j)$. The first one includes the variables in $\mathbf{x}_\mathrm{r}$ and the last one contains the variables in $\mathbf{x}_{\mathrm{g-r}}$.

2. If some rows of $\mathbf{T}^{-1}$ have more than one nonzero element, there are linear combinations between variables in $\mathbf{x}_\mathrm{r}$ and variables in $\mathbf{x}_{\mathrm{g-r}}$. Thus, the estimable portion of the system is of dimension ob less than j $(ob < j)$ and the nonestimable one is of dimension $(g - ob)$.

EXAMPLE 2.3

In this case we consider that the total flowrates of streams 1, 7, and 8 are measured for the process flowsheet presented in Example 2.1, so the matrix **M** is defined as follows:

$$\mathbf{M} = \begin{bmatrix} 1 & -1 & 0 & 0 & 0 & 0 & -1 & 0 \\ 0 & 1 & -1 & 0 & 0 & 0 & 0 & 1 \\ 0 & 0 & 0 & 0 & 0 & -1 & 1 & -1 \\ 0 & 0 & 1 & 1 & -1 & 0 & 0 & 0 \\ 1 & 0 & 0 & 0 & 0 & 0 & 0 & 0 \\ 0 & 0 & 0 & 0 & 0 & 0 & 1 & 0 \\ 0 & 0 & 0 & 0 & 0 & 0 & 0 & 1 \end{bmatrix}.$$

The rank of matrix **M** is 7. As the system is rank deficient, it admits a decomposition into two subsystems, one estimable and the other nonestimable. To determine which variables are observable, the column echelon form of **M** is obtained and $\mathbf{T}^{-1}$

is inspected:

$$
\mathbf{MT} = \begin{array}{c} \begin{matrix} f_2 & f_3 & f_6 & f_4 & f_1 & f_7 & f_8 & f_5 \end{matrix} \\ \begin{bmatrix} -1 & 0 & 0 & 0 & 1 & -1 & 0 & 0 \\ 1 & -1 & 0 & 0 & 0 & 0 & 1 & 0 \\ 0 & 0 & -1 & 0 & 0 & 1 & -1 & 0 \\ 0 & 1 & 0 & 1 & 0 & 0 & 0 & 0 \\ 0 & 0 & 0 & 0 & 1 & 0 & 0 & 0 \\ 0 & 0 & 0 & 0 & 0 & 1 & 0 & 0 \\ 0 & 0 & 0 & 0 & 0 & 0 & 1 & 0 \end{bmatrix} \end{array},
$$

$$
\mathbf{T}^{-1} = \begin{bmatrix} 0 & 1 & 0 & 0 & 0 & 0 & 0 & 0 \\ 0 & 0 & 1 & 0 & 0 & 0 & 0 & 0 \\ 0 & 0 & 0 & 0 & 0 & 1 & 0 & 0 \\ 0 & 0 & 0 & 1 & -1 & 0 & 0 & 0 \\ 1 & 0 & 0 & 0 & 0 & 0 & 0 & 0 \\ 0 & 0 & 0 & 0 & 0 & 0 & 1 & 0 \\ 0 & 0 & 0 & 0 & 0 & 0 & 0 & 1 \\ 0 & 0 & 0 & 0 & 1 & 0 & 0 & 0 \end{bmatrix}.
$$

From the decomposition of $\mathbf{M}$ into its column echelon form, it can be seen that f_5 is nonestimable. The inspection of $\mathbf{T}^{-1}$ indicates that f_4 is also nonestimable, because its calculation depends on f_5.

Let us now extend the results to dealing with systems where the estimability and redundancy conditions are satisfied. A measurement is considered redundant if its removal causes no loss of estimability. If we consider that the rank of $\mathbf{M} = g$ and $(m + l) > g$, that is, more information is available than is necessary for a unique determination, the following can be stated (Stanley and Mah, 1981a).

THEOREM 2.4
If the system of equations (2.14) and (2.15) is

1. *estimable and*
2. *redundant, i.e., $(m + l > g)$, with $(l - i)$ redundant measurements*

and if the rows of $\mathbf{C}$ are permuted so that the first $(l - i)$ rows correspond to the redundant measurements $(\mathbf{y}_1)$, i.e.,

$$
\mathbf{C} = \begin{bmatrix} \mathbf{C}_1 \\ \mathbf{C}_2 \end{bmatrix} \quad \textit{and} \quad i > \mathbf{0}, \tag{2.24}
$$

then there exists a nonsingular $(g \times g)$ matrix $\mathbf{F}$ such that

$$
\mathbf{MF} = \left[\begin{array}{c:c} \mathbf{A}_{\mathrm{U}} & \mathbf{0} \\ \hdashline \mathbf{C}_{1\mathrm{U}} & \mathbf{0} \\ \hdashline \mathbf{C}_{21} & \mathbf{C}_{22} \end{array}\right] \tag{2.25}
$$

and

$$
\operatorname{rank} \mathbf{C}_{22} = i, \quad \operatorname{rank}[\mathbf{A}_{1\mathrm{U}}] = \operatorname{rank}\begin{bmatrix} \mathbf{A}_{\mathrm{U}} \\ \mathbf{C}_{1\mathrm{U}} \end{bmatrix} = g - i \tag{2.26}
$$

with every measurement in the system $\mathbf{A}_{1\mathrm{U}} = \left[\begin{smallmatrix} \mathbf{A}_{\mathrm{U}} \\ \mathbf{C}_{1\mathrm{U}} \end{smallmatrix}\right]$ being redundant.

From the results of the previous theorem, we conclude that any system that is estimable and redundant ($r > 0$) admits a decomposition into its redundant ($\mathbf{x}_1$) and nonredundant parts ($\mathbf{x}_2$). This conclusion is of paramount importance when applied within the framework of the overall estimation problem. Such a decomposition then allows a new equivalent two-problem formulation of the general least squares problem:

PROBLEM 1
Least squares problem:

$$\underset{\mathbf{x}_1}{\text{Minimize}}\ (\mathbf{z}_1 - \mathbf{A}_{1\text{U}}\mathbf{x}_1)^{\text{T}}\mathbf{W}_1(\mathbf{z}_1 - \mathbf{A}_{1\text{U}}\mathbf{x}_1) \tag{2.27}$$

where

$$\mathbf{z}_1 = \begin{bmatrix} \mathbf{0} \\ \mathbf{y}_1 \end{bmatrix}. \tag{2.28}$$

PROBLEM 2
Since the decomposition allows $\mathbf{x}_1$ to be determined first, calculate $\mathbf{x}_2$ using the already known value of $\mathbf{x}_1$ and $\mathbf{y}_2$.

This two-problem formulation results in a significant reduction in the size of the original least squares problem.

EXAMPLE 2.4
If the flowrates of streams 1, 2, 4, 7, and 8 are measured for the process flowsheet of Fig. 1, the matrix $\mathbf{M}$ for this system can be represented as follows:

$$\mathbf{M} = \begin{bmatrix} 1 & -1 & 0 & 0 & 0 & 0 & -1 & 0 \\ 0 & 1 & -1 & 0 & 0 & 0 & 0 & 1 \\ 0 & 0 & 0 & 0 & 0 & -1 & 1 & -1 \\ 0 & 0 & 1 & 1 & -1 & 0 & 0 & 0 \\ 1 & 0 & 0 & 0 & 0 & 0 & 0 & 0 \\ 0 & 1 & 0 & 0 & 0 & 0 & 0 & 0 \\ 0 & 0 & 0 & 0 & 0 & 0 & 1 & 0 \\ 0 & 0 & 0 & 1 & 0 & 0 & 0 & 0 \\ 0 & 0 & 0 & 0 & 0 & 0 & 0 & 1 \end{bmatrix},$$

where matrix $\mathbf{C}$ has been permuted so that the last two rows correspond to flowrate measurements of streams 4 and 8. A nonsingular matrix $\mathbf{F}$ exists such that

$$\mathbf{MF} = \begin{array}{c} \begin{matrix} f_1 & f_2 & f_3 & f_7 & f_5 & f_6 & f_4 & f_8 \end{matrix} \\ \left[\begin{array}{cccccc|cc} 1 & -1 & 0 & -1 & 0 & 0 & 0 & 0 \\ 0 & 1 & -1 & 0 & 0 & 0 & 0 & 0 \\ 0 & 0 & 0 & 1 & 0 & -1 & 0 & 0 \\ 0 & 0 & 1 & 0 & -1 & 0 & 0 & 0 \\ 1 & 0 & 0 & 0 & 0 & 0 & 0 & 0 \\ 0 & 1 & 0 & 0 & 0 & 0 & 0 & 0 \\ 0 & 0 & 0 & 1 & 0 & 0 & 0 & 0 \\ \hline 0 & 0 & 0 & 0 & 0 & 0 & 1 & 0 \\ 0 & 0 & 0 & 0 & 0 & 0 & 0 & 1 \end{array}\right] \end{array},$$

where

$$\mathbf{F} = \begin{bmatrix} 1 & 0 & 0 & 0 & 0 & 0 & 0 & 0 \\ 0 & 1 & 0 & 0 & 0 & 0 & 0 & 0 \\ 0 & 0 & 1 & 0 & 0 & 0 & 0 & 1 \\ 0 & 0 & 0 & 0 & 0 & 0 & 1 & 0 \\ 0 & 0 & 0 & 0 & 1 & 0 & 1 & 1 \\ 0 & 0 & 0 & 0 & 0 & 1 & 0 & -1 \\ 0 & 0 & 0 & 1 & 0 & 0 & 0 & 0 \\ 0 & 0 & 0 & 0 & 0 & 0 & 0 & 1 \end{bmatrix},$$

rank $\mathbf{C}_{22} = 2$, and rank $[\mathbf{A}_{1U}] = 6$. The total flowrate measurements of streams 1, 2, and 7, which are included in the system $\mathbf{A}_{1U}$, are redundant, whereas the total flowrate measurements of streams 4 and 8, which are contained in $\mathbf{C}_{22}$, are nonredundant.

2.4. STRUCTURAL ANALYSIS

From a physical point of view, system parameter values are never known precisely with the exception of zeros. They are fixed, for example, by the absence of physical connections between certain parts of a system. Also, in computing solutions, computers work with "true zeros" and "fuzzy parameters."

Accordingly, let us assume that the entries in the matrices **A** and **C** are either zeros or arbitrary nonzero parameters. For nonlinear systems, their description will be accurate in an infinitesimal region around the point of linearization. Many of the elements of matrices **A** and **C** will vary from one linearization point to another, and some elements will always be zero.

Because of the imprecise knowledge of nonzero system parameters, it is of interest to study system properties that rely on the internal connections of the process under study, and not on the specific numerical values of the system parameters. Among these system properties, structural observability makes the meaning of observability (in the usual sense) more complete from the physical point of view, because the real system involves parameters that are only approximately determined. Indeed, structural observability is a stronger property, as can be demonstrated following the proposition in Lin (1974).

In order to analyze estimability utilizing such ideas, we first include some notions related to "structure." Then we define the concepts of structural observability and the generic rank of a matrix.

DEFINITION 2.3

- A structural matrix **B** is a matrix having fixed zeros in some locations and arbitrary, independent entries in the remaining ones.
- A structured system

$$\mathbf{L} = \begin{pmatrix} \mathbf{B} \\ \mathbf{D} \end{pmatrix} \tag{2.29}$$

is an ordered pair of structured matrices.
- The two systems $\mathbf{L} = \left[\begin{smallmatrix} \mathbf{B} \\ \mathbf{D} \end{smallmatrix}\right]$; $\mathbf{L}' = \left[\begin{smallmatrix} \mathbf{B}' \\ \mathbf{D}' \end{smallmatrix}\right]$ are structurally equivalent if there is a

one-to-one correspondence between the locations of their fixed zero and nonzero entries.

- A matrix $\tilde{\mathbf{B}}$ is called admissible (with respect to $\mathbf{B}$) if it can be obtained by fixing the free parameters of $\mathbf{B}$ at some particular value. The symbol ($\sim$) denotes matrices with fixed elements (matrix in the usual sense).

  For example, consider

$$\mathbf{B} = \begin{bmatrix} 0 & x \\ x & x \end{bmatrix}, \quad \tilde{\mathbf{B}} = \begin{bmatrix} 0 & 1 \\ 2 & 0 \end{bmatrix}. \tag{2.30}$$

  We say $\tilde{\mathbf{B}}$ is admissible with respect to $\mathbf{B}$.

DEFINITION 2.4
A system

$$\mathbf{L} = \begin{bmatrix} \mathbf{B} \\ \mathbf{D} \end{bmatrix} \tag{2.31}$$

is called structurally estimable if there exists a system $\mathbf{L}'$ that satisfies the following conditions:

1. $\mathbf{L}'$ is structurally equivalent to $\mathbf{L}$
2. $\mathbf{L}'$ has an admissible pair

$$\tilde{\mathbf{L}}' = \begin{bmatrix} \tilde{\mathbf{B}}' \\ \tilde{\mathbf{D}}' \end{bmatrix} \tag{2.32}$$

   that is estimable in the usual sense.

It follows directly that any globally estimable system is also structurally estimable.

DEFINITION 2.5
The generic rank of a structured matrix $\mathbf{B}$ is defined to be the maximal rank that $\mathbf{B}$ achieves as a function of its free parameters.

For example, the matrix

$$\mathbf{B} = \begin{bmatrix} x & 0 \\ x & x \end{bmatrix} \tag{2.33}$$

has a generic rank of 2 despite the fact that one diagonal element could be zero as a special case, resulting in a rank of 1.

The maximal rank of an $(m \times g)$ matrix having no specified structure is equal to min (m, g). The inclusion of the structure into the problem makes it possible for matrices to have less than full rank, independent of the values of the free parameters, as was shown by Schields and Pearson (1976). Therefore, a structured matrix $\mathbf{B}$ has full generic rank if, and only if, there exists an admissible matrix $\tilde{\mathbf{B}}$ with full rank.

2.5. CONCLUSIONS

In this chapter, similar arguments to that of dynamic observability were extended to establish the conditions for estimability in steady-state processes when the redundancy condition is satisfied. This concept allows decomposition of the general estimation problem into two smaller subproblems: the estimation of redundant measurements and the calculation of unmeasured observable variables. Also, an easy classification of the process variables is achieved, which is the topic of the following chapter.

The concepts of structural observability are the basic tools for developing variable classification strategies. Some approaches presented in Chapter 3 are based on the fact that the classification of process variables results from the topology of the system and the placement of instruments and has nothing to do with the functional form of the balance equations. Thus, the linearity restriction will be removed and efficient reduction of the large-scale problem will be accomplished.

NOTATION

$\mathbf{A}$	Jacobian matrix of the process constraints
$\mathbf{B}$	general structural matrix
$\mathbf{C}$	Jacobian matrix of the measurement functions
$\mathbf{D}$	general structural matrix
$\mathbf{F}$	matrix defined by Eq. (2.25)
g	number of state variables
i	number of nonredundant measurements
j	rank of $\mathbf{M}$
J	objective function of the least square estimation technique
l	number of measurements
$\mathbf{L}$	general structured system
$\mathbf{M}$	matrix defined by Eq. (2.14)
m	number of process model functions
ob	number of observable variables
r	redundancy
$\mathbf{T}$	matrix defined by Eq. (2.20)
$\mathbf{x}$	vector of state variables
$\mathbf{x}_c$	new coordinates for vector $\mathbf{x} = [\mathbf{x}_r \ \mathbf{x}_{g-r}]$
$\mathbf{x}_1$	vector of state variables that are included in the redundant portion of the system
$\mathbf{x}_2$	vector of state variables that are included in the nonredundant portion of the system
$\mathbf{w}$	vector of expected degree of modeling errors
$\mathbf{W}$	weighting matrix
$\mathbf{z}$	vector defined by Eq. (2.14)
$\mathbf{y}$	vector of measurements

Greek Symbols

ε	measurement random errors
ϕ	measurement model functions
φ	process model functions

Superscripts

$\sim$	admissible matrix

REFERENCES

Deutsch, R. (1973). "Estimation Theory." Prentice-Hall, Englewood Cliffs, NJ.

Kalman, R. E. (1960). Contributions to the theory of optimal control. *Bol. Soc. Matr. Mex.* **5**, 102–119.

Lin, C. T. (1974). Structural controllability. *IEEE Trans. Autom. Control* **AC-19**, 201–208.
Madron, F. (1985). A new approach to the identification of gross errors in chemical engineering measurements. *Chem. Eng. Sci.* **40**, 1855–1860.
Noble, B. (1969). "Applied Linear Algebra." Prentice-Hall, Englewood Cliffs, NJ.
Rao, C. R. (1973). "Linear Statistical Inference and its Applications." Wiley, New York, 1973.
Schields, R., and Pearson, J. B. (1976). Structural controllability of multiinput linear systems. *IEEE Trans. Autom. Control* **AC-21**, 203–212.
Stanley, G., and Mah, R. S. H. (1981a). Observability and redundancy in process data estimation. *Chem. Eng. Sci.* **36**, 259–272.
Stanley, G., and Mah, R. S. H. (1981b). Observability and redundancy classification in process networks—Theorems and algorithms. *Chem. Eng. Sci.* **36**, 1941–1954.

APPENDIX A

Some Results on Matrix Algebra

The utility of matrices in the applied sciences is, in many cases, connected with the fact that they provide a convenient method for the formulation of physical problems in terms of a set of equations. It is therefore important to become familiar with the manipulation of the equations, or equivalently with the manipulation of rows and columns of the corresponding matrix. First, we will be concerned with some basic tools such as column-echelon form and elementary matrices. Let us introduce some definitions (Noble, 1969).

DEFINITION A.1
A matrix is said to be in column echelon normal form if:

1. Certain rows numbered $c_1, c_2, \ldots, c_k$ are precisely the unit vectors $\mathbf{e}_1, \mathbf{e}_2, \ldots, \mathbf{e}_k$, where $\mathbf{e}_i$ is defined to be the $(1 \times m)$ row vector whose ith element is unity with all the other elements being zero.
2. $c_1 < c_2 < \cdots < c_k$.
3. If a row lies above c_1, then it is a row of zeros. If the cth row lies between the rows numbered c_i and c_{i+1}, the last $(m - i)$ elements of the cth row must be zero. If the cth row lies below the row numbered c_k, then the last $(g - k)$ elements of the cth row must be zero.

Note that this definition implies the following:

1. The last $(m - k)$ columns of a column-echelon form are zero. The first k columns of the column echelon form are nonzero.
2. The upper triangle of elements in the (i, j)th positions where $j > i$ are all zero.
3. The first nonzero element in each column is 1. The first c_{i-1} elements of the ith column are zero. The c_kth element of the ith column is zero for $i \neq k$.

Although there is a considerable degree of freedom in the sequence of calculations, when reducing a matrix to a column-echelon form, this is unique and the rank of the matrix is equal to the number of nonzero columns in the column echelon.

DEFINITION A.2
Any matrix $\mathbf{E}$ obtained by performing a single elementary operation on the unit matrix $\mathbf{I}$ is known as an elementary matrix.

For example, $\mathbf{E}_{pq}$ is the elementary matrix obtained by interchanging the pth and the qth rows of $\mathbf{I}$. It can be shown that the elementary matrices possess inverses, and these are also elementary matrices. Now we are in position to recall the following matrix theorem (Noble, 1969).

THEOREM A.1

If $\mathbf{G}$ *is an* $(m \times g)$ *matrix of rank* k *and* $\mathbf{U}_G$ *denotes the column-echelon form of* $\mathbf{G}$*, then a nonsingular matrix* $\mathbf{E}_G$ *exists such that*

1. $\mathbf{G}\mathbf{E}_G = \mathbf{U}_G$ *and* $\mathbf{G} = \mathbf{U}_G\mathbf{E}_G^{-1}$*, where* $\mathbf{E}_G$ *and* $\mathbf{E}_G^{-1}$ *are products of elementary matrices*
2. *A nonsingular matrix can be expressed as a product of elementary matrices*

Proof

The column-echelon form of $\mathbf{G}$ is obtained by performing a sequence of elementary column operations on this matrix. This means that we can find a sequence of elementary matrices $\mathbf{E}_p\mathbf{E}_{p-1}\ldots\mathbf{E}_1$ corresponding to the elementary column operations, such that

$$\mathbf{G}\mathbf{E}_p\mathbf{E}_{p-1}\ldots\mathbf{E}_1 = \mathbf{U}_G, \tag{A2.1}$$

where the elementary matrices are nonsingular. If we multiply through by $\mathbf{E}_p^{-1}\mathbf{E}_{p-1}^{-1}\ldots\mathbf{E}_1^{-1}$ in succession, we obtain

$$\mathbf{G} = \mathbf{U}_G\mathbf{E}_p^{-1}\mathbf{E}_{p-1}^{-1}\ldots\mathbf{E}_1^{-1}. \tag{A2.2}$$

If we denote $\mathbf{E}_p\mathbf{E}_{p-1}\ldots\mathbf{E}_1 = \mathbf{E}_G$, these results give (1).

Also, (2) follows immediately since the column-echelon form of a nonsingular matrix is the unit matrix. ■

Proof of Theorem 2.3

The proof of this theorem follows readily from Theorem A.1, since by definition

$$\mathbf{M} = \underset{g}{\begin{bmatrix}\mathbf{A}\\ \mathbf{C}\end{bmatrix}}\begin{matrix} m \\ \overline{l}\end{matrix}.$$

Then $\mathbf{M}$ is an $(m + l \times g)$ matrix, and by the hypothesis, the rank of $\mathbf{M} = j$. Thus, matrix $\mathbf{M}$ verifies the conditions of Theorem A.1; therefore, there exists a nonsingular $\mathbf{E}_M$ such that

$$\mathbf{M}\mathbf{E}_M = \mathbf{U}_M, \tag{A2.3}$$

where $\mathbf{U}_M$ is the column-echelon form of $\mathbf{M}$. Since by definition (1) the last $(g - j)$ columns of the column-echelon form are zero and the first j columns are nonzero, and (2) the rank of $\mathbf{M}$ equals the number of nonzero columns in the column-echelon, that is,

$$\mathbf{U}_M = [\mathbf{U}_1 \quad \mathbf{0}], \tag{A2.4}$$

then by defining $\mathbf{E}_M = \mathbf{T}$ and $\mathbf{M}_U = \mathbf{U}_1$, we have

$$\mathbf{M}\mathbf{T} = [\mathbf{M}_U \quad \mathbf{0}], \tag{A2.5}$$

or similarly,

$$\begin{bmatrix} \mathbf{A} \\ \mathbf{C} \end{bmatrix} \mathbf{T} = \begin{bmatrix} \mathbf{A}_{\mathrm{U}} & \mathbf{0} \\ \mathbf{C}_{\mathrm{U}} & \mathbf{0} \end{bmatrix} \tag{A2.6}$$
$$\qquad\qquad j \quad g-j$$

That is, $\mathbf{A}_{\mathrm{U}}$ and $\mathbf{C}_{\mathrm{U}}$ have columns and rank $\begin{bmatrix} \mathbf{A}_{\mathrm{U}} \\ \mathbf{C}_{\mathrm{U}} \end{bmatrix} = j$. ■

Proof of Theorem 2.4
Since the system is estimable, then

$$\operatorname{rank} \mathbf{M} = g \quad \text{or} \quad \operatorname{rank} \begin{bmatrix} \mathbf{A} \\ \mathbf{C} \end{bmatrix} = g.$$

Let us partition the matrix $\mathbf{M}$ in the following manner:

$$\mathbf{M} = \begin{bmatrix} \mathbf{A}_1 \\ \mathbf{C}_2 \end{bmatrix}, \quad \text{where} \quad \mathbf{A}_1 = \begin{bmatrix} \mathbf{A} \\ \mathbf{C}_1 \end{bmatrix} \tag{A2.7}$$

and $\mathbf{A}_1$ is a $(m+l-i \times g)$ matrix of rank $(g-i)$. Thus, from the previous theorem, there exists a nonsingular matrix $\mathbf{F}$, such that

$$\mathbf{A}_1\mathbf{F} = [\mathbf{A}_{1\mathrm{U}} \quad \mathbf{0}] \quad m+l-i. \tag{A2.8}$$
$$g-i \quad i$$

Partitioning matrix $\mathbf{C}_2$ accordingly, that is,

$$\mathbf{C}_2 = [\mathbf{C}_{21} \quad \mathbf{C}_{22}], \tag{A2.9}$$
$$g-i \quad i$$

we have, finally,

$$\operatorname{rank} \mathbf{MF} = \operatorname{rank} \begin{bmatrix} \mathbf{A}_{\mathrm{U}} & \mathbf{0} \\ \mathbf{C}_{1\mathrm{U}} & \mathbf{0} \\ \mathbf{C}_{21} & \mathbf{C}_{22} \end{bmatrix}. \tag{A2.10}$$

Since $\mathbf{MF}$ is of full rank g, there is no dependent column and

$$\operatorname{rank}[\mathbf{C}_{22}] = i, \tag{A2.11}$$

and since

$$\operatorname{rank}[\mathbf{C}_{21} \quad \mathbf{C}_{22}] = i, \tag{A2.12}$$

we have

$$\operatorname{rank} \begin{bmatrix} \mathbf{A}_{\mathrm{U}} \\ \mathbf{C}_{1\mathrm{U}} \end{bmatrix} = g-i. \tag{A2.13}$$

Also, since each measurement corresponding to $\mathbf{y}_1$ was redundant for the original system, each row of $\mathbf{C}_1$ was linearly dependent on some other rows of $\begin{bmatrix} \mathbf{A} \\ \mathbf{C} \end{bmatrix}$. But, any row in $\mathbf{C}_1$ must be linearly independent of any row in $\mathbf{C}_2$; otherwise $\mathbf{y}_2$ would be redundant. Hence, any row in $\mathbf{C}_1$ is dependent on the other rows in $\begin{bmatrix} \mathbf{A} \\ \mathbf{C}_1 \end{bmatrix}$. The dependency is unchanged by the transformation $\mathbf{F}$; thus, $\mathbf{y}_1$ is also redundant in the system $\begin{bmatrix} \mathbf{A}_{\mathrm{U}} \\ \mathbf{C}_{1\mathrm{U}} \end{bmatrix}$. ■

3 CLASSIFICATION OF THE PROCESS VARIABLES FOR CHEMICAL PLANTS

In this chapter, the mathematical formulation of the variable classification problem is stated and some structural properties are discussed in terms of graphical techniques. Different strategies are available for carrying out process-variable classification. Both graph-oriented approaches and matrix-based techniques are briefly analyzed in the context of their usefulness for performing variable categorization. The use of output set assignment procedures for variable classification is described and illustrated.

3.1. INTRODUCTION

Steady-state process variables are related by mass and energy conservation laws. Although, for reasons of cost, convenience, or technical feasibility, not every variable is measured, some of them can be estimated using other measurements through balance calculations. Unmeasured variable estimation depends on the structure of the process flowsheet and on the instrument placement. Typically, there is an incomplete set of instruments; thus, unmeasured variables are divided into determinable or estimable and indeterminable or inestimable. An unmeasured variable is determinable, or estimable, if its value can be calculated using measurements. Measurements are classified into redundant and nonredundant. A measurement is redundant if it remains determinable when the observation is deleted.

Because of the complexity of integrated processes and the large volume of available data in highly automated plants, classification algorithms are increasingly used nowadays. They are applied to the design of monitoring systems and to reduce the dimension of the data reconciliation problem.

Variable classification is the essential tool for the design or revamp of monitoring systems. After fixing the degree of required knowledge of the process, that is to say, the subset of variables that must be known, this technique is repeated until the selected set of instruments allows us to obtain the desired information about the process. There is a great economic incentive for robust classification because a deficient procedure will require the installation of extra instrumentation.

For measurement adjustment, a constrained optimization problem with model equations as constraints is resolved at a fixed interval. In this context, variable classification is applied to reduce the set of constraints, by eliminating the unmeasured variables and the nonredundant measurements. The dimensional reduction of the set of constraints allows an easier and quicker mathematical resolution of the problem.

The idea of process variable classification was presented by Vaclavek (1969) with the purpose of reducing the size of the reconciliation problem for linear balances. In a later work Vaclavek and Loucka (1976) covered the case of multicomponent balances (bilinear systems).

A similar approach was undertaken by Mah *et al.* (1976) in their attempt to organize the analysis of process data and to systematize the estimation and measurement correction problem. In this work, a simple graph-theoretic procedure for single component flow networks was developed. They then extended their treatment to multicomponent flow networks (Kretsovalis and Mah, 1987), and to generalized process networks, including bilinear energy balances and chemical reactions (Kretsovalis and Mah, 1988a,b).

Romagnoli and Stephanopoulos (1980) proposed an equation-oriented approach. Solvability of the nodal equations was examined and an output set assignment algorithm (Stadtherr *et al.*, 1974) was employed to simultaneously classify measured and unmeasured variables. These ideas were modified to take into account special situations and a computer implementation (PLADAT) was done by Sánchez *et al.* (1992).

An elegant classification strategy using projection matrices was proposed by Crowe *et al.* (1983) for linear systems and extended later (Crowe, 1986, 1989) to bilinear ones. Crowe suggested a useful method for decoupling the measured variables from the constraint equations, using a projection matrix to eliminate the unmeasured process variables.

Joris and Kalitventzeff (1987) proposed a classification procedure for nonlinear systems, which is based on row and column permutation of the occurrence matrix corresponding to the Jacobian matrix of the linearized model.

Another procedure for variable classification was presented by Madron (1992). The categorization is performed by converting the matrix associated with the linear or linearized model equations to its canonical form.

In this chapter the classification of measurements and unmeasured variables of chemical processes is discussed. After the statement of the problem, variable categorization is posed in terms of a structural analysis of the flowsheet. Then graph- and matrix-based strategies are briefly described and discussed. Illustratives examples of application are included.

3.2. MODELING ASPECTS

Let us consider a process containing K units denoted by $k = 1, \dots, K$, and J oriented streams $j = 1, \dots, J$, with C components $c = 1, \dots, C$. Plant topology is represented by the incidence matrix $\mathbf{L}$, with rows corresponding to units and columns to streams. Then

$$\begin{aligned} L_{jk} &= 1 && \text{if stream } j \text{ enters node } k \\ L_{jk} &= -1 && \text{if stream } j \text{ leaves node } k \\ L_{jk} &= 0 && \text{otherwise} \end{aligned}$$

The balance constraints for a process unit without chemical reactions and heat transfer can be expressed as follows.

Total mass balances:

$$\sum_j L_{j,k} f_j = 0. \tag{3.1}$$

Component mass balances:

$$\sum_j L_{j,k} f_j M_{c,j} = 0. \tag{3.2}$$

Enthalpy balances:

$$\sum_j L_{j,k} f_j h_j = 0. \tag{3.3}$$

Normalization equations:

$$\sum_c f_j M_{c,j} - f_j = \mathbf{0}, \tag{3.4}$$

where f_j is the total flow of stream j, $M_{c,j}$ is the mass fraction of component c in stream j, and h_j represents the specific enthalpy of stream j.

In Appendix 3-A the balance equations for the most common chemical process units are set out.

In general, the model of a plant operating under steady-state conditions is made up of a system of nonlinear algebraic equations of the form

$$\varphi(\mathbf{x}, \mathbf{u}) = \mathbf{0}, \quad \varphi \in \Re^m, \tag{3.5}$$

where φ is a nonlinear vector-valued function, and $\mathbf{x}$ and $\mathbf{u}$ are the vectors of measured and unmeasured process variables, respectively. For linear mass balances, Eq. (3.5) becomes

$$\mathbf{A}_1\mathbf{x} + \mathbf{A}_2\mathbf{u} = \mathbf{0}, \quad \mathbf{x} \in \Re^g, \quad \mathbf{u} \in \Re^n, \tag{3.6}$$

where $\mathbf{A}_1$, and $\mathbf{A}_2$ are compatible matrices of dimension $(m \times g)$ and $(m \times n)$, respectively.

If the state of the system is directly measured, then the measurement model is represented by

$$\mathbf{y} = \mathbf{x} + \varepsilon. \tag{3.7}$$

In this case the Jacobian matrix of the measurements functions $\mathbf{C}$ is equal to the identity matrix, and the vector of random measurement errors is

$$\varepsilon = \mathbf{y} - \mathbf{x}. \tag{3.8}$$

It should be noted here that this formulation of the problem is totally equivalent to that of previous chapter, since the data reconciliation problem is only a special case of the general estimation problem, where we directly measure the process variables.

3.3. CLASSIFICATION OF PROCESS VARIABLES

The unmeasured process variables can be classified into determinable and indeterminable (Fig. 1).

DETERMINABLE
An unmeasured variable, belonging to the subset **u**, is determinable if it can be evaluated from the available measurements using the balance equations.

INDETERMINABLE
An unmeasured variable, belonging to the subset **u**, is indeterminable if it cannot be evaluated from the available measurements using the balance equations.

Similarly, some of the elements of vector **x** of measured variables can be classified into redundant and non-redundant measured process variables (Fig. 2).

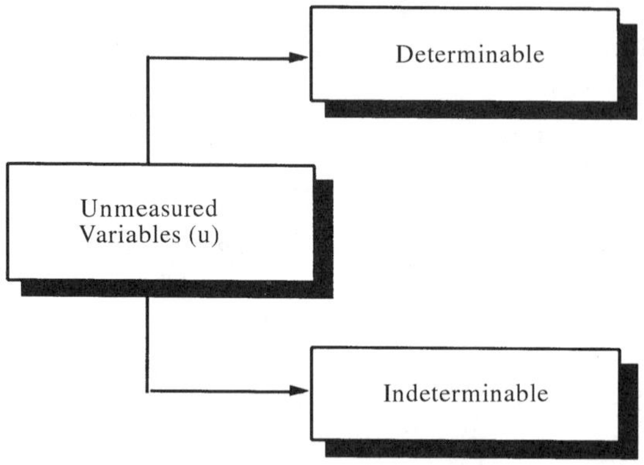

FIGURE 1 Classification of the unmeasured process variables.

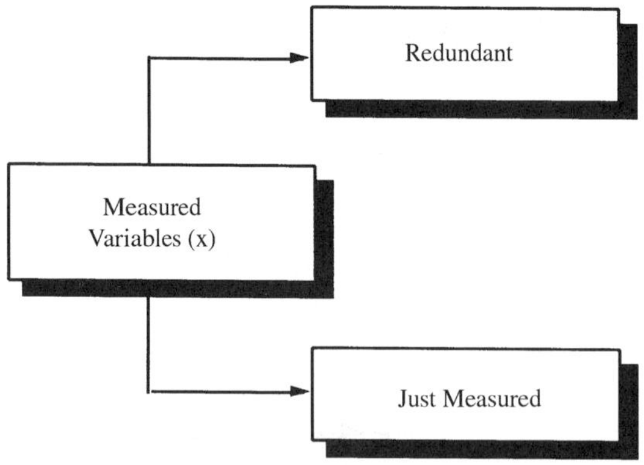

FIGURE 2 Classification of the measured process variables.

Redundant

A measured process variable, belonging to subset **x**, is called redundant (overdetermined) if it can also be computed from the balance equations and the rest of the measured variables.

Nonredundant

A measured process variable, belonging to subset **x**, is called nonredundant (just-measured) if it cannot be computed from the balance equations and the rest of the measured variables.

Based on the preceding formulation, the following problems can be defined:

1. Classify the unmeasured variables
2. Define the subset of redundant equations to be used for the adjustment of measurements
3. Classify the measured variables

In the following sections, the basic tools for the structural evaluation of the process equations are briefly discussed. They allow us to systematically analyze the topological structure of the balance equations and to solve the three problems defined earlier.

3.4. ANALYSIS OF THE PROCESS TOPOLOGY

In the previous section we showed that process variables could be divided into vectors **x** and **u**, corresponding to measured and unmeasured variables, respectively. Accordingly, linear systems of balance equations can be represented in terms of compatible matrices $\mathbf{A}_1$ and $\mathbf{A}_2$ by Eq. (3.6).

We use a linear system for simplicity, but it is not restrictive, since $\mathbf{A}_1$ and $\mathbf{A}_2$ may arise from the linearization of the nonlinear balances. This suggests a structural representation of the system where matrices $\mathbf{A}_1$ and $\mathbf{A}_2$ consist of some elements that are generally nonzero and others that are always zero.

The system matrices $\mathbf{A}_1$ and $\mathbf{A}_2$ describe the structural topology of streams and units in terms of variables and equations. We can associate a graph with the system, which shows the mutual influences of the variables in a more pictorial way.

Definition 3.1 (Signal Flow Graph)

Let the nodes of the graph represent process variables and the edges the relationships (balance equations) between them. There is a directed edge from node a to node i, if a belongs to the interval of i, i.e., if we need a to evaluate i.

Example 3.1

Let us take the trivial example in Fig. 3 to show the signal flow graph concept. In this example, the flowrates of streams 1, 2, and 4 are considered measured. Performing a total mass balance around each unit and according to Eq. (3.1)

$$f_1 + f_2 - f_3 = 0$$
$$f_3 + f_4 - f_5 = 0$$
$$f_5 + f_6 - f_7 = 0$$

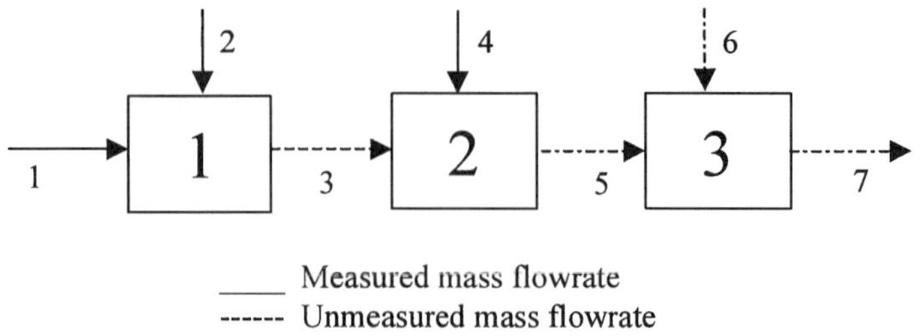

FIGURE 3 The flowsheet diagram for a simple serial system (from Romagnoli and Stephanopoulos, 1980).

We have three equations, one for each unit. Solving these equations with respect to the variables f_3, f_5, and f_6 results in the associated signal flow graph shown in Figure 4.

In the previous chapter we have defined the generic rank as a property of structural matrices. Let us now introduce some new concepts in connection with structural systems and their associated graphs.

DEFINITION 3.2 (NONACCESSIBILITY)

We define a node i to be nonaccessible from node a if there is no possibility of reaching node i by starting from node a (which corresponds to a measured variable) and going to node i in the direction of the arrows, along a path in the signal graph.

DEFINITION 3.3 (DETERMINABILITY)

We define a node i as determinable if any path going to node i always starts in a measured node.

EXAMPLE 3.2

Consider the system used in Example 3.1. Applying the definition of accessibility:

- Nodes 3, 5, and 6 are accessible
- Node 7 is nonaccessible

Also, from the concept of determinability:

- Nodes 3 and 5 are determinable
- Nodes 6 and 7 are nondeterminable

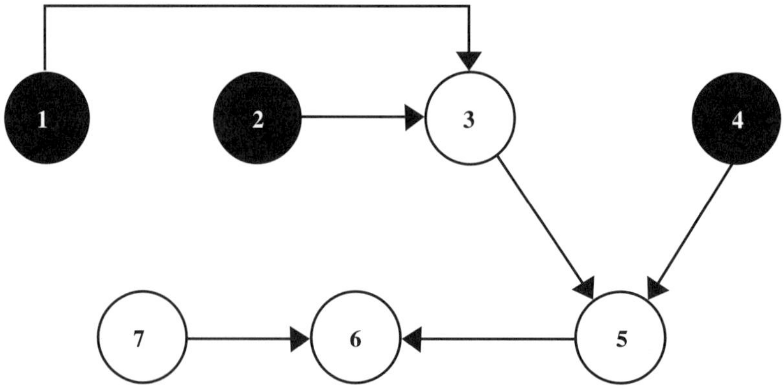

FIGURE 4 The signal flow graph for the system in Figure 3 (from Romagnoli and Stephanopoulos, 1980).

Consequently, f_3 and f_5 are determinable, unmeasured process variables and f_6 and f_7 are indeterminable.

By a natural extension of the concepts developed in the previous chapter (structural estimability), if the generic rank of the composite matrix $(\mathbf{A}_1; \mathbf{A}_2)$ is not less than n (n: number of unmeasured variables), then the system does not include structural singularities. Furthermore, if all the unmeasured nodes are determinable, then there are no isolated variables, which cannot be computed from the balance equations.

Consequently, the following can be stated. The structural pair $(\mathbf{A}_1; \mathbf{A}_2)$ is completely solvable with respect to the unmeasured variables, if the following two conditions are satisfied:

- The generic rank of the composite matrix $(\mathbf{A}_1; \mathbf{A}_2)$ is not less than n
- Each of the unmeasured nodes is accessible.

These two conditions stated for determinability correspond to those for the existence of an output set, given by Steward (1962). The first condition warrants that the number of equations is at least equal to the number of unmeasured variables, while the second condition of accessibility takes into account the existence of a subset of equations containing fewer variables than equations. We have shown that if either of the above two conditions is not satisfied, the structural pair $(\mathbf{A}_1; \mathbf{A}_2)$ admits a decomposition analogous to that given in the previous section. Thus the same results are still valid when only the structural aspects are considered. A graphical interpretation of these two conditions is instructive.

EXAMPLE 3.3

Let us consider $\mathbf{A}_1$ and $\mathbf{A}_2$ for the system of Example 3.1:

$$\mathbf{A}_1 = \begin{bmatrix} x & x & 0 \\ 0 & 0 & x \\ 0 & 0 & 0 \end{bmatrix}, \quad \mathbf{A}_2 = \begin{bmatrix} x & 0 & 0 & 0 \\ x & x & 0 & 0 \\ 0 & x & x & x \end{bmatrix}.$$

The generic rank of $(\mathbf{A}_1; \mathbf{A}_2)$ is 3 and the number of unmeasured process variables $n = 4$, so the system exhibits generic rank deficiency. The signal flow graph is given in Fig. 4.

Let us now consider the case where stream 3 is also measured. For this new situation we have

$$\mathbf{A}_1 = \begin{bmatrix} x & x & x & 0 \\ 0 & 0 & x & x \\ 0 & 0 & 0 & 0 \end{bmatrix}, \quad \mathbf{A}_2 = \begin{bmatrix} 0 & 0 & 0 \\ x & 0 & 0 \\ x & x & x \end{bmatrix}.$$

Now, the generic rank of $(\mathbf{A}_1; \mathbf{A}_2)$ is equal to the number of unmeasured process variable and the system does not exhibit generic rank deficiency. However, from the corresponding signal graph of Fig. 5, we can see that node 7 is nonaccessible.

On the other hand, when stream 6 is considered measured instead of stream 3, all of the unmeasured nodes are accessible and the generic rank is equal to 3.

This example is particularly instructive, especially for the case shown in Fig. 5, since in this case we have no generic rank deficiency but one node is nonaccessible. In such cases we will have overmeasured process variables. This can be seen from Fig. 5, where variable f_2 is measured but can also be computed from the balance around unit 1. Note that Equation 1 could also be assigned to solve for variables

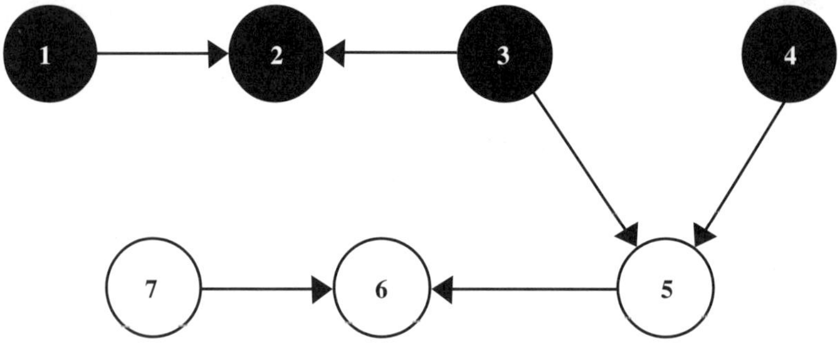

FIGURE 5 Signal flow graph for Example 3.3 under new measurement structure (from Romagnoli and Stephanopoulos, 1980).

f_1 or f_3. Consequently, in this example the process variables f_1, f_2, and f_3 are overmeasured or redundant.

3.5. DIFFERENT APPROACHES FOR SOLVING THE CLASSIFICATION PROBLEM

During the past three decades, several strategies have been formulated for performing process variable classification. These strategies may be divided into two major groups. One group applies the concepts of graph theory to achieve the categorization; the other makes use of matrix ordering techniques and computations. In this section the main features of both approaches are briefly presented and some aspects relating to their ranges of application are discussed.

3.5.1. Graph-Oriented Techniques

Given the process topology, an unoriented graph is built where nodes correspond to units and arcs represent process streams. The process graph contains an environment node from which the process receives feeds, and to which it supplies products.

The main contributions to the development of graph-oriented techniques are due to the following authors.

Vaclavek

Vaclavek (1969) first defined the concepts of observability and redundancy. He formulated two rules for achieving variable categorization for linear plant models:

1. Aggregate two nodes connected with an unmeasured stream. The resulting Reduced Balance Scheme contains only redundant measurements;
2. Delete all measured streams and search for cycles on the reduced graph. The cycles in the resulting graph represent indeterminable flows.

Vaclavek and Loucka (1976) extended the approach to multicomponent processes with the assumption that, for any stream, either all or none of the mass fractions are measured. Chemical reactions are taken into account by adding fictitious streams to the graph. Splitter units are not considered in his formulation.

Mah and Co-workers

These workers have presented a comprehensive theory and algorithms for the design of measured and unmeasured variable classification. For single-component process networks (mass balances only), Mah *et al.* (1976) derived a simple classification procedure based on graph theory. In a later work (Kretsovalis and Mah, 1987) they described the categorization of variables for multicomponent flow networks without assumptions in the location of the sensors. Chemical reactions and splitters were not taken into account. Kretsovalis and Mah (1988a,b) extended their treatment to include reactors, splitters, and units where pure energy flows take place. The following stream variables were accounted for in their analysis: mass flows, mass fractions, component and energy flows, and temperatures. The set of measurements is restricted to mass flows, mass fractions, and temperatures. It was also assumed that there is a one-to-one correspondence between temperature and enthalpy per unit mass.

The technique requires an extensive analysis of the process graph and its derived subgraphs (16 + number of components). They are tested against a set of 19 theorems on observability and redundancy. These subgraphs are updated during the execution of the procedure. The classification of unmeasured variables is accomplished using rules derived only from graph theory and matrix algebra.

Meyer *et al.*

The authors (Meyer *et al.*, 1993) introduced a variant method derived from Kretsovalis and Mah (1987) that allows chemical reactions and splitters to be treated. It leads to a decrease in the size of the data reconciliation problem as well as a partitioning of the equations for unmeasured variable classification.

3.5.2. Equation-Oriented Approaches

Given the topology of the process and a measurement set, these strategies generate the system of model equations for the plant first. Different kinds of rearrangements and calculations, involving matrixes and nonlinear equations, are then performed to classify process variables. The main contributions to this line of work are considered in the following paragraphs.

Romagnoli and Stephanopoulos

Romagnoli and Stephanopoulos (1980) proposed a classification procedure based on the application of an output set assignment algorithm to the occurrence submatrix of unmeasured variables, associated with linear or nonlinear model equations. An assigned unmeasured variable is classified as determinable, after checking that its calculation may be possible through the resolution of the corresponding equation or subset of equations.

A set of redundant equations is constructed using unassigned equations without indeterminable variables and specific balance equations around disjoint systems of units. The measurements involved in this set are classified as redundant.

The procedure was originally applied to variable classification for bilinear systems of equations. In the case of multicomponent balances, it was considered that the composition of a stream is either completely measured, or not measured at all.

A more detailed description of an update strategy based on the use of output set assignments will be presented in the next main section.

Crowe

For linear plant models Crowe *et al.* (1983) used a projection matrix to obtain a reduced system of equations that allows the classification of measured variables. They identified the unmeasured variables by column reduction of the submatrix corresponding to these variables.

Crowe (1986) extended this methodology to the classification of variables involved in bilinear component balances. The model is modified to linear form using a knowledge of process topology, instrument locality, and a set of measurements that must be consistent with process constraints. Then Crowe (1989) proposed a variable classification algorithm based on a set of lemmas. In this formulation, bilinear energy balances are included in the model equations, assuming there is a one-to-one correspondence between temperature and enthalpy per unit mass. The procedure allows the inclusion of arbitrary placement of measurements, chemical reactions, flow splitters, and pure energy flows.

The strategies developed by Crowe are described further in the next chapter, where other matrix computations for variable classification are analyzed.

Joris and Kalitventzeff

The procedure developed by Joris and Kalitventzeff (1987) aims to classify the variables and measurements involved in any type of plant model. The system of equations that represents plant operation involves state variables (temperature, pressure, partial molar flowrates of components, extents of reactions), measurements, and link variables (those that relate certain measurements to state variables). This system is made up of material and energy balances, liquid–vapor equilibrium relationships, pressure equality equations, link equations, etc.

The classification of unmeasured variables and measurements is accomplished by permuting rows and columns of the occurrence matrix corresponding to the Jacobian matrix of the model equations.

In most cases, the structural procedure is able to determine whether the measurements can be corrected and whether they enable the computation of all of the state variables of the process. In some configurations this technique, used alone, fails in the detection of indeterminable variables. This situation arises when the Jacobian matrix used for the resolution is singular.

Madron

The classification procedure developed by Madron is based on the conversion, into the canonical form, of the matrix associated with the linear or linearized plant model equations. First a composed matrix, involving unmeasured and measured variables and a vector of constants, is formed. Then a Gauss–Jordan elimination, used for pivoting the columns belonging to the unmeasured quantities, is accomplished. In the next phase, the procedure applies the elimination to a resulting submatrix which contains measured variables. By rearranging the rows and columns of the macro-matrix, the final canonical form is obtained, which allows the classification of both types of variables. Initial estimates for all variables should be supplied by the user. This strategy is extensively described in the monograph by Madron (1992).

3.6. USE OF OUTPUT SET ASSIGNMENTS FOR VARIABLE CLASSIFICATION

Generally, a chemical plant is composed of several units with several streams and components. The set of material and energy balances constitutes a set of linear/nonlinear equations, which can be represented by an undirected graph. However, when the number of units and streams is large, the graphical representation becomes cumbersome. An alternative representation of the topological structure of the balances in a chemical process is achieved using the *occurrence matrix*.

DEFINITION 3.4 (OCCURRENCE MATRIX)

The rows of the occurrence matrix correspond to the balance equations and the columns to the process variables, both measured and unmeasured. An element of the matrix O_{ij} is a Boolean 1 or 0, that is,

$$O_{ij} = \begin{cases} 1 & \text{if variable } j \text{ appears in equation } i \\ 0 & \text{otherwise} \end{cases}$$

To classify the variables, one must first establish what information each equation is to supply, that is, to obtain an *output set assignment* for the balance equations.

DEFINITION 3.5 (OUTPUT SET ASSIGNMENT)

The output set assignment assigns to any unmeasured process variable one equation, or to two or more variables the same number of equations. This is equivalent to transforming the original undirected graph to a directed one.

EXAMPLE 3.4

Consider again the system used in the previous examples. For f_1, f_2, and f_4 measured, the corresponding occurrence matrix is given in Table 1. The following assignments can be done:

- Assign Equation 1 to f_3
- Assign Equation 2 to f_5

Equation 3 cannot be assigned, because it contains two unmeasured process variables. Since f_3 and f_5 can be calculated from the available information, they are unmeasured but determinable variables. On the other hand, f_6 and f_7 cannot be calculated from the available information; thus, they are indeterminable.

TABLE 1 Occurrence Matrix for Example 3.4

	f_3	f_5	f_6	f_7	f_1	f_2	f_4
Equation 1	1				1	1	
Equation 2	1	1					1
Equation 3		1	1	1			

TABLE 2 New Occurrence Matrix for Example 3.4

	f_5	f_6	f_7	f_3	f_1	f_2	f_4
Equation 1				1	1	1	
Equation 2	1			1			1
Equation 3	1	1	1				

Now, let us consider the case where f_3 is measured. The corresponding occurrence matrix is in Table 2.

In this case we assign Equation 2 to f_5 (determinable), leaving two unassigned equations, 1 and 3. Variables f_6 and f_7 are still not determinable, but now Equation 1 is not assigned and contains only measured variables; thus, this is a redundant equation and the associated variables are also redundant or overmeasured.

Finally, let us consider f_6 measured instead of f_3. The new occurrence matrix is in Table 3. Now, we can assign all the equations by assigning Equations 1, 2, and 3 to f_3, f_5, and f_7, respectively. In this case, all the unmeasured process variables are determinable from the available information; however, there are no redundant measured variables.

The output set assignment is not unique; however, this does not affect the result of the classification. As Steward (1962) has shown, if there is no structural singularity, the determinable unmeasured variables are always assigned independently of the obtained output set assignment. The classification of the unmeasured variables allows us to define the sequence of calculation for these variables. That is, expressions are obtained to solve them as functions of the measurements. The expressions are also used in the classification of the measured variables and in the formulation of the reconciliation equations. After the reconciliation procedure is applied to the measurements, these equations are used to find an estimate of the unmeasured determinable variables in terms of the reconciled measurements.

After the classification of the unmeasured variables is completed, we need to classify the measured ones. First, the set of equations is divided into two groups:

1. Assigned equations
2. Unassigned equations

The latter can be further divided into three groups:

- Equations that contain only measured variables (*NA1*)

TABLE 3 New Occurrence Matrix for Example 3.4

	f_5	f_7	f_3	f_6	f_1	f_2	f_4
Equation 1			1		1	1	
Equation 2	1		1				1
Equation 3	1	1		1			

- Equations that contain measured and unmeasured determinable variables (*NA2*)
- Equations that contain unmeasured indeterminable variables (*NA3*)

The unmeasured determinable variables in set *NA2* are then substituted by their corresponding expressions as function of the measured variables and set *NA2′* is obtained. After this is accomplished, sets *NA1* and *NA2′* contain only measured variables, which are then redundant. The corresponding equations constitute the set of constraints in the reconciliation problem.

3.6.1. Balances Around a Set of Units

When the balance equations are formulated around individual units only, it is possible that the classification by output set assignment may not be satisfactory. Some variables classified as indeterminable may actually be determinable if we consider additional balances around groups of units. An erroneous measurement classification is also possible. The problem is in the system of equations used in the classification rather than in the assignment method. The most common problem arises because of the presence of parallel streams between two units.

In order to analyze the existence of parallel streams, let us consider the case shown in Fig. 6 when total mass balances are included in the set of process constraints. The flowrate of stream 1 is assumed to be measured. By performing a material balance around each unit we have

$$f_1 + f_4 - f_2 = 0$$
$$f_2 - f_4 - f_3 = 0.$$

According to the output set assignment approach, flowrates f_2, f_3, and f_4 are indeterminable from this set of equations. However, if one of these equations is substituted by a balance around units 1 and 2, the result is different. In this case the set of balances is given by

$$f_1 + f_4 - f_2 = 0$$
$$f_1 - f_3 = 0 \rightarrow f_3.$$

Now flowrate f_3 is determinable since it can be assigned to one of the equations that contains only measured variables.

A similar situation arises for measurement categorization when f_1 and f_3 are measured. Although f_2 and f_4 constitute an output set assignment for the individual

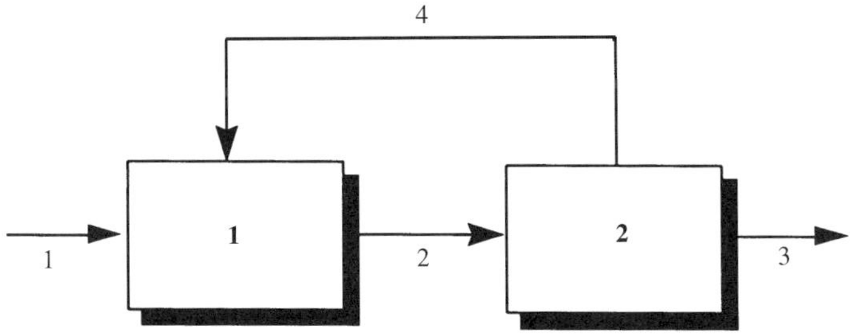

FIGURE 6 Flow diagram for parallel streams.

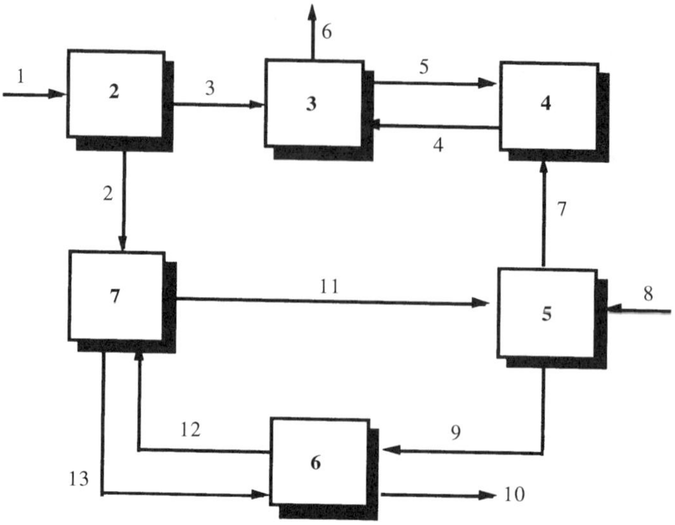

FIGURE 7 Flow diagram for Example 3.5 (adapted from Kretsovalis and Mah, 1987).

mass balances, there is a structural singularity in the unmeasured variable occurrence submatrix. Consequently, unmeasured and measured flowrates are classified as nondeterminable and nonredundant, respectively. The categorization of observations is erroneous since the mass balance around units 1 and 2 contains only measured variables, so they are redundant.

To avoid these situations, we need to check for the presence of parallel streams in the flow diagram of the process. When this is the case, one of the individual balances is substituted by a combined balance around the units involved.

EXAMPLE 3.5

Consider the flow diagram in Fig. 7 (Krestovalis and Mah, 1987), with 6 units and 13 streams. All the streams have three components and we have considered total and component balances around the units. There is no limit to the number of compositions measured in each stream; thus, normalization equations are also included in the classification. Additional information regarding the status of each variable is given in Table 4.

First, balances around the individual units were considered, indicated as b2, b3, b4, b5, b6, and b7. With this set of equations, only four variables are classified as determinable from the unmeasured process variables. They are f_2, f_4, $M_{1,2}$, $M_{3,2}$. It

TABLE 4 Measured Variables for Example 3.5

Measurement	Stream
Flowrate (f_j)	1 3 5 6 7 9
Molar fraction 1 $(M_{1,j})$	1 3 6 9 12
Molar fraction 2 $(M_{2,j})$	2 6 7 9 12
Molar fraction 3 $(M_{3,j})$	4 6 8 9 10 12

TABLE 5 Unmeasured Variable Classification for Example 3.5

Classification	Flowrates	Molar fractions
Determinable	f_2 f_4 f_8 f_{10} f_{11}	$M_{1,2}$ $M_{1,7}$ $M_{2,3}$ $M_{3,2}$ $M_{3,3}$ $M_{3,7}$ $M_{2,1}$ $M_{3,1}$ $M_{3,5}$ $M_{3,11}$
Nondeterminable	f_{12} f_{13}	$M_{1,4}$ $M_{2,4}$ $M_{1,5}$ $M_{2,5}$ $M_{1,8}$ $M_{2,8}$ $M_{1,10}$ $M_{2,10}$ $M_{1,11}$ $M_{2,11}$ $M_{1,13}$ $M_{2,13}$ $M_{3,13}$

is evident that the presence of parallel streams and the use of balances for individual units does not allow a correct classification.

However, analyzing the flow diagram we can see that units 3 and 4 are connected by parallel streams; thus, the balance around unit 4 (b4) is substituted by a balance around unit 3 and 4 (b(3 + 4)). The new balances are now

$$\text{b2, b3, b}(3+4)\text{, b5, b6, b7.}$$

In the same way b7 is substituted by b(6 + 7), leaving

$$\text{b2, b3, b}(3+4)\text{, b5, b6, b}(6+7).$$

By obtaining the output set assignment on the previous set of balances we can classify the unmeasured process variables as determinable or nondeterminable. The results are given in Table 5; this classification is coincident with those from other works cited in the literature.

3.7. THE SOLUTION OF SPECIAL PROBLEMS

Once the process variables have been classified, a great deal of information about the process topology is also available. The question now is how to use the classification and this information to attack other problems. In a real process we will have different kind of problems to solve and the goals will vary from one process to another. Among the possible situations that may be encountered are the following:

1. Using a classification algorithm we can determine the measured variables that are overmeasured, that is, the measurements that may also be obtained from mathematical relationships using other measured variables. In certain cases we are not interested in all of them, but rather in some that for some reason (control, optimization, reliability) are required to be known with good accuracy. On the other hand, there are unmeasured variables that are also required and whose intervals are composed of over measured parameters. Then we can state the following problem: *Select the set of measured variables that are to be corrected in order to improve the accuracy of the required measured and unmeasured process variables.*

2. Consider a system that after the classification has all the unmeasured variables determinable. Suppose also that the system under study has some overmeasured variables. Then we want to select which of the overmeasured variables need not be measured, while preserving the condition of determinability for the unmeasured variables. That is, we want to minimize the number of measurements in such a way that all the unmeasured variables are determinable. This problem can be stated as

follows: *Select the minimum number of measurements so that all the unmeasured variables are determinable.*

3. In some cases, we do not want all the variables to be determinable: only those that are required. Consequently, we must identify which of the measurable variables have to be measured. Let **p** be the set of variables that for various reasons should be known correctly; **p** may be composed of measured and unmeasured variables. Sometimes we are not interested in the whole system being determinable, so we want to select which of the process variables have to be measured to have complete determinability of the variables in set **p**. This problem can be stated as follows: *Select the necessary measurements for the subset of required variables to be determinable.*

3.8. A COMPLETE CLASSIFICATION EXAMPLE

In this section an example, taken from Kretsovalis and Mah (1988b), is given to illustrate how the output set assignment strategy combined with the use of symbolic mathematics can be successfully used to classify measured and unmeasured variables. The process flowsheet for this example (Example 3.6) is given in Fig. 8. There are five components involved in the process streams. The sixth component corresponds to the refrigerant fluid (streams 10 and 11). Component 5 is not present in stream 1. The process units are a mixer (MX), a reactor (RX) where two exothermic reactions take place, a divisor (DIV), a separator (SEP), and three heat exchangers (HX1, HX2, and HX3). The following two reactions can be formulated using the available information:

$$\text{Reaction 1:} \quad s_1^1 c_1 + s_3^1 c_3 \rightarrow s_4^1 c_4 + s_5^1 c_5$$

$$\text{Reaction 2:} \quad s_1^2 c_1 + s_5^2 c_5 \rightarrow s_2^2 c_2 + s_3^2 c_3$$

It is assumed that enthalpy is a function of temperature only. Measurements for this process are included in Table 6.

In order to classify measurements and unmeasured variables for the process flowsheet in Fig. 8, the following tasks are performed:

1. Analysis of the process flowsheet to identify the presence of parallel streams.
2. Formulation of balances and normalization equations for each type of unit

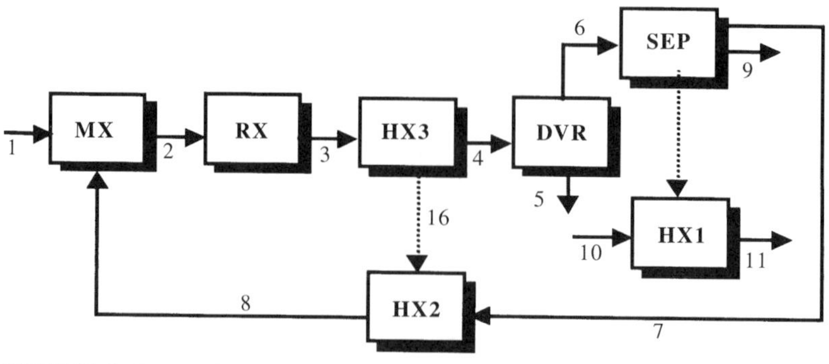

FIGURE 8 Process flowsheet for Example 3.6 (adapted from Kretsovalis and Mah, 1988b).

TABLE 6 Measurements for Example 3.6

Measurement	Stream	Measurement	Stream
Total flowrate	1 2 4 8 10	Mass fraction 4	2 4 8 9
Mass fraction 1	1 7	Mass fraction 5	3 5 8 9
Mass fraction 2	1 3 5 7 9	Temperature	1 4 6 7 8 9 10
Mass fraction 3	7 8	—	—

TABLE 7 Measured Variable Classification for Example 3.6[a]

	Streams						
Category	Flowrate	Mass fraction 1	Mass fraction 2	Mass fraction 3	Mass fraction 4	Mass fraction 5	Temperature
R	1 2 4 8	7	1 3 5 7 9	7 8	2 4 8 9	3 5 8 9	1 4 6 7
NR	10	1	—	—	—	—	8 9 10

[a] *R*, redundant measurement; *NR*, nonredundant measurement.

or for clusters of units. The set of equations used for classification is presented in Appendix 3-B.

3. Application of output set assignment algorithms to classify the unmeasured variables. For the process under study, the type and placement of instruments is such that all unmeasured variables are determinable.

4. Formulation of expressions for the unmeasured variables in terms of the measured ones (see Appendix 3-B), using the sequence of calculations that is obtained as a by-product of the assignment procedure.

5. Categorization of nonassigned equations into types *NA1*, *NA2*, and *NA3*.

6. Substitution of the determinable unmeasured variables in *NA2* by the corresponding expressions in terms of measurements to obtain the set *NA2′*.

7. Classification of the measured variables included in *NA1* and *NA2′* as redundant. The other measurements are categorised as nonredundant. Measured variable classification results for this example are in Table 7.

8. Analysis of the set of equations (*NA1* + *NA2′*) to eliminate dependencies.

3.9. FORMULATION OF A REDUCED RECONCILIATION PROBLEM

Let us consider the system of linear balance equations described by Eq. (3.8). In the presence of measurement errors the balance equations are not satisfied exactly, and any general data reconciliation procedure must solve the following least squares problem:

$$\begin{aligned} &\text{Minimize } (\mathbf{y}-\mathbf{x})^{\mathrm{T}}\mathbf{W}(\mathbf{y}-\mathbf{x}) \\ &\text{s.t.} \qquad \mathbf{A}_1\mathbf{x}+\mathbf{A}_2\mathbf{u}=\mathbf{0}, \end{aligned} \tag{3.9}$$

where $\mathbf{W}$ is a weighting matrix.

In the previous sections the structural topology of the balance equations was exploited to classify the operational variables into four categories. Accordingly, we can define

$\mathbf{x}_1$: set of over measured (redundant) variables
$\mathbf{x}_2$: set of just determined measurements
$\mathbf{u}_1$: set of unmeasured determinable variables
$\mathbf{u}_2$: set of unmeasured indeterminable variables

In the same way, the system matrices $\mathbf{A}_1$ and $\mathbf{A}_2$ are also partitioned into the following matriccs: $\mathbf{A}_{11}$ and $\mathbf{A}_{12}$ from $\mathbf{A}_1$, and $\mathbf{A}_{21}$ and $\mathbf{A}_{22}$ from $\mathbf{A}_2$. Following this partitioning, the balance equations can now be written

$$\mathbf{A}_{11}\mathbf{x}_1 + \mathbf{A}_{12}\mathbf{x}_2 + \mathbf{A}_{21}\mathbf{u}_1 + \mathbf{A}_{22}\mathbf{u}_2 = \mathbf{0}. \tag{3.10}$$

If $\mathbf{A}_{22} \neq \mathbf{0}$, the system possesses unmeasured variables that cannot be determined from the available information (measurements and equations). In such cases the system is indeterminable and additional information is needed. This can be provided by additional balances that may be overlooked, or by making additional measurements (placing a measurement device to an unmeasured process variable). Also, from the classification strategy we can identify those equations that contain only measured variables, i.e., the redundant equations. Thus, we can define the reduced subsystem of equations

$$\mathbf{A}_{01}\mathbf{x}_1 = \mathbf{0}, \tag{3.11}$$

where $\mathbf{A}_{01}$ is the corresponding system matrix of the redundant subsystem and $\mathbf{x}_1$ belongs to the subset of overmeasured process variables. Similar arguments can be extended to nonlinear systems, which arise from component and energy balances, since the classification algorithm depends on the structural characteristics of the balances.

The constrained least squares problem for the overall plant can now be replaced by the equivalent two-problem formulation.

PROBLEM 1
Least squares estimation of redundant measurements:

$$\begin{aligned} &\text{Minimize } (\mathbf{y}_1 - \mathbf{x}_1)^T \mathbf{W} (\mathbf{y}_1 - \mathbf{x}_1) \\ &\text{s.t.} \qquad \mathbf{A}_{01}\mathbf{x}_1 = \mathbf{0} \end{aligned} \tag{3.12}$$

The solution of this problem is given by

$$\hat{\mathbf{x}}_1 = \mathbf{y}_1 - \mathbf{W}^{-1}\mathbf{A}_{01}^T \left(\mathbf{A}_{01}\mathbf{W}^{-1}\mathbf{A}_{01}^T\right)^{-1} \mathbf{A}_{01}\mathbf{y}_1 \tag{3.13}$$

and is discussed in Chapter 5.

PROBLEM 2
Calculate $\mathbf{u}_1$ using our knowledge of $\mathbf{y}_1$, $\mathbf{y}_2$, and the balance equations.

3.10. CONCLUSIONS

This chapter has shown that the analysis of the topological structure of the balance equations allows classification of the measured and unmeasured process variables, finally leading to system decomposition.

Various strategies are available for performing process variable classification. These have been briefly presented. Their application for different types of plant model (linear, bilinear, nonlinear) and the sets of process variables concerned with them are also discussed.

Furthermore, a variable classification strategy based on an output set assignment algorithm and the symbolic manipulation of process constraints is discussed. It manages any set of unmeasured variables and measurements, such as flowrates, compositions, temperatures, pure energy flows, specific enthalpies, and extents of reaction. Although it behaves successfully for any relationship between variables, it is well suited to nonlinear systems, which are the most common in process industries.

NOTATION

a	graph node index
$\mathbf{A}_1$	submatrix corresponding to measured variables for linear model equations
$\mathbf{A}_2$	submatrix corresponding to unmeasured variables for linear model equations
$\mathbf{A}_{01}$	redundant subsystem of equations
c	index of components
C	number of components
f_j	total mass flowrate of stream j
g	number of measured variables
h_j	specific enthalpy of stream j
i	graph node index
j	index of streams
J	number of streams
k	index of units
K	number of units
$\mathbf{L}$	incidence matrix
m	number of process model functions
$M_{c,j}$	mass fraction of component c in stream j
n	number of unmeasured variables
$NA1$	nonassigned equations with only measured variables
$NA2$	nonassigned equations with measured and unmeasured determinable variables
$NA3$	nonassigned equations with unmeasured nondeterminable variables
$\mathbf{O}$	occurrence matrix
$\mathbf{p}$	vector of required process variables
$\mathbf{s}$	vector of stoichiometric coefficient for c in reaction r
$\mathbf{u}$	vector of unmeasured variables
$\mathbf{u}_1$	vector of unmeasured determinable variables
$\mathbf{u}_2$	vector of unmeasured indeterminable variables
$\mathbf{W}$	weighting matrix
$\mathbf{x}$	vector of measured variables
$\mathbf{x}_1$	vector of overmeasured (redundant) variables
$\mathbf{x}_2$	vector of just-determined measurements
$\mathbf{y}$	measurement vector

Greek Symbols

ε	measurement random errors
φ	process model functions

Superscripts

$\widehat{\ }$	estimated value

REFERENCES

Crowe, C. M. (1986). Reconciliation of process flow rates by matrix projection. Part II: The non-linear case. *AIChE J.* **32**, 616–623.

Crowe, C. M. (1989). Observability and redundancy of process data for steady state reconciliation. *Chem. Eng. Sci.* **44**, 2909–2917.

Crowe, C. M., García Campos, Y. A., and Hrymak, A. (1983). Reconciliation of process flow rates by matrix projection. Part I: Linear case. *AIChE J.* **29**, 881–888.

Joris, P., and Kalitventzeff, B. (1987). Process measurements analysis and validation. *Proc. CEF'87: Use Comput. Chem. Eng.*, Italy, pp. 41–46.

Kretsovalis, A., and Mah, R. S. H. (1987). Observability and redundancy classification in multicomponent process networks. *AIChE J.* **33**, 70–82.

Kretsovalis, A., and Mah, R. S. H. (1988a). Observability and redunancy classification in generalised process networks. I: Theorems. *Comput. Chem. Eng.* **12**, 671–687.

Kretsovalis, A., and Mah, R. S. H. (1988b). Observability and redundancy classification in generalised process networks. II. Algorithms. *Comput. Chem. Eng.* **12**, 689–703.

Madron, F. (1992). "Process Plant Performance. Measurement and Data Processing for Optimisation and Retrofits." Ellis Horwood, Chichester, England.

Mah, R. S. H., Stanley, G., and Downing, D. (1976). Reconciliation and rectification of process flow and inventory data. *Ind. Eng. Chem. Process Des. Dev.* **15**, 175–183.

Meyer, M., Koehret, B., and Enjalbert, M. (1993). Data reconciliation on multicomponent network process. *Comput. Chem. Eng.* **17**, 807–817.

Romagnoli, J., and Stephanopoulos, G. (1980). On the rectification of measurement errors for complex chemical plants. *Chem. Eng. Sci.* **35**, 1067–1081.

Sánchez, M., Bandoni, A., and Romagnoli, J. (1992). PLADAT—A package for process variable classification and plant data reconciliation. *Comput. Chem. Eng.* **S16**, 499–506.

Stadtherr, M., Gifford, W., and Scriven, L. (1974). Efficient solution of sparse sets of design equations. *Chem. Eng. Sci.* **29**, 1025–1034.

Steward, D. (1962). On an approach to techniques for the analysis of the structure of large systems of equations. *SIAM Rev.* **4**, 321–342.

Václavek, V. (1969). Studies on system engineering. III. Optimal choice of the balance measurements in complicated chemical engineering systems. *Chem. Eng. Sci.* **24**, 947–955.

Václavek, V., and Loucka, M. (1976). Selection of measurements necessary to achieve multicomponent mass balances in chemical plants. *Chem. Eng. Sci.* **31**, 1199–1205.

APPENDIX A: BALANCE EQUATIONS FOR COMMON CHEMICAL PROCESS UNITS

General Unit

This is equipment where chemical reactions and heat or work transfer do not take place.

Total mass balance:

$$\sum_j L_{jk} f_j = \mathbf{0}. \tag{A3.1}$$

Component mass balance:

$$\sum_j L_{j,k} f_j M_{c,j} = \mathbf{0}. \tag{A3.2}$$

Energy balance:

$$\sum_j L_{jk} f_j h_j = \mathbf{0}. \tag{A3.3}$$

Normalization equations:

$$\sum_{c} f_j M_{c,j} - f_j = \mathbf{0}, \tag{A3.4}$$

where f_j is the total flow of stream j, $M_{c,j}$ is the mass fraction of component c in stream j, and h_j represents the specific enthalpy of stream j.

Heat Exchanger

This is equipment where heat transfer through a solid wall without chemical reactions takes place. The heat transfer between two fluids without phase change is considered as an example.

The unit is divided in two pseudounits, each one corresponding to a different fluid. Pseudounits are interconnected by a pure energy flow that represents the heat transfer between them (see Fig. A.1). The balance equations for each pseudounit k' can be stated as follows:

Total mass balance:

$$\sum_{j} L_{jk'} f_j = \mathbf{0}. \tag{A3.5}$$

Component mass balance:

$$\sum_{j} L_{jk'} f_j M_{c,j} = \mathbf{0}. \tag{A3.6}$$

Energy balance:

$$\sum_{j} L_{jk'} f_j h_j + q_{k'} = \mathbf{0}, \tag{A3.7}$$

where $q_{k'}$ is the vector of pure energy flows for unit k'.

Reactor

This is a unit where chemical reactions, and possibly heat transfer with the environment, take place.

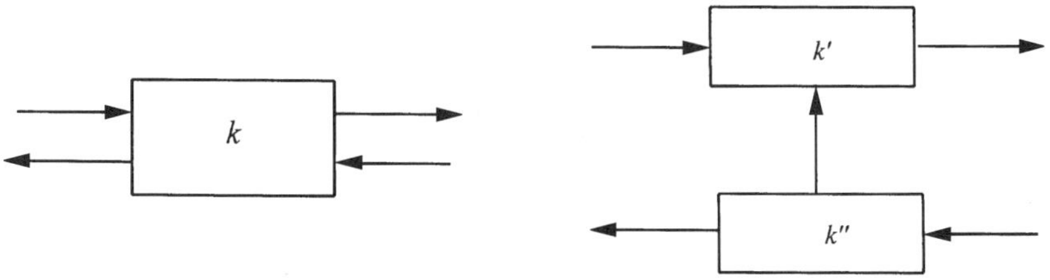

FIGURE A.1 Scheme of pseudounits for a heat exchanger.

Total mass balance:

$$\sum_j L_{jk} f_j = 0. \tag{A3.8}$$

Component mass balance:

$$\sum_j L_{j,k} f_j M_{c,j} + \sum_r S_{k,rc} \gamma_{kr} = 0. \tag{A3.9}$$

Energy balance:

$$\sum_j L_{jk} f_j h_j + \sum_r \chi_{k,r} + q_k = 0, \tag{A3.10}$$

where $S_{k,rc}$ is the coefficient of the stoichiometric matrix for component c of reaction r of unit k (Crowe *et al.*, 1983) times the corresponding molecular weight; $\gamma_{k,r}$ is the extent of reaction r of unit k (moles/time); and $\chi_{k,r}$ is the total heat involved in reaction r of unit k.

Stream Divisor

An input stream is divided into two or more output streams in this equipment. All streams related with the divisor have the same intensive properties (composition, temperature, pressure). For the case of one input stream ($j1$) and two output stream ($j2$, $j3$), the balance equations are as follows:

Total mass balance:

$$f_{j1} - f_{j2} - f_{j3} = 0. \tag{A3.11}$$

Intensive equality constraints:

$$M_{c,j1} - M_{c,j2} = 0. \tag{A3.12}$$

$$M_{c,j1} - M_{c,j3} = 0 \tag{A3.13}$$

$$t_{j1} - t_{j2} = 0 \tag{A3.14}$$

$$t_{j1} - t_{j3} = 0 \tag{A3.15}$$

$$p_{j1} - p_{j2} = 0 \tag{A3.16}$$

$$p_{j1} - p_{j3} = 0. \tag{A3.17}$$

It is necessary to include the normalization equation of only one stream because of composition equalities.

Pump, Compressor, Turbine

Work transfer to a process fluid is accomplished in these types of equipment. Their balance equations are as follows:

Total mass balance:

$$\sum_j L_{jk} f_j = 0. \tag{A3.18}$$

Component mass balance:

$$\sum_j L_{jk} f_j M_{c,j} = 0. \tag{A3.19}$$

Energy balance:

$$\sum_j L_{jk} f_j h_j + \mathbf{e}_k^{\mathrm{w}} = 0, \tag{A3.20}$$

where $\mathbf{e}_k^{\mathrm{w}}$ is the vector of energy flows, as work, that are exchanged in unit k.

APPENDIX B: ADDITIONAL INFORMATION FOR EXAMPLE 3.6

Process Model Equations

Normalization Equations

$M_{1,1} + M_{2,1} + M_{3,1} + M_{4,1} - \mathbf{1} = \mathbf{0}$	*A-19* $\Rightarrow M_{3,1}$
$M_{1,2} + M_{2,2} + M_{3,2} + M_{4,2} + M_{5,2} - 1 = 0$	*NA2*
$M_{1,3} + M_{2,3} + M_{3,3} + M_{4,3} + M_{5,3} - 1 = 0$	*NA2*
$M_{1,4} + M_{2,4} + M_{3,4} + M_{4,4} + M_{5,4} - 1 = 0$	*NA2*
$M_{1,7} + M_{2,7} + M_{3,7} + M_{4,7} + M_{5,7} - 1 = 0$	*NA2*
$M_{1,8} + M_{2,8} + M_{3,8} + M_{4,8} + M_{5,8} - 1 = 0$	*NA2*
$M_{1,9} + M_{2,9} + M_{3,9} + M_{4,9} + M_{5,9} - 1 = 0$	*NA2*

Equality of Intensive Variables for the Divisor

$h_4 - h_5 = 0$	*A-1* $\Rightarrow h_5$
$h_4 - h_6 = 0$	*NA1*
$M_{1,4} - M_{1,5} = 0$	*A-35* $\Rightarrow M_{1,5}$
$M_{2,4} - M_{2,5} = 0$	*A-2* $\Rightarrow M_{2,4}$
$M_{3,4} - M_{3,5} = 0$	*A-36* $\Rightarrow M_{3,5}$
$M_{4,4} - M_{4,5} = 0$	*A-3* $\Rightarrow M_{4,5}$
$M_{5,4} - M_{5,5} = 0$	*A-4* $\Rightarrow M_{5,4}$
$M_{1,4} - M_{1,6} = 0$	*A-37* $\Rightarrow M_{1,6}$
$M_{2,4} - M_{2,6} = 0$	*A-20* $\Rightarrow M_{2,6}$
$M_{3,4} - M_{3,6} = 0$	*A-38* $\Rightarrow M_{3,6}$
$M_{4,4} - M_{4,6} = 0$	*A-5* $\Rightarrow M_{4,6}$
$M_{5,4} - M_{5,6} = 0$	*A-6* $\Rightarrow M_{5,6}$

Total Mass Balance Equations

$f_1 + f_8 - f_2 = 0$	*NA1*
$f_2 - f_3 = 0$	*A-7* $\Rightarrow f_3$
$f_3 - f_4 = 0$	*NA2*
$f_4 - f_5 - f_6 = 0$	*A-30* $\Rightarrow f_5$
$f_6 - f_7 - f_9 = 0$	*A-27* $\Rightarrow f_6, f_9$
$f_7 - f_8 = 0$	*A-8* $\Rightarrow f_7$
$f_{10} - f_{11} = 0$	*A-9* $\Rightarrow f_{11}$

Component and Energy Balance Equations

$f_1M_{1,1} + f_8M_{1,8} - f_2M_{1,2} = 0$	*A-21* $\Rightarrow M_{1,2}$
$f_1M_{2,1} + f_8M_{2,8} - f_2M_{2,2} = 0$	*A-22* $\Rightarrow M_{2,2}$
$f_1M_{3,1} + f_8M_{3,8} - f_2M_{3,2} = 0$	*A-23* $\Rightarrow M_{3,2}$
$f_1M_{4,1} + f_8M_{4,8} - f_2M_{4,2} = 0$	*A-10* $\Rightarrow M_{4,1}$
$f_8M_{5,8} - f_2M_{5,2} = 0$	*A-11* $\Rightarrow M_{5,2}$
$f_1h_1 + f_8h_8 - f_2h_2 = 0$	*A-12* $\Rightarrow h_2$
$f_2M_{1,2} + v_1^1PM_1\gamma_1 + v_1^2PM_1\gamma_2 - f_3M_{1,3} = 0$	*A-31* $\Rightarrow M_{1,3}$
$f_2M_{2,2} + v_2^2PM_2\gamma_2 - f_3M_{2,3} = 0$	*A-24* $\Rightarrow \gamma_2$
$f_2M_{3,2} + v_3^1PM_3\gamma_1 + v_3^2PM_3\gamma_2 - f_3M_{3,3} = 0$	*A-32* $\Rightarrow M_{3,3}$
$f_2M_{4,2} + v_4^1PM_4\gamma_1 - f_3M_{4,3} = 0$	*A-25* $\Rightarrow \gamma_1$
$f_2M_{5,2} + v_5^1PM_5\gamma_1 + v_5^2PM_5\gamma_2 - f_3M_{5,3} = 0$	*NA2*
$f_2h_2 + \chi_r^1\gamma_1 + \chi_r^2\gamma_2 - f_3h_3 = 0$	*NA2*
$f_3M_{1,3} - f_4M_{1,4} = 0$	*A-33* $\Rightarrow M_{1,4}$
$f_3M_{2,3} - f_4M_{2,4} = 0$	*NA2*
$f_3M_{3,3} - f_4M_{3,4} = 0$	*A-34* $\Rightarrow M_{3,4}$
$f_3M_{4,3} - f_4M_{4,4} = 0$	*A-13* $\Rightarrow M_{4,3}$
$f_3M_{5,3} - f_4M_{5,4} = 0$	*NA2*
$f_3h_3 - q_{16} - f_4h_4 = 0$	*A-26* $\Rightarrow h_3$
$f_6M_{1,6} - f_7M_{1,7} - f_9M_{1,9} = 0$	*A-39* $\Rightarrow M_{1,9}$
$f_6M_{2,6} - f_7M_{2,7} - f_9M_{2,9} = 0$	*NA2*
$f_6M_{3,6} - f_7M_{3,7} - f_9M_{3,9} = 0$	*A-40* $\Rightarrow M_{3,9}$
$f_6M_{4,6} - f_7M_{4,7} - f_9M_{4,9} = 0$	*A-27* $\Rightarrow f_6, f_9$
$f_6M_{5,6} - f_7M_{5,7} - f_9M_{5,9} = 0$	*NA2*
$f_6h_6 - q_{17} - f_7h_7 - f_9h_9 = 0$	*A-28* $\Rightarrow q_{17}$
$f_7M_{1,7} - f_7M_{1,8} = 0$	*A-14* $\Rightarrow M_{1,8}$
$f_7M_{2,7} - f_7M_{2,8} = 0$	*A-15* $\Rightarrow M_{2,8}$
$f_7M_{3,7} - f_7M_{3,8} = 0$	*NA2*
$f_7M_{4,7} - f_7M_{4,8} = 0$	*A-16* $\Rightarrow M_{4,7}$
$f_7M_{5,7} - f_7M_{5,8} = 0$	*A-17* $\Rightarrow M_{5,7}$
$f_7h_7 + q_{16} - f_8h_8 = 0$	*A-18* $\Rightarrow q_{16}$
$f_{10}h_{10} + q_{17} - f_{11}h_{11} = 0$	*A-29* $\Rightarrow h_{11}$

A-X $\Rightarrow$ *J* Assigned equation for the estimation of unmeasured variable *J*; it belongs to the output set assignment obtained in order *X*
NA1 $\Rightarrow$ Nonassigned equation of type 1
NA2 $\Rightarrow$ Nonassigned equation of type 2

$$v_c^r = \frac{S_c^r}{S_R^r} \quad \text{where } r = 1, 2; \quad c = 1, \ldots 5; \quad R = \text{reference component}$$

Determinable Variables Expressed as Functions of Measurements

1. $h_5 = h_4$
2. $M_{2,4} = M_{2,5}$
3. $M_{4,5} = M_{4,4}$

4. $M_{5,4} = M_{5,5}$
5. $M_{4,6} = M_{4,4}$
6. $M_{5,6} = M_{5,5}$
7. $f_3 = f_2$
8. $f_7 = f_8$
9. $f_{11} = f_{10}$
10. $M_{4,1} = \dfrac{[f_2 M_{4,2} - f_8 M_{4,8}]}{f_1}$
11. $M_{5,2} = \dfrac{f_8 M_{5,8}}{f_2}$
12. $h_2 = \dfrac{f_1 h_1 + f_8 h_8}{f_2}$
13. $M_{4,3} = \dfrac{f_4 M_{4,4}}{f_2}$
14. $M_{1,8} = M_{1,7}$
15. $M_{2,8} = M_{2,7}$
16. $M_{4,7} = M_{4,8}$
17. $M_{5,7} = M_{5,8}$
18. $q_{16} = f_8 h_8 - f_8 h_7$
19. $M_{3,1} = 1 - M_{1,1} - M_{2,1} - \dfrac{[f_2 M_{4,2} - f_8 M_{4,8}]}{f_1}$
20. $M_{2,6} = M_{2,5}$
21. $M_{1,2} = \dfrac{[f_1 M_{1,1} + f_8 M_{1,7}]}{f_2}$
22. $M_{2,2} = \dfrac{[f_1 M_{2,1} + f_8 M_{2,7}]}{f_2}$
23. $M_{3,2} = \dfrac{f_1 - f_1 M_{1,1} - f_1 M_{2,1} - f_2 M_{4,2} + f_8 M_{4,8} + f_8 M_{3,8}}{f_2}$
24. $\gamma_2 = \dfrac{f_2 M_{2,3} - f_1 M_{2,1} - f_8 M_{2,7}}{v_2^2 P M_2}$
25. $\gamma_1 = \dfrac{f_4 M_{4,4} - f_2 M_{4,2}}{v_4^1 P M_4}$
26. $h_3 = \dfrac{f_4 h_4 + f_8 h_8 - f_8 h_7}{f_2}$
27. $f_9 = f_8 \dfrac{(M_{4,8} - M_{4,4})}{(M_{4,4} - M_{4,9})}, \quad f_6 = f_8 \left[1 + \dfrac{(M_{4,8} - M_{4,4})}{(M_{4,4} - M_{4,9})}\right]$
28. $q_{17} = f_8(h_6 - h_7) + \dfrac{f_8(M_{4,8} - M_{4,4})(h_6 - h_9)}{(M_{4,4} - M_{4,9})} - f_8 h_7$
29. $h_{11} = \left[f_{10} h_{10} + f_8(h_6 - h_7) + \dfrac{f_8(M_{4,8} - M_{4,4})(h_6 - h_9)}{(M_{4,4} - M_{4,9})} - f_8 h_7\right] \Big/ f_{10}$
30. $f_5 = f_4 - f_8 \left[1 + \dfrac{(M_{4,8} - M_{4,4})}{(M_{4,4} - M_{4,9})}\right]$

31. $$M_{1,3} = \left[f_1 M_{1,1} + f_8 M_{1,7} + \frac{v_1^1 P M_1 (f_4 M_{4,4} - f_2 M_{4,2})}{v_4^1 P M_4} + \frac{v_1^2 P M_1 (f_2 M_{2,3} - f_1 M_{2,1} - f_8 M_{2,7})}{v_2^2 P M_2} \right] \Big/ f_2$$

32. $$M_{3,3} = \left[f_1 - f_1 M_{1,1} - f_1 M_{2,1} - f_2 M_{4,2} + f_8 M_{4,8} + f_8 M_{3,8} + \frac{v_3^1 P M_3 (f_4 M_{4,4} - F_2 M_{4,2})}{v_4^1 P M_4} + \frac{v_3^2 P M_3 (f_2 M_{2,3} - f_1 M_{2,1} - f_8 M_{2,7})}{v_2^2 P M_2} \right] \Big/ f_2$$

33. $$M_{1,4} = \left[f_1 M_{1,1} + f_8 M_{1,7} + \frac{v_1^1 P M_1 (f_4 M_{4,4} - f_2 M_{4,2})}{v_4^1 P M_4} + \frac{v_1^2 P M_1 (f_2 M_{2,3} - f_1 M_{2,1} - f_8 M_{2,7})}{v_2^2 P M_2} \right] \Big/ f_4$$

34. $$M_{3,4} = \left[f_1 - f_1 M_{1,1} - f_1 M_{2,1} - f_2 M_{4,2} + f_8 M_{4,8} + f_8 M_{3,8} + \frac{v_3^1 P M_3 (f_4 M_{4,4} - f_2 M_{4,2})}{v_4^1 P M_4} + \frac{v_3^2 P M_3 (f_2 M_{2,3} - f_1 M_{2,1} - f_8 M_{2,7})}{v_2^2 P M_2} \right] \Big/ f_4$$

35. $$M_{1,5} = \left[f_1 M_{1,1} + f_8 M_{1,7} + \frac{v_1^1 P M_1 (f_4 M_{4,4} - f_2 M_{4,2})}{v_4^1 P M_4} + \frac{v_1^2 P M_1 (f_2 M_{2,3} - f_1 M_{2,1} - f_8 M_{2,7})}{v_2^2 P M_2} \right] \Big/ f_4$$

36. $$M_{3,5} = \left[f_1 - f_1 M_{1,1} - f_1 M_{2,1} - f_2 M_{4,2} + f_8 M_{4,8} + f_8 M_{3,8} + \frac{v_3^1 P M_3 (f_4 M_{4,4} - F_2 M_{4,2})}{v_4^1 P M_4} + \frac{v_3^2 P M_3 (f_2 M_{2,3} - f_1 M_{2,1} - f_8 M_{2,7})}{v_2^2 P M_2} \right] \Big/ f_4$$

37. $$M_{1,6} = \left[f_1 M_{1,1} + f_8 M_{1,7} + \frac{v_1^1 P M_1 (f_4 M_{4,4} - f_2 M_{4,2})}{v_4^1 P M_4} + \frac{v_1^2 P M_1 (f_2 M_{2,3} - f_1 M_{2,1} - f_8 M_{2,7})}{v_2^2 P M_2} \right] \Big/ f_4$$

38. $$M_{3,6} = \left[f_1 - f_1 M_{1,1} - f_1 M_{2,1} - f_2 M_{4,2} + f_8 M_{4,8} + f_8 M_{3,8} + \frac{v_3^1 P M_3 (f_4 M_{4,4} - f_2 M_{4,2})}{v_4^1 P M_4} + \frac{v_3^2 P M_3 (f_2 M_{2,3} - f_1 M_{2,1} - f_8 M_{2,7})}{v_2^2 P M_2} \right] \Big/ f_4$$

39. $$M_{1,9} = \frac{\left[1+\frac{(M_{4,8}-M_{4,4})}{(M_{4,4}-M_{4,9})}\right]\left[\frac{1}{f_4}\left[\begin{array}{l} f_1M_{1,1}+f_8M_{1,7}+\frac{v_1^1PM_1(f_4M_{4,4}-f_2M_{4,2})}{v_4^1PM_4} \\ +\frac{v_1^2PM_1(f_2M_{2,3}-f_1M_{2,1}-f_8M_{2,7})}{v_2^2PM_2}\end{array}\right]\right]-M_{1,7}}{\frac{(M_{4,8}-M_{4,4})}{(M_{4,4}-M_{4,9})}}$$

40. $$M_{3,9} = \frac{\left[1+\frac{(M_{4,8}-M_{4,4})}{(M_{4,4}-M_{4,9})}\right]\frac{1}{f_4}\left[\begin{array}{l} f_1-f_1M_{1,1}-f_1M_{2,1}-f_2M_{4,2}+f_8M_{4,8}+f_8M_{3,8} \\ +\frac{v_3^1PM_3(f_4M_{4,4}-f_2M_{4,2})}{v_4^1PM_4}+\frac{v_3^2PM_3(f_2M_{2,3}-f_1M_{2,1}-f_8M_{2,7})}{v_2^2PM_2}\end{array}\right]-M_{3,7}}{\frac{(M_{4,8}-M_{4,4})}{(M_{4,4}-M_{4,9})}}$$

Reconciliation Equations

1. $f_1 + f_8 - f_2 = 0$
2. $f_2 - f_4 = 0$
3. $h_4 - h_6 = 0$
4. $M_{3,7} - M_{3,8} = 0$
5. $f_2M_{2,3} - f_4M_{2,5} = 0$
6. $f_2M_{5,3} - f_4M_{5,5} = 0$
7. $$\left[1+\frac{(M_{4,8}-M_{4,4})}{(M_{4,4}-M_{4,9})}\right]M_{2,5} - M_{2,7} - \frac{(M_{4,8}-M_{4,4})}{(M_{4,4}-M_{4,9})}M_{2,9} = 0$$
8. $$\left[1+\frac{(M_{4,8}-M_{4,4})}{(M_{4,4}-M_{4,9})}\right]M_{5,5} - M_{5,7} - \frac{(M_{4,8}-M_{4,4})}{(M_{4,4}-M_{4,9})}M_{5,9} = 0$$
9. $$f_8M_{5,8} + \frac{v_5^1PM_5[f_4M_{4,4}-f_2M_{4,2}]}{v_4^1PM_4} + \frac{v_5^2PM_5[f_2M_{2,3}-f_1M_{2,1}-f_8M_{2,7}]}{v_2^2PM_2} - f_2M_{5,3} = 0$$
10. $$f_8(M_{1,7}+M_{4,8}+M_{3,8}) + \left[\frac{v_1^1PM_1}{v_4^1PM_4}+\frac{v_3^1PM_3}{v_4^1PM_4}\right][f_4M_{4,4}-f_2M_{4,2}] + \left[\frac{v_1^2PM_1}{v_2^2PM_2}+\frac{v_3^2PM_3}{v_2^2PM_2}\right][f_2M_{2,3}-f_1M_{2,1}-f_8M_{2,7}] + f_1 - f_1M_{2,1} + f_4M_{4,4} + f_2(M_{2,3}+M_{5,3}-M_{4,2}-1) = 0$$
11. $$f_4(M_{2,5}+M_{4,4}+M_{5,5}-1) + f_1(1-M_{2,1}) + f_8(M_{1,7}+M_{4,8}+M_{3,8}) + \left[\frac{v_1^1PM_1}{v_4^1PM_4}+\frac{v_3^1PM_3}{v_4^1PM_4}\right][f_4M_{4,4}-f_2M_{4,2}] + \left[\frac{v_1^2PM_1}{v_2^2PM_2}+\frac{v_3^2PM_3}{v_2^2PM_2}\right] \times [F_2M_{2,3}-F_1M_{2,1}-F_8M_{2,7}] - F_2M_{4,2} = 0$$
12. $M_{1,7} + M_{2,7} + M_{3,7} + M_{4,8} + M_{5,8} - 1 = 0$

13. $$\left[1+\frac{(M_{4,8}-M_{4,4})}{(M_{4,4}-M_{4,9})}\right]\frac{1}{F_4}\left[f_8(M_{1,7}+M_{4,8}+M_{3,8})+\left[\frac{v_1^1 PM_1}{v_4^1 PM_4}+\frac{v_3^1 PM_3}{v_4^1 PM_4}\right]\right.$$
$$\times[f_4M_{4,4}-f_2M_{4,2}]+\left[\frac{v_1^2 PM_1}{v_2^2 PM_2}+\frac{v_3^2 PM_3}{v_2^2 PM_2}\right][f_2M_{2,3}-f_1m_{2,1}-f_8m_{2,7}]$$
$$\left.+f_1(1-m_{2,1})-f_2m_{4,2}\right]-M_{1,7}-M_{3,7}+\frac{(M_{4,8}-M_{4,4})}{(M_{4,4}-M_{4,9})}$$
$$\times[M_{2,9}+M_{4,9}+M_{5,9}-1]=0$$

14. $$f_1h_1-f_4h_4+f_8h_7+\chi_r^1\left[\frac{f_4M_{4,4}-f_2M_{4,2}}{v_4^1 PM_4}\right]$$
$$+\chi_r^2\left[\frac{f_2M_{2,3}-f_1M_{2,1}-f_8M_{2,7}}{v_2^2 PM_2}\right]=0$$

4 DECOMPOSITION USING ORTHOGONAL TRANSFORMATIONS

This chapter is devoted to the analysis of variable classification and the decomposition of the data reconciliation problem for linear and bilinear plant models, using the so-called matrix projection approach. The use of orthogonal factorizations, more precisely the Q-R factorization, to solve the aforementioned problems is discussed and its range of application is determined. Several illustrative examples are included to show the applicability of such techniques in practical applications.

4.1. INTRODUCTION

Crowe *et al.* (1983) proposed an elegant strategy for decoupling measured variables from the linear constraint equations. This procedure allows both the reduction of the data reconciliation problem and the classification of process variables. It is based on the use of a projection matrix to eliminate unmeasured variables. Crowe later extended this methodology (Crowe, 1986, 1989) to bilinear systems.

An equivalent decomposition can be performed using the Q-R orthogonal transformation (Sánchez and Romagnoli, 1996). Orthogonal factorizations were first used by Swartz (1989), in the context of successive linearization techniques, to eliminate the unmeasured variables from the constraint equations.

In this chapter the use of Q-R factorizations with the purpose of system decomposition and instrumentation analysis, for linear and bilinear plant models, is thoroughly investigated. Simple expressions are provided using subproducts of Q-R factorizations for application in data reconciliation. Furthermore, the use of factorization procedures when energy balances are included in the set of process constraints is discussed, in order to establish the limitations of this technique.

4.2. LINEAR MASS BALANCES

4.2.1. Crowe's Projection Matrix Approach

A method for decomposing the unmeasured process variables from the measured ones was proposed by Crowe *et al.* (1983) for linear constraints. This strategy is based on the use of a projection matrix.

Let us represent the operation of a process under steady-state conditions by the following set of linear equations:

$$\mathbf{A}_1\mathbf{x} + \mathbf{A}_2\mathbf{u} = \mathbf{0}, \quad \mathbf{x} \in \Re^g; \quad \mathbf{u} \in \Re^n, \tag{4.1}$$

where $\mathbf{x}$ is the $(g \times 1)$ vector of measured variables and $\mathbf{u}$ is the $(n \times 1)$ vector of unmeasured variables. $\mathbf{A}_1$ and $\mathbf{A}_2$ are compatible matrices of dimension $(m \times g)$ and $(m \times n)$, respectively.

A projection matrix $\mathbf{P}$ was defined by Crowe, such that premultiplying the Jacobian matrix $\mathbf{A}_2$ with $\mathbf{P}$ yields

$$\mathbf{P}\mathbf{A}_2 = \mathbf{0}. \tag{4.2}$$

The columns of $\mathbf{P}$ span the null space of $\mathbf{A}_2$, and thus the unmeasured variables are eliminated.

In order to obtain the projection matrix $\mathbf{P}$, Crowe proposed the following procedure:

1. Column reduce $\mathbf{A}_2$ to obtain a matrix $\mathbf{X}$ with linearly independent columns

$$\mathbf{A}_2\mathbf{A}_3 = [\mathbf{X} \quad \mathbf{0}], \tag{4.3}$$

where $\mathbf{A}_3$ represents the nonsingular matrix that performs the necessary operations on the columns of $\mathbf{A}_2$.

2. Partition $\mathbf{X}$ such that

$$\mathbf{A}_4\mathbf{A}_2\mathbf{A}_3 = \begin{bmatrix} \mathbf{X}_1 & \mathbf{0} \\ \mathbf{X}_2 & \mathbf{0} \end{bmatrix} \tag{4.4}$$

with $\mathbf{X}_1$ square and nonsingular. Then $\mathbf{P}$ is calculated by the following expression:

$$\mathbf{P} = \begin{bmatrix} -\mathbf{X}_2\mathbf{X}_1^{-1} & \mathbf{I} \end{bmatrix} \mathbf{A}_4. \tag{4.5}$$

The reduced problem is now formulated as

$$\mathbf{G}\mathbf{x} = \mathbf{0}, \tag{4.6}$$

where

$$\mathbf{G} = \mathbf{P}\mathbf{A}_1. \tag{4.7}$$

Data reconciliation can now be performed on the reduced subsystem containing only measured variables. We can now state the general problem as the equivalent two-problem formulation discussed in the previous chapter.

4.2.2. The Q-R Approach

An alternative decomposition can be performed using a Q-R factorization of matrix $\mathbf{A}_2$ to decouple the unmeasured variables from the measured ones (Sánchez and Romagnoli, 1996). Let us state the following theorem (Dahlquist and Bjork, 1974).

THEOREM 4.1 (Q-R THEOREM)

Let $\mathbf{A}$ *be a given* $(m \times n)$ *matrix with* $m \geq n$ *and* n *linearly independent columns. Then there exists a unique* $(m \times m)$ $\mathbf{Q}$*,*

$$\mathbf{Q}^{\mathrm{T}}\mathbf{Q} = \mathbf{D}_{\mathrm{i}}, \quad \mathbf{D}_i = \mathrm{diag}(d_1, \ldots, d_n); \quad d_k > \mathbf{0}, k = 1, \ldots, n \tag{4.8}$$

and a unique $(m \times n)$ *upper-triangular matrix* $\mathbf{R}$*, with* $R_{kk} = 1, k = 1, \ldots, n$*, such that*

$$\mathbf{A} = \mathbf{QR}. \tag{4.9}$$

Now, if $\mathbf{A}$ is rank deficient and $\mathbf{A} = \mathbf{QR}$ is the Q-R factorization of $\mathbf{A}$, then at least one diagonal entry in $\mathbf{R}$ is zero. Let us examine why the Q-R factorization approach can fail in the case when $\mathrm{R}(\mathbf{A}) = \mathrm{rank}(\mathbf{A}) = r < n$.

The mission of any orthogonalization method is to compute the orthonormal basis for $\mathrm{R}(\mathbf{A})$. Indeed, if $\mathrm{R}(\mathbf{A}) = \mathrm{R}(\mathrm{Q}_1)$, where $\mathbf{Q}_1 = [\mathbf{q}_1, \ldots, \mathbf{q}_r]$ has orthonormal columns, then $\mathbf{A} = \mathrm{Q}_1\mathrm{S}$ for some $\mathbf{S} \in \Re^{r\times n}$.

Unfortunately, if $\mathrm{rank}(\mathbf{A}) < n$, then the Q-R factorization does not necessarily produce an orthonormal basis for $\mathrm{R}(\mathbf{A})$. However, the Q-R decomposition can be modified in a simple way so as to produce an orthonormal basis for $\mathbf{A}$'s range. The modified algorithm computes the factorization

$$\mathrm{A}\Pi = \begin{bmatrix}\mathbf{Q}_1 & \mathbf{Q}_2\end{bmatrix}\begin{bmatrix}\mathbf{R}_{11} & \mathbf{R}_{12}\\ \mathbf{0} & \mathbf{0}\end{bmatrix}. \tag{4.10}$$

where $r = \mathrm{rank}(\mathbf{A})$, $\mathbf{Q}$ is orthogonal, $\mathbf{R}_{11}$ is upper triangular, and Π is a permutation. If $\mathrm{A}\Pi = [\mathbf{a}_{c_1}, \ldots, \mathbf{a}_{c_n}]$ and $\mathbf{Q} = [\mathbf{q}_1, \ldots, \mathbf{q}_m]$, then for $k = 1, \ldots, n$, we have

$$\mathbf{a}_{ck} = \sum_{i=1}^{\min\{r,k\}} \mathbf{r}_{ik}\mathbf{q}_i \in \mathrm{span}\{\mathbf{q}_1, \ldots, \mathbf{q}_{\mathrm{r}}\}. \tag{4.11}$$

Also, it follows that for any vector satisfying $\mathbf{Ax} = \mathbf{b}$,

$$\mathbf{\Pi}^{\mathrm{T}}\mathbf{x} = \begin{bmatrix}\mathbf{s}\\ \mathbf{z}\end{bmatrix} \quad \text{and} \quad \mathbf{Q}^{\mathrm{T}}\mathbf{b} = \begin{bmatrix}\mathbf{i}\\ \mathbf{l}\end{bmatrix}, \tag{4.12}$$

where $\mathbf{s}$ and $\mathbf{i}$ are vectors of dimension r, $\mathbf{z}$ is an $(n-r)$-dimensional vector, and $\mathbf{l}$ is a vector of dimension $(m-r)$.

Returning to our reconciliation problem, the Q-R decomposition of matrix $\mathbf{A}_2$ allows us to obtain $\mathbf{Q}_{\mathrm{u}}$ and $\mathbf{R}_{\mathrm{u}}$ matrices and the permutation matrix $\mathbf{\Pi}_{\mathrm{u}}$, such that

$$\mathbf{A}_2\mathbf{\Pi}_{\mathrm{u}} = \mathbf{Q}_{\mathrm{u}}\mathbf{R}_{\mathrm{u}}, \tag{4.13}$$

where $\mathbf{Q}_{\mathrm{u}}$ and $\mathbf{R}_{\mathrm{u}}$ can be divided into

$$\mathbf{Q}_{\mathrm{u}} = \begin{bmatrix}\mathbf{Q}_{\mathrm{u}1} & \mathbf{Q}_{\mathrm{u}2}\end{bmatrix}, \quad \mathbf{R}_{\mathrm{u}} = \begin{bmatrix}\mathbf{R}_{\mathrm{u}1} & \mathbf{R}_{\mathrm{u}2}\\ \mathbf{0} & \mathbf{0}\end{bmatrix}, \tag{4.14}$$

with $r_u = \text{rank}(\mathbf{A}_2) = \text{rank}(\mathbf{R}_{u1})$. Notice that $\mathbf{Q}_u$ is an orthogonal matrix and $\mathbf{R}_{u1}$ is a nonsingular upper triangular matrix of dimension r_u.

In the same way the unmeasured process variables can be partitioned into two subsets,

$$\boldsymbol{\Pi}_u^T \mathbf{u} = \begin{bmatrix} \mathbf{u}_{r_u} \\ \mathbf{u}_{n-r_u} \end{bmatrix} \tag{4.15}$$

Now, premultiplying the linearized constraints by $\mathbf{Q}_u^T = \mathbf{Q}_u^{-1}$, we obtain

$$\begin{bmatrix} \mathbf{Q}_{u1}^T \mathbf{A}_1 & \mathbf{R}_{u1} & \mathbf{R}_{u2} \\ \mathbf{Q}_{u2}^T \mathbf{A}_1 & \mathbf{0} & \mathbf{0} \end{bmatrix} \begin{bmatrix} \mathbf{x} \\ \mathbf{u}_{r_u} \\ \mathbf{u}_{n-r_u} \end{bmatrix} = \mathbf{0}. \tag{4.16}$$

The first r_u equations for $\mathbf{u}_{r_u}$ can be written in terms of the other variables:

$$\mathbf{u}_{r_u} = -\mathbf{R}_{u1}^{-1} \mathbf{Q}_{u1}^T \mathbf{A}_1 \mathbf{x} - \mathbf{R}_{u1}^{-1} \mathbf{R}_{u2} \mathbf{u}_{n-r_u}. \tag{4.17}$$

Since the unmeasured variables do not appear in the remaining equations, the first reduced subproblem becomes the following problem.

PROBLEM 1

Solve

$$\begin{aligned} &\min_x \, (\mathbf{y} - \mathbf{x})^T \Psi_x^{-1} (\mathbf{y} - \mathbf{x}) \\ &\text{s.t. } \mathbf{G}_x \mathbf{x} = \mathbf{0}, \end{aligned} \tag{4.18}$$

where

$$\mathbf{G}_x = \mathbf{Q}_{u2}^T \mathbf{A}_1.$$

PROBLEM 2

Estimate the unmeasured variables, $\mathbf{u}$, by solving Eq. (4.17) where the components $\mathbf{u}_{n-r_u}$ were arbitrarily set. The uniqueness of $\mathbf{u}$ is related to the system estimability and will be discussed later.

The following illustrative examples will allow us to understand the implications of the previous results on the reconciliation problem.

EXAMPLE 4.1

Consider the simple serial system of Fig. 1. A total mass balance around each unit is considered.

The matrices $\mathbf{A}_1$ and $\mathbf{A}_2$ for this example are given by

$$\mathbf{A}_1 = \begin{bmatrix} 1 & 1 & 0 & 0 & 0 \\ 0 & 0 & 1 & 0 & 0 \\ 0 & 0 & 0 & 1 & -1 \end{bmatrix}, \quad \mathbf{A}_2 = \begin{bmatrix} -1 & 0 \\ 1 & -1 \\ 0 & 1 \end{bmatrix}.$$

Performing the Q-R decomposition on $\mathbf{A}_2$, we have

$$\mathbf{Q}_u = \begin{bmatrix} -0.7071 & -0.4082 & 0.5774 \\ 0.7071 & -0.4082 & 0.5774 \\ 0 & 0.8165 & 0.5774 \end{bmatrix}, \quad \mathbf{R}_u = \begin{bmatrix} 1.4142 & -0.7071 \\ 0 & 1.2247 \\ 0 & 0 \end{bmatrix}.$$

Now matrix $\mathbf{Q}_{u2}^T$ in the previous development is given by

$$\mathbf{Q}_{u2}^T = [0.5754 \quad 0.5754 \quad 0.5754].$$

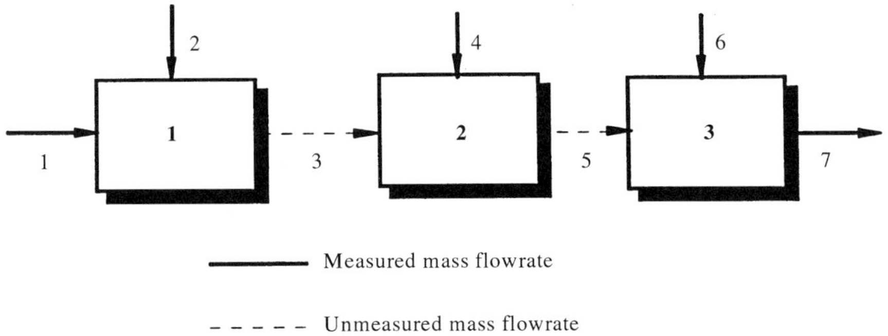

FIGURE 1 Simple serial system.

Consequently, the reduced set containing only measured variables is

$$\mathbf{G}_x = \mathbf{Q}_{u2}^T \mathbf{A}_1 = [0.5754 \quad 0.5754 \quad 0.5754 \quad 0.5754 \quad -0.5754],$$

which in this specific situation corresponds to a global mass balance around all units. Note also that in this case $\text{rank}(\mathbf{A}_2) = 2$ and there are two unmeasured process variables. Consequently, the subset $\mathbf{u}_{n-r_u}$ is empty and all the unmeasured variables are determinable.

EXAMPLE 4.2
Consider now the modified serial system of Fig. 2.
In this case,

$$\mathbf{A}_1 = \begin{bmatrix} 1 & 0 \\ 0 & -1 \\ 0 & 1 \end{bmatrix}; \quad \mathbf{A}_2 = \begin{bmatrix} 1 & -1 & 0 & 0 \\ -1 & 1 & 0 & 0 \\ 0 & 0 & 1 & -1 \end{bmatrix}.$$

Applying the Q-R decomposition, we have

$$\mathbf{Q}_u = \begin{bmatrix} -0.7071 & -0.7071 & 0 \\ 0.7071 & -0.7071 & 0 \\ 0 & 0 & 1 \end{bmatrix}; \quad \mathbf{R}_u = \begin{bmatrix} -1.4142 & 1.4142 & 0 & 0 \\ 0 & 0 & 0 & 0 \\ 0 & 0 & 1 & -1 \end{bmatrix}.$$

Note that in this case one diagonal entry in $\mathbf{R}_u$ is zero; thus, the modified decomposition

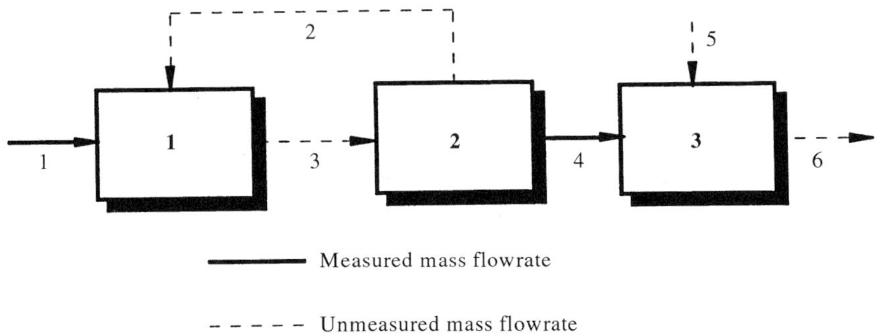

FIGURE 2 Alternative serial system.

should be used. In this case, using a permutation matrix Π_u we have

$$\Pi_u = \begin{bmatrix} 1 & 0 & 0 & 0 \\ 0 & 0 & 1 & 0 \\ 0 & 1 & 0 & 0 \\ 0 & 0 & 0 & 1 \end{bmatrix}, \quad \mathbf{R}_u = \begin{bmatrix} -1.4142 & 0 & 1.4142 & 0 \\ 0 & -1 & 0 & 1 \\ 0 & 0 & 0 & 0 \end{bmatrix}.$$

Matrices $\mathbf{Q}_{u2}^T$ and $\mathbf{G}_x$ are now

$$\mathbf{Q}_{u2}^T = [-0.7071 \quad -0.7071 \quad 0]$$

$$\mathbf{G}_x = \mathbf{Q}_{u2}^T \mathbf{A}_1 = [-0.7071 \quad 0.7071],$$

which corresponds to a balance around units 1 and 2. Also by permuting the vector of unmeasured process variables using $\mathbf{\Pi}_u$, we have

$$\Pi^T \mathbf{u} = \begin{bmatrix} f_2 \\ f_5 \\ f_3 \\ f_6 \end{bmatrix},$$

meaning that in this case, f_2 and f_5 can be rewritten as function of f_3 and f_6. The unmeasured process variables f_3 and f_6 are said to be nonestimable. In this case f_2 and f_5 are also nonestimable, since they cannot be calculated from the measured variables because they depend on the assumed values f_3 and f_6.

EXAMPLE 4.3

Now, consider the same system under a new arrangement of the measurements, as shown in Fig. 3. In this case, the flowrate of stream number 2 is considered measured. The matrices $\mathbf{A}_1$ and $\mathbf{A}_2$ are now

$$\mathbf{A}_1 = \begin{bmatrix} 1 & 1 & 0 \\ 0 & -1 & -1 \\ 0 & 0 & 1 \end{bmatrix}, \quad \mathbf{A}_2 = \begin{bmatrix} -1 & 0 & 0 \\ 1 & 0 & 0 \\ 0 & 1 & -1 \end{bmatrix}.$$

The Q-R decomposition gives

$$\mathbf{Q}_u = \begin{bmatrix} -0.7071 & 0 & -0.7071 \\ 0.7071 & 0 & -0.7071 \\ 0 & -1 & 0 \end{bmatrix}, \quad \mathbf{R}_u = \begin{bmatrix} 1.4142 & 0 & 0 \\ 0 & -1 & 1 \\ 0 & 0 & 0 \end{bmatrix}.$$

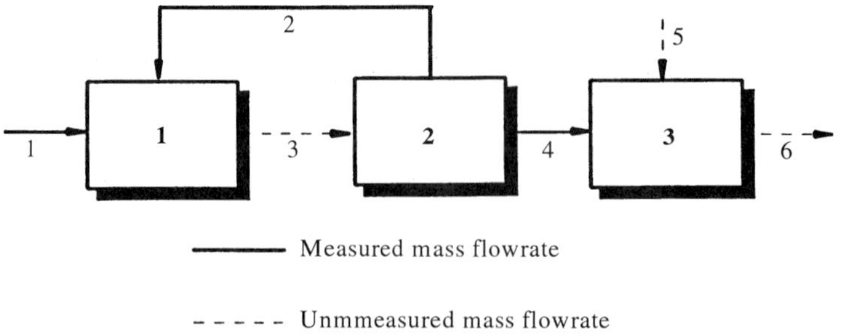

FIGURE 3 Alternative system under new measurement configuration.

Finally, matrices $\mathbf{Q}_{u2}^{T}$ and $\mathbf{G}_{x}$ are

$$\mathbf{Q}_{u2}^{T} = [-0.7071 \quad -0.7071 \quad 0]$$

$$\mathbf{G}_{x} = \mathbf{Q}_{u2}^{T}\mathbf{A}_{1} = [-0.7071 \quad 0 \quad 0.7071],$$

which again corresponds to a global mass balance around units 1 and 2. Note that the flowrate of stream 2 does not appear in the reduced set. Actually, the flowrate of stream 2 corresponds to a just-measured variable (nonredundant) as defined before. In terms of the unmeasured variables, we have that rank $(\mathbf{A}_2) = 2$ and there are three unmeasured variables; f_5 and f_6 are nonestimable.

From the previous discussions several remarks are in order:

Remark 1. Matrix $\mathbf{R}_u$ in the Q-R factorization of the $\mathbf{A}_2$ matrix contains the topological information about the system in terms of the available measurements.

1. If rank $(\mathbf{R}_u) = r_u = n$, where n is the number of unmeasured variables, then all unmeasured process variables are estimable from the available information.
2. If rank $(\mathbf{R}_u) = r_u < n$, then at least $(n - r_u)$ variables cannot be calculated from the available information.

Remark 2. The permutation matrix $\mathbf{\Pi}_u$, obtained as a by-product of the Q-R factorization procedure of $\mathbf{A}_2$, enables an easy classification of the unmeasured process variables, as is indicated by Eq. (4.15). The variables in subset $\mathbf{u}_{n-r_u}$ correspond to the minimum number and the location of measurements needed for the system to satisfy the estimability condition, that is, that all unmeasured variables be determinable.

Remark 3. The $\mathbf{Q}_{u2}^{T}$ matrix in the Q-R factorization of $\mathbf{A}_2$ is such that its columns span the null space of $\mathbf{A}_2$. That is,

$$\mathbf{Q}_{u2}^{T}\mathbf{A}_{2} = \mathbf{0}. \tag{4.19}$$

The Q-R factorization algorithm provides the information about the estimability conditions of the variables allowing a direct classification and decomposition. We already have shown [Eq. (4.15)] that, in general, the unmeasured process variables can be divided into two subsets. The subset $\mathbf{u}_{n-r_u}$ corresponds to the indeterminable unmeasured process variables. Regarding the subset $\mathbf{u}_{r_u}$, nothing can be said since some of these variables can be calculated directly from the available measurements and some depend on the assumption of the $\mathbf{u}_{n-r_u}$ variables. Further information for classifying the variables in subset $\mathbf{u}_{r_u}$ can be obtained from the Q-R factorization. Note that in Eq. (4.17), if the last term in the RHS is zero, then all the corresponding $\mathbf{u}_{r_u}$ variables can be calculated from the available information. To further classify the remaining $\mathbf{u}_{r_u}$ variables, we need to look at the rows of the matrix

$$\mathbf{R}_{IU} = \mathbf{R}_{u1}^{-1}\mathbf{R}_{u2}. \tag{4.20}$$

The following can be stated:

1. A variable in subset $\mathbf{u}_{r_u}$ is said to be estimable if the corresponding row in the $\mathbf{R}_{IU}$ matrix is zero.
2. A variable in subset $\mathbf{u}_{r_u}$ is said to be nonestimable otherwise.

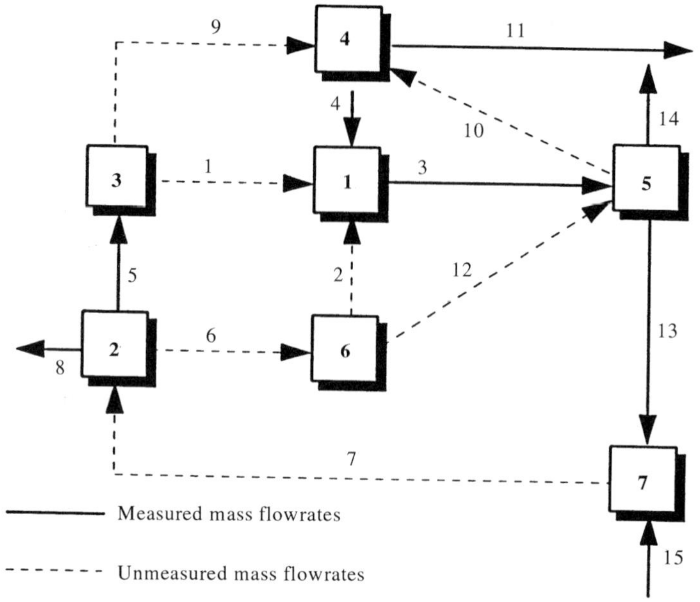

FIGURE 4 Flow diagram for Example 4.4.

Remark 4. As indicated by Crowe *et al.* (1983), measured variable classification is performed by examining the matrix associated with the reconciliation equations. The zero columns of **G** or $\mathbf{G}_x$ correspond to variables that do not participate in the reconciliation, so they are nonredundant. The remaining columns correspond to redundant measurements.

Let us now apply the Q-R factorization approach to a larger and more complicated example.

Example 4.4

A subsystem comprising seven units is shown in Fig. 4. In this case we have eight measured process variables and seven unmeasured ones. Matrices $\mathbf{A}_1$ and $\mathbf{A}_2$ are stated as follows:

$$\mathbf{A}_1 = \begin{bmatrix} -1 & 1 & 0 & 0 & 0 & 0 & 0 & 0 \\ 0 & 0 & -1 & -1 & 0 & 0 & 0 & 0 \\ 0 & 0 & 1 & 0 & 0 & 0 & 0 & 0 \\ 0 & 0 & 0 & 0 & -1 & 0 & 0 & 0 \\ 1 & 0 & 0 & 0 & 0 & -1 & -1 & 0 \\ 0 & 0 & 0 & 0 & 0 & 0 & 0 & 0 \\ 0 & 0 & 0 & 0 & 0 & 1 & 0 & 1 \end{bmatrix},$$

$$\mathbf{A}_2 = \begin{bmatrix} 1 & 1 & 0 & 0 & 0 & 0 & 0 \\ 0 & 0 & -1 & 1 & 0 & 0 & 0 \\ -1 & 0 & 0 & 0 & -1 & 0 & 0 \\ 0 & 0 & 0 & 0 & 1 & 1 & 0 \\ 0 & 0 & 0 & 0 & 0 & -1 & 1 \\ 0 & -1 & 1 & 0 & 0 & 0 & -1 \\ 0 & 0 & 0 & -1 & 0 & 0 & 0 \end{bmatrix}.$$

By applying the orthogonal projection approach, the following $\mathbf{Q}_{u1}$, $\mathbf{Q}_{u2}$, $\mathbf{R}_{u1}$, and $\mathbf{R}_{u2}$ matrices are obtained:

$$\mathbf{Q}_{u1} = \begin{bmatrix} -0.7071 & 0 & 0 & 0 & -0.5 & 0.3273 \\ 0 & -0.7071 & 0 & -0.4082 & 0 & -0.4364 \\ 0.7071 & 0 & 0 & 0 & -0.5 & 0.3273 \\ 0 & 0 & -0.7071 & 0 & 0.5 & 0.3273 \\ 0 & 0 & 0.7071 & 0 & 0.5 & 0.3273 \\ 0 & 0.7071 & 0 & -0.4082 & 0 & -0.4364 \\ 0 & 0 & 0 & 0.8165 & 0 & -0.4364 \end{bmatrix},$$

$$\mathbf{Q}_{u2} = \begin{bmatrix} 0.378 \\ 0.378 \\ 0.378 \\ 0.378 \\ 0.378 \\ 0.378 \\ 0.378 \end{bmatrix}.$$

$$\mathbf{R}_{u1} = \begin{bmatrix} -1.4142 & 0 & 0 & 0 & -0.7071 & 0 \\ 0 & 1.4142 & 0 & -0.7071 & 0 & -0.7071 \\ 0 & 0 & -1.4142 & 0 & -0.7071 & 0.7071 \\ 0 & 0 & 0 & -1.2247 & 0 & 0.4082 \\ 0 & 0 & 0 & 0 & 1. & 0.5 \\ 0 & 0 & 0 & 0 & 0 & 0.7638 \end{bmatrix},$$

$$\mathbf{R}_{u2} = \begin{bmatrix} -0.7071 \\ -0.7071 \\ 0 \\ 0.4082 \\ -0.5 \\ 0.7638 \end{bmatrix}.$$

Matrix $\mathbf{G}_x$ now becomes

$$\mathbf{G}_x = [0 \quad 0.378 \quad 0 \quad -0.378 \quad -0.378 \quad 0 \quad -0.378 \quad 0.378].$$

Thus, we have identified the subset of redundant equations containing only the redundant process variables f_4, f_8, f_{11}, f_{14}, and f_{15}. Furthermore, the rank of $\mathbf{R}_u$ is equal to 6, which means that at least one of the unmeasured variables is indeterminable. The remaining ones can be written in terms of it, as indicated by Eq. (4.15). In this case, from the orthogonal transformation, the subsets of $\mathbf{u}$ are defined as

$$\mathbf{u}_{r_u} = [f_1 \quad f_6 \quad f_{10} \quad f_7 \quad f_9 \quad f_{12}], \quad \mathbf{u}_{n-r_u} = [f_2].$$

Furthermore,

$$\mathbf{R}_{IU} = \mathbf{R}_{u1}^{-1}\mathbf{R}_{u2} = \begin{bmatrix} 1 \\ 0 \\ 1 \\ 0 \\ -1 \\ 1 \end{bmatrix}$$

There are two zero rows in matrix $\mathbf{R}_{IU}$; thus, f_6 and f_7 are determinable. The other unmeasured flowrates are indeterminable since they depend on the assumed values of vector $\mathbf{u}_{n-r_u}$.

4.3. BILINEAR MULTICOMPONENT AND ENERGY BALANCES

In the following sections, the use of the Q-R decomposition approach is discussed within the framework of the general multicomponent and energy (bilinear) reconciliation problem. In this case the classification of the measured and unmeasured process variables involves a sequence of steps.

4.3.1. Modification of Bilinear Constraints

Component mass and energy balances and normalization equations are first rewritten using the method proposed by Crowe (1986) for the bilinear terms. Streams are divided into three categories depending on the combination of total flowrates (**f**), concentration (**M**), and temperature (**t**) measurements as shown in Table 1.

Component mass/energy balances:

$$\mathbf{B}_1\mathbf{f}_{ch} + \mathbf{B}_2\mathbf{V}\mathbf{d} + \mathbf{B}_3\mathbf{v} = \mathbf{0}. \tag{4.21}$$

Normalization equations:

$$\mathbf{E}_1\mathbf{f}_{ch} + \mathbf{E}_2\mathbf{V}\mathbf{d} + \mathbf{E}_3\mathbf{v} + \mathbf{E}_4\mathbf{f}_M + \mathbf{E}_5\mathbf{f}_U = \mathbf{0}, \tag{4.22}$$

where

$\mathbf{f}_{ch}$ is the vector of component or enthalpy flows for streams in Category 1
$\mathbf{d}$ is the vector of measured concentrations or temperatures for streams in Category 2
$\mathbf{v}$ is the vector of component or enthalpy flows for streams in Category 3, extents of reaction, unknown pure energy flows
$\mathbf{f}_M$ stands for measured total flowrates and $\mathbf{f}_U$ for the unmeasured ones
$\mathbf{V}$ represents the diagonal matrix of unmeasured total flowrates of Category 2 The number of entries for a stream in $\mathbf{V}$ is equal to the number of elements of $\mathbf{d}$ corresponding to this stream

TABLE 1 Categories of the Streams[a]

Category	f	M/t
1	M	M
2	U	M
3	M/U	U

[a] M and U indicate measured and unmeasured variables, respectively (from Sánchez and Romagnoli, 1986).

The measured variable $\mathbf{d}$ is replaced by a consistent measured value plus the correction term ε_{d},

$$\mathbf{d} = (\tilde{\mathbf{d}} + \varepsilon_{\mathrm{d}}), \tag{4.23}$$

and a new variable $\boldsymbol{\theta}$ is defined as

$$\boldsymbol{\theta} = \mathbf{V}\varepsilon_{\mathrm{d}}). \tag{4.24}$$

The terms that contain variable $\mathbf{d}$ in Eqs. (4.21) and (4.22) are replaced by

$$\begin{aligned} \mathbf{B}_2\mathbf{V}\mathbf{d} &= \mathbf{B}_2\boldsymbol{\theta} + \mathbf{B}_2\mathbf{V}\tilde{\mathbf{d}} \\ \mathbf{E}_2\mathbf{V}\mathbf{d} &= \mathbf{E}_2\boldsymbol{\theta} + \mathbf{E}_2\mathbf{V}\tilde{\mathbf{d}} \end{aligned} \tag{4.25}$$

In order to display the unmeasured total flow rates of a stream with specific flowrates of category 2, $\mathbf{B}_4$ and $\mathbf{E}_6$ matrices are defined as

$$\begin{aligned} \mathbf{B}_4(\tilde{\mathbf{d}})\mathbf{f}_{\mathrm{U2}} &= \mathbf{B}_2\mathbf{V}\tilde{\mathbf{d}} \\ \mathbf{E}_6(\tilde{\mathbf{d}})\mathbf{f}_{\mathrm{U2}} &= \mathbf{E}_2\mathbf{V}\tilde{\mathbf{d}} \end{aligned} \tag{4.26}$$

In order to group all unmeasured total flowrates, zero columns are added to $\mathbf{B}_4$ and $\mathbf{E}_6$ if necessary. New matrices $\mathbf{B}_5$ and $\mathbf{E}_7$ are obtained such that:

$$\begin{aligned} \mathbf{B}_5(\tilde{\mathbf{d}})\mathbf{f}_{\mathrm{U}} &= \mathbf{B}_2\mathbf{V}\tilde{\mathbf{d}} \\ \mathbf{E}_7(\tilde{\mathbf{d}})\mathbf{f}_{\mathrm{U}} &= \mathbf{E}_2\mathbf{V}\tilde{\mathbf{d}}. \end{aligned} \tag{4.27}$$

The set of component/energy balances and normalization equations after modification of bilinear terms can now be written as

$$\begin{bmatrix} \mathbf{0} & \mathbf{B}_1 & \mathbf{B}_2 & \mathbf{B}_5 & \mathbf{B}_3 \\ \mathbf{E}_4 & \mathbf{E}_1 & \mathbf{E}_2 & \mathbf{E}_8 & \mathbf{E}_3 \end{bmatrix} \begin{bmatrix} \mathbf{f}_{\mathrm{M}} \\ \mathbf{f}_{\mathrm{ch}} \\ \boldsymbol{\theta} \\ \mathbf{f}_{\mathrm{U}} \\ \mathbf{v} \end{bmatrix} = \mathbf{0} \tag{4.28}$$

where $\mathbf{E}_8 = \mathbf{E}_7 + \mathbf{E}_5$.

If we considered the adjustments of total flow rates (ε_{f}) and the component and enthalpy flows ($\varepsilon_{\mathrm{f_{ch}}}$), the previous equations are finally modified as follows:

$$[\mathbf{B}_{11} \quad \mathbf{B}_{22} \quad \mathbf{B}_{33}] \begin{bmatrix} \mathbf{a} \\ \mathbf{f}_{\mathrm{U}} \\ \mathbf{v} \end{bmatrix} = -\begin{bmatrix} \mathbf{0}_1 & \mathbf{B}_1 \\ \mathbf{E}_4 & \mathbf{E}_1 \end{bmatrix} \begin{bmatrix} \tilde{\mathbf{f}}_{\mathrm{M}} \\ \tilde{\mathbf{f}}_{\mathrm{ch}} \end{bmatrix} = \mathbf{e}, \tag{4.29}$$

where

$$\mathbf{a} = \begin{bmatrix} \varepsilon_{\mathrm{f_M}} \\ \varepsilon_{\mathrm{f_{ch}}} \\ \theta \end{bmatrix}, \quad \mathbf{B}_{11} = \begin{bmatrix} \mathbf{0}_1 & \mathbf{B}_1 & \mathbf{B}_2 \\ \mathbf{E}_4 & \mathbf{E}_1 & \mathbf{E}_2 \end{bmatrix}, \quad \mathbf{B}_{22} = \begin{bmatrix} \mathbf{B}_5 \\ \mathbf{E}_8 \end{bmatrix}, \quad \mathbf{B}_{33} = \begin{bmatrix} \mathbf{B}_3 \\ \mathbf{E}_3 \end{bmatrix}. \tag{4.30}$$

Then the general reconciliation problem can be written as

$$\begin{gathered}\min_{\delta,\theta}\left(\varepsilon_{\mathrm{f_M}}^{\mathrm{T}}\boldsymbol{\Psi}_{\mathrm{f_M}}^{-1}\varepsilon_{\mathrm{f_M}}+\varepsilon_{\mathrm{f_{ch}}}^{\mathrm{T}}\boldsymbol{\Psi}_{\mathrm{f_{ch}}}^{-1}\varepsilon_{\mathrm{f_{ch}}}+\boldsymbol{\theta}^{\mathrm{T}}\boldsymbol{\Psi}_{\theta}^{-1}\boldsymbol{\theta}\right)\\ \text{s.t.}\\ [\mathbf{B}_{11}\quad \mathbf{B}_{22}\quad \mathbf{B}_{33}]\begin{bmatrix}\mathbf{a}\\ \mathbf{f}_{\mathrm{U}}\\ \mathbf{v}\end{bmatrix}=-\begin{bmatrix}\mathbf{0}_1 & \mathbf{B}_1\\ \mathbf{E}_4 & \mathbf{E}_1\end{bmatrix}\begin{bmatrix}\tilde{\mathbf{f}}_{\mathrm{M}}\\ \tilde{\mathbf{f}}_{\mathrm{ch}}\end{bmatrix}=\mathbf{e}.\end{gathered} \tag{4.31}$$

$\boldsymbol{\Psi}_{\mathrm{f_M}}$, $\boldsymbol{\Psi}_{\mathrm{f_{ch}}}$, $\boldsymbol{\Psi}_{\theta}$, and $\boldsymbol{\Psi}_{\mathrm{d}}$ are the weighting matrices for $\mathbf{f}_{\mathrm{M}}, \mathbf{f}_{\mathrm{ch}}, \boldsymbol{\theta}$, and $\mathbf{d}$. $\boldsymbol{\Psi}_{\theta}$ is defined as

$$\boldsymbol{\Psi}_{\theta}=\mathbf{V}\boldsymbol{\Psi}_{\mathrm{d}}\mathbf{V}. \tag{4.32}$$

4.3.2. Decomposition Using Orthogonal Transformations

A method for decomposing unmeasured process variables from the measured ones, using the Q-R orthogonal transformation, was discussed before for the linear case. A similar procedure is applied twice in order to resolve the nonlinear reconciliation problem.

Step 1

A Q-R decomposition of $(mb \times nb)$ matrix $\mathbf{B}_{33}$ is accomplished, then

$$\mathbf{B}_{33}\boldsymbol{\Pi}_{\mathrm{v}}=[\mathbf{QB}]\,[\mathbf{RB}]=[\mathbf{QB}_1\quad \mathbf{QB}_2]\begin{bmatrix}\mathbf{RB}_1 & \mathbf{RB}_2\\ \mathbf{0} & \mathbf{0}\end{bmatrix}, \tag{4.33}$$

where $r_{\mathrm{v}}=\mathrm{rank}(\mathbf{RB}_1)$ and $\mathbf{QB}_2^{\mathrm{T}}$ is such that its columns span the null space of $\mathbf{B}_{33}$. That is,

$$\mathbf{QB}_2^{\mathrm{T}}\mathbf{B}_{33}=\mathbf{0}. \tag{4.34}$$

If Eq. (4.26) is multiplied by $\mathbf{QB}_2^{\mathrm{T}}$, the unmeasured variables $\mathbf{v}$ are eliminated, and the process constraints are defined as

$$\mathbf{QB}_2^{\mathrm{T}}\mathbf{B}_{11}\mathbf{a}+\mathbf{QB}_2^{\mathrm{T}}\mathbf{B}_{22}\mathbf{f}_{\mathrm{u}}=\mathbf{QB}_2^{\mathrm{T}}\mathbf{e}. \tag{4.35}$$

Step 2

A new $(md \times nd)$ matrix $\mathbf{D}$ is defined as

$$\mathbf{D}=\mathbf{QB}_2^{\mathrm{T}}\cdot\mathbf{B}_{22}. \tag{4.36}$$

Then Eq. (4.35) is rewritten as

$$\mathbf{QB}_2^{\mathrm{T}}\mathbf{B}_{11}\mathbf{a}+\mathbf{D}\mathbf{f}_{\mathrm{U}}=\mathbf{QB}_2^{\mathrm{T}}\mathbf{e}. \tag{4.37}$$

A Q-R orthogonal transformation is performed on matrix $\mathbf{D}$:

$$\mathbf{D}\boldsymbol{\Pi}_{\mathrm{f_U}}=[\mathbf{QD}]\,[\mathbf{RD}]=[\mathbf{QD}_1\quad \mathbf{QD}_2]\begin{bmatrix}\mathbf{RD}_1 & \mathbf{RD}_2\\ \mathbf{0} & \mathbf{0}\end{bmatrix}, \tag{4.38}$$

where $rf=\mathrm{rank}(\mathbf{RD}_1)$ and $\mathbf{QD}_2^{\mathrm{T}}$ spans the column space of $\mathbf{D}$. Then the process constraints can be reduced to

$$\mathbf{QD}_2^{\mathrm{T}}\mathbf{QB}_2^{\mathrm{T}}\mathbf{B}_{11}\mathbf{a}=\mathbf{G}_{\mathrm{a}}\mathbf{a}=\mathbf{QD}_2^{\mathrm{T}}\mathbf{QB}_2^{\mathrm{T}}\mathbf{e}. \tag{4.39}$$

All unmeasured variables are eliminated from the constraints by using simple Q-R transformations and a linear reconciliation problem results. The zero columns of $\mathbf{G}_{\mathrm{a}}$

correspond to nonredundant measurements. The remaining ones are associated with redundant measured variables.

4.3.3. Reconciliation of Measured Variables and Estimation of Unmeasured Total Flowrates

After eliminating unmeasured variables, the reconciliation of measured variables and the estimation of unmeasured total flowrates are accomplished, again by an iterative procedure.

Step I

(a) Using an estimation of unmeasured total flow rates, the weighting matrix for variable θ is evaluated.

(b) The following linear reconciliation problem needs to be resolved:

$$\begin{aligned} &\min_{w} \mathbf{a}^{\mathrm{T}} \boldsymbol{\Psi}_{\mathrm{a}}^{-1} \mathbf{a} \\ &\text{s.t.} \quad \mathbf{G}_{\mathrm{a}} \mathbf{a} = \mathbf{b}, \end{aligned} \tag{4.40}$$

where

$$\begin{aligned} \mathbf{G}_{\mathrm{a}} &= \mathbf{Q}\mathbf{D}_2^{\mathrm{T}} \mathbf{Q}\mathbf{B}_2^{\mathrm{T}} \mathbf{B}_{11} \\ \mathbf{b} &= \mathbf{Q}\mathbf{D}_2^{\mathrm{T}} \mathbf{Q}\mathbf{B}_2^{\mathrm{T}} \mathbf{e}, \end{aligned} \tag{4.41}$$

with the solution given by

$$\hat{\mathbf{a}} = \boldsymbol{\Psi}_{\mathrm{a}} \mathbf{G}_{\mathrm{a}}^{\mathrm{T}} \left(\mathbf{G}_{\mathrm{a}} \boldsymbol{\Psi}_{\mathrm{a}} \mathbf{G}_{\mathrm{a}}^{\mathrm{T}} \right)^{-1} \mathbf{b}. \tag{4.42}$$

Step 2

The estimation of unmeasured total flow rates is done by using the Q-R orthogonal decomposition of matrix $\mathbf{D}$. The equation can be written as

$$\mathbf{Q}\mathbf{B}_2^{\mathrm{T}} \mathbf{B}_{11} \hat{\mathbf{a}} + [\mathbf{Q}\mathbf{D}_1 \quad \mathbf{Q}\mathbf{D}_2] \begin{bmatrix} \mathbf{R}\mathbf{D}_1 & \mathbf{R}\mathbf{D}_2 \\ \mathbf{0} & \mathbf{0} \end{bmatrix} \begin{bmatrix} \mathbf{f}_{\mathrm{U_{rf}}} \\ \mathbf{f}_{\mathrm{U_{nd-rf}}} \end{bmatrix} = \mathbf{Q}\mathbf{B}_2^{\mathrm{T}} \mathbf{e}, \tag{4.43}$$

where

$$\begin{bmatrix} \mathbf{f}_{\mathrm{U_{rf}}} \\ \mathbf{f}_{\mathrm{U_{nd-rf}}} \end{bmatrix} = \boldsymbol{\Pi}_{\mathrm{f_U}}^{\mathrm{T}} \mathbf{f}_{\mathrm{U}}. \tag{4.44}$$

The subset $\mathbf{f}_{\mathrm{U_{nd-rf}}}$ corresponds to the indeterminable total flowrates. Regarding the subset $\mathbf{f}_{\mathrm{U_{rf}}}$, nothing can be said, since some of these variables can be calculated directly from the measurements and some depend on $\mathbf{f}_{\mathrm{U_{nd-rf}}}$, as is explained in Remark 3.

Further information for classifying the variables in subset $\mathbf{f}_{\mathrm{U_{rf}}}$ can be obtained by premultiplying Eq. (4.37) by $\mathbf{QD}^{\mathrm{T}}$ and writing the vector $\mathbf{f}_{\mathrm{U_{rf}}}$ in terms of the other variables:

$$\begin{bmatrix} \mathbf{Q}\mathbf{D}_1^{\mathrm{T}} \mathbf{Q}\mathbf{B}_2^{\mathrm{T}} \mathbf{B}_{11} & \mathbf{R}\mathbf{D}_1 & \mathbf{R}\mathbf{D}_2 \\ \mathbf{Q}\mathbf{D}_2^{\mathrm{T}} \mathbf{Q}\mathbf{B}_2^{\mathrm{T}} \mathbf{B}_{11} & \mathbf{0} & \mathbf{0} \end{bmatrix} \begin{bmatrix} \mathbf{a} \\ \mathbf{f}_{\mathrm{U_{rf}}} \\ \mathbf{f}_{\mathrm{U_{nd-rf}}} \end{bmatrix} = \begin{bmatrix} \mathbf{Q}\mathbf{D}_1^{\mathrm{T}} \mathbf{Q}\mathbf{B}_2^{\mathrm{T}} \mathbf{e} \\ \mathbf{Q}\mathbf{D}_2^{\mathrm{T}} \mathbf{Q}\mathbf{B}_2^{\mathrm{T}} \mathbf{e} \end{bmatrix} \tag{4.45}$$

$$\mathbf{f}_{\mathrm{u_{rf}}} = \mathbf{R}\mathbf{D}_1^{-1} \mathbf{Q}\mathbf{D}_1^{\mathrm{T}} \mathbf{Q}\mathbf{B}_2^{\mathrm{T}} \mathbf{e} - \mathbf{R}\mathbf{D}_1^{-1} \mathbf{Q}\mathbf{D}_1^{\mathrm{T}} \mathbf{Q}\mathbf{B}_2^{\mathrm{T}} \mathbf{B}_{11} \hat{\mathbf{a}} - \mathbf{R}\mathbf{D}_1^{-1} \mathbf{R}\mathbf{D}_2 \mathbf{f}_{\mathrm{U_{nd-rf}}}. \tag{4.46}$$

Note that if the last term in the RHS of Eq. (4.46) is zero, all of $\mathbf{f}_{\mathrm{U_{rf}}}$ can be calculated from the available information.

In order to classify the variables in $\mathbf{f}_{U_{rf}}$, a matrix $\mathbf{R}_{IF}$ is defined as

$$\mathbf{R}_{IF} = \mathbf{RD}_1^{-1}\mathbf{RD}_2, \tag{4.47}$$

and the following can be stated:

1. A variable in subset $\mathbf{f}_{U_{rf}}$ is estimable if the corresponding row in $\mathbf{R}_{IF}$ is zero.
2. A variable in subset $\mathbf{f}_{U_{rf}}$ is nonestimable otherwise.

At this point the vector $\mathbf{f}_U$ can be divided into

$$\mathbf{f}_U = \begin{bmatrix} \mathbf{f}_{Ud} \\ \mathbf{f}_{Ui} \end{bmatrix}, \tag{4.48}$$

where

$\mathbf{f}_{Ud}$ is the fe-dimensional vector of determinable total flowrates $(fe \leq rf)$
$\mathbf{f}_{Ui}$ is the $(nd - fe)$-dimensional vector of indeterminable total flowrates

$\mathbf{f}_{Ud}$ contains the fe variables in subset $f_{U_{rf}}$ that satisfy condition (1), whereas $\mathbf{f}_{Ui}$ includes those that satisfy condition (2) plus the variables in subset $\mathbf{f}_{U_{nd-rf}}$.

After updating the values of determinable total flowrates, the procedure is reinitiated until convergence is achieved.

Estimation of Vector v

In order to estimate the unmeasured variables contained in $\mathbf{v}$, the matrix $\mathbf{B}_{22}$ is divided into two parts by column permutation. The first fe columns correspond to the determinable total flowrates, and the $(nd - fe)$ remaining ones belong to indeterminable total flowrates:

$$\mathbf{B}_{22} = [\mathbf{B}_{2d} \quad \mathbf{B}_{2i}]. \tag{4.49}$$

The set of process constraints (4.29) is then expressed as

$$[\mathbf{B}_{11} \quad \mathbf{B}_{2d} \quad \mathbf{B}_{2i} \quad \mathbf{B}_{33}] \begin{bmatrix} \hat{\mathbf{a}} \\ \mathbf{f}_{Ud} \\ \mathbf{f}_{Ui} \\ \mathbf{v} \end{bmatrix} = \mathbf{e}, \tag{4.50}$$

where $\mathbf{f}_{Ud}$ and $\mathbf{f}_{Ui}$ are obtained by the solution of the previous subproblem.

Using the Q-R decomposition of matrix $\mathbf{B}_{33}$, the set of constraints (4.50) is rewritten as

$$\mathbf{B}_{11}\hat{\mathbf{a}} + \mathbf{B}_{2d}\mathbf{f}_{Ud} + \mathbf{B}_{2i}\mathbf{f}_{Ui} + [\mathbf{QB}_1 \quad \mathbf{QB}_2] \begin{bmatrix} \mathbf{RB}_1 & \mathbf{RB}_2 \\ \mathbf{0} & \mathbf{0} \end{bmatrix} \begin{bmatrix} \mathbf{v}_{r_v} \\ \mathbf{v}_{nb-r_v} \end{bmatrix} = \mathbf{e}, \tag{4.51}$$

where

$$\mathbf{\Pi}_v^T \mathbf{v} = \begin{bmatrix} \mathbf{v}_{r_v} \\ \mathbf{v}_{nb-r_v} \end{bmatrix}. \tag{4.52}$$

Consequently,

$$\begin{aligned} \mathbf{v}_{rv} = {} & \mathbf{RB}_1^{-1}\mathbf{QB}_1^T\mathbf{e} - \mathbf{RB}_1^{-1}\mathbf{QB}_1^T\mathbf{B}_{11}\hat{\mathbf{a}} - \mathbf{RB}_1^{-1}\mathbf{QB}_1^T\mathbf{B}_{2d}\mathbf{f}_{Ud} \\ & - \mathbf{RB}_1^{-1}\mathbf{QB}_1^T\mathbf{B}_{2i}\mathbf{f}_{Ui} - \mathbf{RB}_1^{-1}\mathbf{RB}_2\mathbf{v}_{nb-r_v} \end{aligned} \tag{4.53}$$

The first three terms of the previous equation are known, so if the last ones are zero, all the variables in $\mathbf{v}_{r_v}$ can be evaluated using the available information. In order to classify the variables in $\mathbf{v}_{r_v}$, two new matrixes are defined,

$$\mathbf{R}_{\mathrm{IV}} = \mathbf{RB}_1^{-1}\mathbf{RB}_2 \tag{4.54}$$

$$\mathbf{R}_{\mathrm{IFi}} = \mathbf{RB}_1^{-1}\mathbf{QB}_1^{\mathrm{T}}\mathbf{B}_{2i}, \tag{4.55}$$

and the following can be stated:

1. A variable in subset v_{r_v} is estimable if the corresponding rows of $\mathbf{R}_{\mathrm{IV}}$ and $\mathbf{R}_{\mathrm{IFi}}$ are zero.
2. A variable in subset v_{r_v} is nonestimable otherwise.

At this time the vector $\mathbf{v}$ can be divided into

$$\mathbf{v} = \begin{bmatrix} \mathbf{v}_d \\ \mathbf{v}_i \end{bmatrix}, \tag{4.56}$$

where

$\mathbf{v}_d$ is the ve-dimensional vector of determinable variables in $\mathbf{v}$ $(ve \leq rv)$
$\mathbf{v}_i$ is the $(nb - ve)$-dimensional vector of indeterminable variables in $\mathbf{v}$

$\mathbf{v}_d$ contains the ve variables in subset $\mathbf{v}_{r_v}$ that satisfy condition (1), whereas $\mathbf{v}_i$ includes those that satisfy condition (2) plus the variables in subset $\mathbf{v}_{nb-r_v}$.

Unmeasured temperatures or concentrations that correspond to enthalpy or component flows in $\mathbf{v}_d$ are determinable if the total flow rate of the stream is measured. Otherwise, they are indeterminable. Measured total flow rates are nonredundant and unmeasured total flow rates are indeterminable. The analysis of intensive constraints between variables may change previous classification.

4.3.4. Energy Balances in the Context of the Projection Approach

Throughout the preceding discussion, simplified expressions of stream-specific enthalpy as a function of temperature are used. They have to be updated during process operation to consider changes in steady-state compositions.

The application of a more precise expression for enthalpy, at least as a function of temperature and composition, requires a new categorization of enthalpy flowrates. They can be divided into three categories depending on the combination of total flowrate, composition, and temperature measurements, as indicated in Table 2.

The problem arises for the last combination of measurements. It is due to the difficulty of adjusting temperature measurement values for streams whose compositions

TABLE 2 Categories of Enthalpy Flowrates (from Sánchez and Romagnoli, 1986)

Category	f	t	M
1	M	M	M
2	U	M	M
3	M/U	U	M
3	M/U	M	U

are unmeasured or partially measured. In this context, the temperature of a stream j may be adjusted only for the following conditions:

1. All component molar fractions are unmeasured
2. Rule of mixing: $h_j = \sum M_{jc} H_{jc}$
3. H_{jc} is approximated as a linear function of temperature for the steady-state operation range

The following resolution scheme can be implemented:

1. Estimation of unmeasured total flowrates and unmeasured species flowrates for streams with unmeasured temperatures
2. Simultaneous elimination of unmeasured variables; a two-stage decomposition procedure is not advantageous because measurements are involved in Category 3 flowrates
3. Least squares adjustment of measurements
4. Estimation of unmeasured variables
5. Iteration until convergence is achieved

Hence, factorization methods can only be applied to solve particular cases of data reconciliation when energy balances are considered. Other equation-oriented techniques, such as PLADAT (Sánchez *et al.*, 1992), perform better in tackling the more general problem.

EXAMPLE 4.5

Let us consider the well-known Vaclavek's example (Vaclavek and Loucka, 1976); the process flowsheet and measured variables are presented in Fig. 5 and Table 3, respectively. It consists of six units and four components. In this case, the concentrations of a stream are assumed to be either all measured or all unmeasured.

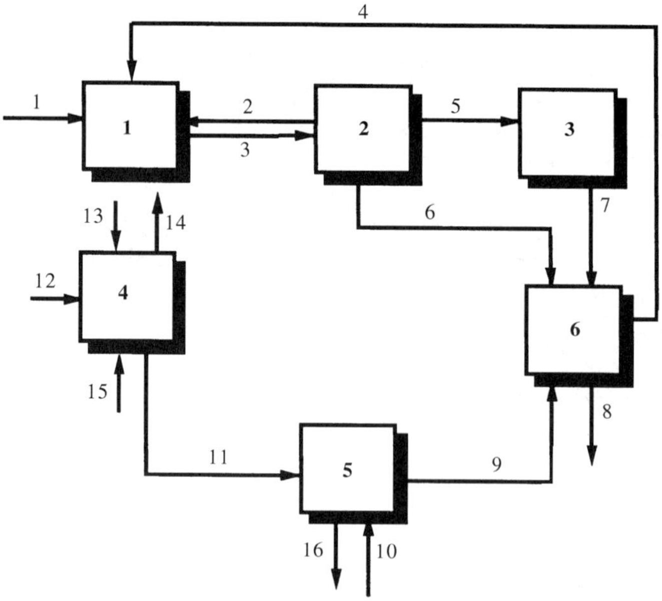

FIGURE 5 Process flowsheet for Example 4.5 (from Vaclavek and Loucka, 1976).

TABLE 3 Measurements for Example 4.5 (from Vaclavek and Loucka, 1976)

Measurement	Stream
Total flowrates	1 2 3 4 6 8 9 12 16
Composition	1 4 5 8 9 10 11 12 13 14 15

The normalization constraints are taken into account by eliminating the measured concentration of one component and calculating this concentration by difference after reconciliation. There are 24 process constraints. Six equations belong to total mass balances and 18 correspond to component mass balances. Enthalpy balances are not included.

The set of process constraints for Vaclavek's example is

$$\begin{bmatrix} \mathbf{A}_1 & \mathbf{O}_1 & \mathbf{O}_2 & \mathbf{A}_2 & \mathbf{O}_3 \\ \mathbf{O}_4 & \mathbf{B}_1 & \mathbf{B}_2 & \mathbf{B}_5 & \mathbf{B}_3 \end{bmatrix} \begin{bmatrix} \mathbf{f}_M \\ \mathbf{f}_{ch} \\ \boldsymbol{\theta} \\ \mathbf{f}_U \\ \mathbf{v} \end{bmatrix} = \mathbf{0},$$

where

$$\mathbf{B}_{11} = \begin{bmatrix} \mathbf{A}_1 & \mathbf{O}_1 & \mathbf{O}_2 \\ \mathbf{O}_4 & \mathbf{B}_1 & \mathbf{B}_2 \end{bmatrix}, \quad \mathbf{B}_{22} = \begin{bmatrix} \mathbf{A}_2 \\ \mathbf{B}_5 \end{bmatrix}, \quad \mathbf{B}_{33} = \begin{bmatrix} \mathbf{O}_3 \\ \mathbf{B}_3 \end{bmatrix}$$

$$\mathbf{A}_1 = \begin{bmatrix} 1 & 1 & -1 & 1 & 0 & 0 & 0 & 0 & 0 \\ 0 & -1 & 1 & 0 & -1 & 0 & 0 & 0 & 0 \\ 0 & 0 & 0 & 0 & 0 & 0 & 0 & 0 & 0 \\ 0 & 0 & 0 & 0 & 0 & 0 & 0 & 1 & 0 \\ 0 & 0 & 0 & 0 & 0 & 0 & -1 & 0 & -1 \\ 0 & 0 & 0 & -1 & 1 & -1 & 1 & 0 & 0 \end{bmatrix}, \quad \mathbf{f}_M = \begin{bmatrix} f_1 \\ f_2 \\ f_3 \\ f_4 \\ f_6 \\ f_8 \\ f_9 \\ f_{12} \\ f_{16} \end{bmatrix}, \quad \mathbf{f}_U = \begin{bmatrix} f_5 \\ f_{10} \\ f_{11} \\ f_{13} \\ f_{14} \\ f_{15} \\ f_7 \end{bmatrix}$$

$$\mathbf{A}_2 = \begin{bmatrix} 0 & 0 & 0 & 0 & 0 & 0 & 0 \\ -1 & 0 & 0 & 0 & 0 & 0 & 0 \\ 1 & 0 & 0 & 0 & 0 & 0 & -1 \\ 0 & 0 & -1 & 1 & -1 & 1 & 0 \\ 0 & 1 & 1 & 0 & 0 & 0 & 0 \\ 0 & 0 & 0 & 0 & 0 & 0 & 1 \end{bmatrix}, \quad \mathbf{B}_1 = \begin{bmatrix} \mathbf{I} & \mathbf{I} & \mathbf{O} & \mathbf{O} & \mathbf{O} \\ \mathbf{O} & \mathbf{O} & \mathbf{O} & \mathbf{O} & \mathbf{O} \\ \mathbf{O} & \mathbf{O} & \mathbf{O} & \mathbf{O} & \mathbf{O} \\ \mathbf{O} & \mathbf{O} & \mathbf{O} & \mathbf{O} & \mathbf{I} \\ \mathbf{O} & \mathbf{O} & \mathbf{O} & -\mathbf{I} & \mathbf{O} \\ \mathbf{O} & -\mathbf{I} & -\mathbf{I} & \mathbf{I} & \mathbf{O} \end{bmatrix}$$

$$\mathbf{B}_2 = \begin{bmatrix} \mathbf{O} & \mathbf{O} & \mathbf{O} & \mathbf{O} & \mathbf{O} & \mathbf{O} \\ -\mathbf{I} & \mathbf{O} & \mathbf{O} & \mathbf{O} & \mathbf{O} & \mathbf{O} \\ \mathbf{I} & \mathbf{O} & \mathbf{O} & \mathbf{O} & \mathbf{O} & \mathbf{O} \\ \mathbf{O} & \mathbf{O} & -\mathbf{I} & \mathbf{I} & -\mathbf{I} & \mathbf{I} \\ \mathbf{O} & \mathbf{I} & \mathbf{I} & \mathbf{O} & \mathbf{O} & \mathbf{O} \\ \mathbf{O} & \mathbf{O} & \mathbf{O} & \mathbf{O} & \mathbf{O} & \mathbf{O} \end{bmatrix}, \quad \mathbf{B}_3 = \begin{bmatrix} \mathbf{I} & -\mathbf{I} & \mathbf{O} & \mathbf{O} & \mathbf{O} \\ -\mathbf{I} & \mathbf{I} & -\mathbf{I} & \mathbf{O} & \mathbf{O} \\ \mathbf{O} & \mathbf{O} & \mathbf{O} & -\mathbf{I} & \mathbf{O} \\ \mathbf{O} & \mathbf{O} & \mathbf{O} & \mathbf{O} & \mathbf{O} \\ \mathbf{O} & \mathbf{O} & \mathbf{O} & \mathbf{O} & -\mathbf{I} \\ \mathbf{O} & \mathbf{O} & \mathbf{I} & \mathbf{I} & \mathbf{O} \end{bmatrix}$$

$$\mathbf{I} = \begin{bmatrix} 1 & 0 & 0 \\ 0 & 1 & 0 \\ 0 & 0 & 1 \end{bmatrix}, \quad \mathbf{O} = \begin{bmatrix} 0 & 0 & 0 \\ 0 & 0 & 0 \\ 0 & 0 & 0 \end{bmatrix}$$

$$\mathbf{B}_5 = \begin{bmatrix} \mathbf{o} & \mathbf{o} & \mathbf{o} & \mathbf{o} & \mathbf{o} & \mathbf{o} & \mathbf{o} \\ -\mathbf{d}_5 & \mathbf{o} & \mathbf{o} & \mathbf{o} & \mathbf{o} & \mathbf{o} & \mathbf{o} \\ \mathbf{d}_5 & \mathbf{o} & \mathbf{o} & \mathbf{o} & \mathbf{o} & \mathbf{o} & \mathbf{o} \\ \mathbf{o} & \mathbf{o} & -\mathbf{d}_{11} & \mathbf{d}_{13} & -\mathbf{d}_{14} & -\mathbf{d}_{15} & \mathbf{o} \\ \mathbf{o} & \mathbf{d}_{10} & \mathbf{d}_{11} & \mathbf{o} & \mathbf{o} & \mathbf{o} & \mathbf{o} \\ \mathbf{o} & \mathbf{o} & \mathbf{o} & \mathbf{o} & \mathbf{o} & \mathbf{o} & \mathbf{o} \end{bmatrix}, \quad \mathbf{o} = \begin{bmatrix} 0 \\ 0 \\ 0 \end{bmatrix},$$

where $\mathbf{O}_1, \mathbf{O}_2, \mathbf{O}_3, \mathbf{O}_4$ are compatible matrixes of zeros.

$$\mathbf{f}_{\mathrm{ch}}^{\mathrm{T}} = [f_{1,1} \quad f_{1,2} \quad f_{1,3} \quad f_{4,1} \quad f_{4,2} \quad f_{4,3} \quad f_{8,1} \quad f_{8,2} \quad f_{8,3} \quad f_{9,1} \quad f_{9,2} \quad f_{9,3} \quad f_{12,1} \quad f_{12,2} \quad f_{12,3}]$$

$$\boldsymbol{\theta}^T = [\theta_{5,1} \quad \theta_{5,2} \quad \theta_{5,3} \quad \theta_{10,1} \quad \theta_{10,2} \quad \theta_{10,3} \quad \theta_{11,1} \quad \theta_{11,2} \quad \theta_{11,3} \quad \theta_{13,1} \quad \theta_{13,2} \quad \theta_{13,3} \quad \theta_{14,1} \quad \theta_{14,2} \quad \theta_{14,3} \quad \theta_{15,1} \quad \theta_{15,2} \quad \theta_{15,3}]$$

$$\mathbf{v}^{\mathrm{T}} = [\upsilon_{2,1} \quad \upsilon_{2,2} \quad \upsilon_{2,3} \quad \upsilon_{3,1} \quad \upsilon_{3,2} \quad \upsilon_{3,3} \quad \upsilon_{61} \quad \upsilon_{62} \quad \upsilon_{63} \quad \upsilon_{71} \quad \upsilon_{72} \quad \upsilon_{73} \quad \upsilon_{16,1} \quad \upsilon_{16,2} \quad \upsilon_{16,3}].$$

In classifying the measured and unmeasured process variables the following procedure is applied:

1. A Q-R decomposition is performed on matrix $\mathbf{B}_{33}$, and $\mathbf{QB}_1, \mathbf{QB}_2, \mathbf{RB}_1, \mathbf{RB}_2, \mathbf{\Pi}_{\mathrm{v}}$ matrices are obtained.

2. Matrix $\mathbf{D}$ is calculated and the application of the orthogonal projection approach leads to $\mathbf{QD}_1, \mathbf{QD}_2, \mathbf{RD}_1, \mathbf{RD}_2, \mathbf{\Pi}_{\mathrm{fU}}$ matrices.

3. With the calculation of matrix $\mathbf{G}_{\mathrm{a}}$ [using Ex. (4.41)], all the necessary information for the measured variables classification is available. The zero columns of matrix $\mathbf{G}_{\mathrm{a}}$ correspond to $f_6, f_{12}, f_{16}, f_{4,1}, f_{4,2}, f_{4,3}, f_{12,1}, f_{12,2}, f_{12,3}$ and all the elements of $\boldsymbol{\theta}$. Consequently, the nonredundant measured variables are:

   (a) f_6, f_{12}, f_{16} total flowrates
   (b) All concentrations that belong to the streams in θ and the concentrations of streams 4 and 12

4. As $\mathbf{RD}_2$ is an empty matrix, the subset $\mathbf{f}_{\mathrm{U_{nd-rf}}}$ is also empty, so all the unmeasured total flowrates are determinable, as can be seen from Eq. (4. 46). The subsets $\mathbf{v}_{\mathrm{r_v}}, \mathbf{v}_{\mathrm{nb-r_v}}$ and matrices $\mathbf{R}_{\mathrm{IB}}, \mathbf{R}_{\mathrm{IFi}}$ are then obtained:

$$\mathbf{v}_{\mathrm{r_v}} = [\upsilon_{2,1}, \upsilon_{2,2}, \upsilon_{2,3}, \upsilon_{7,1}, \upsilon_{7,2}, \upsilon_{7,3}, \upsilon_{6,1}, \upsilon_{6,2}, \upsilon_{6,3}, \upsilon_{16,1}, \upsilon_{16,2}, \upsilon_{16,3}]$$

$$\mathbf{v}_{\mathrm{nb-r_v}} = [\upsilon_{3,1}, \upsilon_{3,2}, \upsilon_{3,3}].$$

Matrix $\mathbf{R}_{\mathrm{IB}}$ is, in this case,

$$\mathbf{R}_{\mathrm{IB}} = \begin{bmatrix} -\mathbf{I} \\ \mathbf{O} \\ \mathbf{O} \\ \mathbf{O} \end{bmatrix},$$

and $\mathbf{R}_{\text{IFi}}$ is empty. As the first three rows of $\mathbf{R}_{\text{IB}}$ are nonzero, the following $\mathbf{v}_{\text{d}}$ and $\mathbf{v}_{\text{i}}$ vectors result:

$$\mathbf{v}_{\text{d}} = [v_{7,1}, v_{7,2}, v_{7,3}, v_{6,1}, v_{6,2}, v_{6,3}, v_{16,1}, v_{16,2}, v_{16,3}]$$
$$\mathbf{v}_{\text{i}} = [v_{2,1}, v_{2,2}, v_{2,3}, v_{3,1}, v_{3,2}, v_{3,3}].$$

As all total flowrates are measured or determinable, the concentrations of streams 6, 7, and 16 are determinable, while those of streams 2 and 3 are indeterminable.

4.4. CONCLUSIONS

In this chapter, the use of projection matrix techniques, more precisely the Q-R factorization, to analyze, decompose, and solve the linear and bilinear data reconciliation problem was discussed. This type of transformation is selected because it provides a very good balance of numerical accuracy, flexibility, and computational cost (Goodall, 1993).

An extension of the strategy for bilinear systems was also discussed, which allows us to separate total flowrates from specific flowrates. It has two advantages:

1. A sequence of simple expressions for application in instrumentation analysis and data reconciliation were obtained using subproducts of Q-R factorizations
2. Assumptions are avoided for the adjustment of total flowrates

Q-R factorization is successful in decomposing linear systems of equations. It is also satisfactory when bilinear systems contain component balances and normalization equations. If energy balances are included in the set of process constraints, the procedure has the drawback that only simple thermodynamic relations for the specific enthalpy of the stream can be considered.

NOTATION

$\mathbf{a}$	vector defined by Eq. (4.30)
$\mathbf{A}$	a general matrix $(m \times n)$
$\mathbf{A}_1$	matrix for measured variables $(m \times g)$
$\mathbf{A}_2$	matrix for unmeasured variables $(m \times n)$
$\mathbf{A}_3$	matrix defined by Eq. (4.3)
$\mathbf{A}_4$	matrix defined by Eq. (4.4)
$\mathbf{B}_i$	matrices for component/enthalpy balances
$\mathbf{B}_{ii}$	matrices defined by Eq. (4.30), $ii = 1, 3$
$\mathbf{b}$	vector defined by Eq. (4.41)
$\mathbf{d}$	measured molar fractions and specific enthalpies vector
$\mathbf{D}_{\text{i}}$	diagonal matix defined by Eq. (4.8)
$\mathbf{D}$	matrix defined by Eq. (4.36)
$\mathbf{E}_{\text{i}}$	matrices for normalization equations
$\mathbf{e}$	vector defined by Eq. (4.29)
$\mathbf{f}$	vector of total flowrate
$\mathbf{f}_{\text{ch}}$	vector of specific flowrates of Category 1
$\mathbf{f}_{\text{U}_{\text{rf}}}, \mathbf{f}_{\text{U}_{\text{nd-rf}}}$	partitions of $\mathbf{f}_{\text{U}}$
$\mathbf{G}$	matrix defined by Eq. (4.7) $[(m - r_u) \times g]$
$\mathbf{G}_{\text{X}}$	matrix defined by Eq. (4.18) $[(m - r_u) \times g]$

$\mathbf{G}_a$	matrix defined by Eq. (4.41)
$\mathbf{h}$	vector of stream enthalpies
$\mathbf{H}$	matrix of specific component enthalpies
$\mathbf{M}$	matrix of mass fractions
m_b, m_d	number of files of $\mathbf{B}_{33}$ and $\mathbf{D}$, respectively
n_b, n_d	number of columns of $\mathbf{B}_{33}$ and $\mathbf{D}$, respectively
$\mathbf{O}$	zero matrix
$\mathbf{P}$	projection matrix
$[\mathbf{Q}, \mathbf{R}, \mathbf{\Pi}]$	QR(**A**)
$[\mathbf{Q}_u, \mathbf{R}_u, \mathbf{\Pi}_u]$	QR($\mathbf{A}_2$)
$[\mathbf{QB}, \mathbf{RB}, \mathbf{\Pi}_v]$	QR($\mathbf{B}_{33}$)
$[\mathbf{QD}, \mathbf{RD}, \mathbf{\Pi}_{f_U}]$	QR(**A**)
r	rank(**D**)
r_u	rank($\mathbf{A}_2$)
r_v	rank($\mathbf{B}_{33}$)
rf	rank(**D**)
$\mathbf{R}_{IU}$	matrix defined by Eq. (4.20) $[r_u \times (n - r_u)]$
$\mathbf{R}_{IF}, \mathbf{R}_{IV}, \mathbf{R}_{IFi}$	inspection matrices
$\mathbf{t}$	vector of stream temperatures
$\mathbf{u}$	vector of unmeasured variables $(n \times 1)$
$\mathbf{u}_{r_u}, \mathbf{u}_{n-r_u}$	partitions of **u**
$\mathbf{v}$	vector of specific flowrates in Category 3
$\mathbf{v}_{r_v}, \mathbf{v}_{nb-r_v}$	partitions of **v**
$\mathbf{V}$	diagonal matrix of $\mathbf{f}_{U2}$
$\mathbf{x}$	vector of measured variables $(g \times 1)$
$\mathbf{X}$	matrix defined by Eq. (4.3)
$\mathbf{X}_1$	matrix defined by Eq. (4.4)
$\mathbf{X}_2$	matrix defined by Eq. (4.4)

Greek

ε_i	vector of i random errors
θ	vector defined by Eq. (4.24)
Ψ_i	variance-covariance matrix of i

Superscripts

$\sim$	with measured values
$\wedge$	with reconciled values

Subscripts

M, U	measured or unmeasured variable
d, i	determinable or indeterminable variable

REFERENCES

Crowe, C. M. (1986). Reconciliation of process flow rates by matrix projection. Part II: The nonlinear case. *AIChE J.* **32**, 616–623.

Crowe, C. M. (1989). Observability and redundancy of process data for steady state reconciliation. *Chem. Eng. Sci.* **44**, 2909–2917.

Crowe, C. M., García Campos, Y. A., and Hrymak, A. (1983). Reconciliation of process flow rates by matrix projection. Part I: Linear case. *AIChE J.* **29**, 881–888.

Dahlquist, G., and Bjork, A. (1974). "Numerical Methods." Prentice-Hall, Englewood Cliffs. NJ.

Goodall, C. R. (1993). "Computation Using the Q-R Decomposition Computational Statistics." Elsevier, Amsterdam.

Sánchez, M., and Romagnoli, J. (1996). Use of orthogonal transformations in data classification—reconciliation. *Comput. Chem. Eng.* **20**, 483–493.

Sánchez, M., Bandoni, A., and Romagnoli, J. (1992). PLADAT—A package for process variable classification and plant data reconciliation. *Comput. Chem. Eng.* **S16**, 499–506.

Swartz, C. L. E. (1989). Data reconciliation for generalised flowsheet applications. *197th Natl. Meet. Am. Chem. Soc.*, Dallas, TX. 1989.

Václavek, V., and Loucka, M. (1976). Selection of measurements necessary to achieve multicomponent mass balances in chemical plants. *Chem. Eng. Sci.* **31**, 1199–1205.

5 STEADY-STATE DATA RECONCILIATION

In this chapter we concentrate on the statement and further solution of the general steady-state data reconciliation problem. Initially, we analyze its resolution for linear plant models, and then the nonlinear case is discussed.

The use of Q-R orthogonal factorizations is presented as an alternative methodology for performing data reconciliation for bilinear systems. Finally, we briefly describe current techniques for tackling the general nonlinear problem.

5.1. INTRODUCTION

Reliable process data are the key to the efficient operation of chemical plants. With the increasing use of on-line digital computers, numerous data are acquired and used for on-line optimization and control. Frequently these activities are based on small improvements in process performance, but it must be noted that errors in process data, or inaccurate and unreliable methods of resolving these errors, can easily exceed or mask actual changes in process performance.

Inadequate knowledge of the process models and poor estimation of process parameters (physical properties, processing constants, etc.) mean that any technique for correcting the measurements should rely on simple, well-known, and indubitable process relationships, which should be satisfied independent of the measurements' accuracy. Such relationships are the multicomponent mass and energy balances.

In chemical engineering, Kuehn and Davidson (1961) were the first to publish an analysis of data reconciliation. Since then, many articles on this subject have appeared in the literature. General reviews of data reconciliation have been published by Hlavacek (1977), Tamhane and Mah (1985), Mah (1982, 1990), Madron (1992), and Crowe (1996).

5.2. PROBLEM FORMULATION

Process measurements are subject to errors. These errors give rise to discrepancies in material and energy balances.

DEFINITION 5.1
Data reconciliation is the process of adjusting or reconciling the process measurements to obtain more accurate estimates of flowrates, temperatures, compositions, etc., that are consistent with material and energy balances.

Let us first define the models to be used in our formulation of the data reconciliation problem.

DEFINITION 5.2 (MEASUREMENTS)
In the absence of gross errors, the measurement vector can be written as

$$\mathbf{y} = \mathbf{x} + \varepsilon, \quad \mathbf{y} \in \Re^g, \quad \mathbf{x} \in \Re^g, \tag{5.1}$$

where $\mathbf{y}$ is the $(g \times 1)$ measurement vector, $\mathbf{x}$ is the $(g \times 1)$ vector of true values of variables, and ε stands for the vector of random measurement errors. The following assumptions are usually made:

1. The expected value of ε is the null vector, i.e., $E(\varepsilon) = \mathbf{0}$
2. The successive vectors of measurements are independent, i.e., $E(\varepsilon_i \varepsilon_j^{\mathrm{T}}) = \mathbf{0}$, for $i \neq j$
3. The covariance matrix of the measurement errors is known and positive definite, i.e., $\mathrm{Cov}(\varepsilon) = \Psi = E(\varepsilon_i \varepsilon_i^{\mathrm{T}})$

Note that $\mathbf{x}$ and $\mathbf{y} \in \Re^g$, that is, we are assuming that all process variables are measured.

DEFINITION 5.3 (CONSTRAINTS)
Additional information must be introduced through the process model equations (constraint equations). They occur in practice when some or all of the process variables must conform to some relationships arising from the physical characteristics of the model. In general we will represent them as a set of nonlinear algebraic equations, such as

$$\varphi(\mathbf{x}, \mathbf{u}) = \mathbf{0}, \quad \mathbf{u} \in \Re^n, \quad \varphi \in \Re^m, \tag{5.2}$$

where $\mathbf{u}$ indicates the $(n \times 1)$ vector of unmeasured process variables.

The data reconciliation problem can be generally stated as the following constrained weighted least-squares estimation problem:

$$\begin{aligned} &\operatorname*{Min}_{\mathbf{x},\mathbf{u}} \; (\mathbf{y} - \mathbf{x})^{\mathrm{T}} \Psi^{-1} (\mathbf{y} - \mathbf{x}) \\ &\text{s.t.} \\ &\qquad \varphi(\mathbf{x}, \mathbf{u}) = \mathbf{0} \\ &\qquad \mathbf{x}^{\mathrm{L}} \leq \mathbf{x} \leq \mathbf{x}^{\mathrm{U}} \\ &\qquad \mathbf{u}^{\mathrm{L}} \leq \mathbf{u} \leq \mathbf{u}^{\mathrm{U}} \end{aligned} \tag{5.3}$$

If it is assumed that the measurement errors are normally distributed, the resolution of problem (5.3) gives maximum likelihood estimates of process variables, so they are minimum variance and unbiased estimators.

Different methodologies are required for solving problem (5.3) depending on whether the constraints are a linear or a nonlinear set of equations. These methods will be discussed in detail in the following sections.

5.3. LINEAR DATA RECONCILIATION

Two situations arise in linear data reconciliation. Sometimes all the variables included in the process model are measured, but more frequently some variables are not measured. Both cases will be separately analyzed.

5.3.1. Linear Data Reconciliation with All Measured Variables

For this case problem (5.3) can be reformulated as

$$\begin{aligned} &\operatorname*{Min}_{\mathrm{x}} J = (\mathbf{y} - \mathbf{x})^{\mathrm{T}} \boldsymbol{\Psi}^{-1} (\mathbf{y} - \mathbf{x}) \\ &\text{s.t.} \\ &\qquad \mathbf{A}_1 \mathbf{x} = \mathbf{0}, \end{aligned} \tag{5.4}$$

where $\mathbf{A}_1$ is an $(m \times g)$ matrix of known constants. It should be noted that all variables are redundant in this case.

Several different resolution methods for this problem have been developed. Of these we will discuss the traditional methodology, called batch resolution, and an alternative methodology that is based on Q-R factorizations.

Batch Solution

Introducing measurement error into the process constraints through Eq. (5.1) gives

$$\mathbf{A}_1(\mathbf{y} - \boldsymbol{\varepsilon}) = \mathbf{0} \tag{5.5}$$

Consequently, the previous optimization problem (5.4) is now

$$\begin{aligned} &\operatorname*{Min}_{\varepsilon} \boldsymbol{\varepsilon}^{\mathrm{T}} \boldsymbol{\Psi}^{-1} \boldsymbol{\varepsilon} \\ &\text{s.t.} \\ &\qquad \mathbf{A}_1 (\mathbf{y} - \boldsymbol{\varepsilon}) = \mathbf{0}. \end{aligned} \tag{5.6}$$

The solution is obtained by means of the Lagrange multipliers method. The Lagrangian for this problem is

$$L = \boldsymbol{\varepsilon}^{\mathrm{T}} \boldsymbol{\Psi}^{-1} \boldsymbol{\varepsilon} - 2\boldsymbol{\lambda}^{\mathrm{T}} (\mathbf{A}_1 \mathbf{y} - \mathbf{A}_1 \boldsymbol{\varepsilon}). \tag{5.7}$$

Since $\boldsymbol{\Psi}$ is positive definite and the constraints are linear, the necessary and sufficient conditions for minimization are

$$\begin{aligned} \frac{\partial L}{\partial \boldsymbol{\varepsilon}} &= 2\boldsymbol{\Psi}^{-1} \boldsymbol{\varepsilon} + 2\mathbf{A}_1^{\mathrm{T}} \boldsymbol{\lambda} = 0 \\ \frac{\partial L}{\partial \boldsymbol{\lambda}} &= \mathbf{A}_1 (\mathbf{y} - \boldsymbol{\varepsilon}) = \mathbf{0}, \end{aligned} \tag{5.8}$$

TABLE 1 Data for Example 5.1 (from Ripps, 1965)

Flowrates	Measured values	True values	Variances
f_1	0.1858	0.1739	0.000289
f_2	4.7935	5.0435	0.0025
f_3	1.2295	1.2175	0.000576
f_4	3.88	4.00	0.04

which yield

$$\varepsilon = -\boldsymbol{\Psi}\mathbf{A}_1^{\mathrm{T}}\boldsymbol{\lambda} \tag{5.9}$$

$$\boldsymbol{\lambda} = -\left(\mathbf{A}_1\boldsymbol{\Psi}\mathbf{A}_1^{\mathrm{T}}\right)^{-1}\mathbf{A}_1\mathbf{y} \tag{5.10}$$

Finally, the estimate of the process variable, $\hat{\mathbf{x}}$, can be obtained as

$$\hat{\mathbf{x}} = \mathbf{y} - \boldsymbol{\Psi}\mathbf{A}_1^{\mathrm{T}}\left(\mathbf{A}_1\boldsymbol{\Psi}\mathbf{A}_1^{\mathrm{T}}\right)^{-1}\mathbf{A}_1\mathbf{y} \tag{5.11}$$

EXAMPLE 5.1

To illustrate the application of data reconciliation to linear systems, we will consider the problem presented by Ripps (1965). Four mass flows are measured, two entering and two leaving a chemical reactor. Three elemental balances are considered:

$$0.1f_1 + 0.6f_2 - 0.2f_3 - 0.7f_4 = 0$$
$$0.8f_1 + 0.1f_2 - 0.2f_3 - 0.1f_4 = 0$$
$$0.1f_1 + 0.3f_2 - 0.6f_3 - 0.2f_4 = 0.$$

The data for the problem are given in Table 1.

From the balance equations and the stochastic characteristics of the measuring devices the matrices $\mathbf{A}_1$ and $\boldsymbol{\Psi}$ are given by

$$\mathbf{A}_1 = \begin{bmatrix} 0.1 & 0.6 & -0.2 & -0.7 \\ 0.8 & 0.1 & -0.2 & -0.1 \\ 0.1 & 0.3 & -0.6 & -0.2 \end{bmatrix}, \quad \boldsymbol{\Psi} = \begin{bmatrix} 0.000289 & & & \\ & 0.0025 & & \\ & & 0.000576 & \\ & & & 0.04 \end{bmatrix}.$$

Thus, by applying Eqs. (5.9) to (5.11), we have the following measurement errors and process variables estimates:

$$\hat{\varepsilon} = \begin{bmatrix} 0.0182 \\ -0.0659 \\ +0.0565 \\ 0.026 \end{bmatrix}, \quad \hat{\mathbf{x}} = \begin{bmatrix} 0.1676 \\ 4.8594 \\ 1.1730 \\ 3.854 \end{bmatrix}.$$

A comparison between the original measurements, and the new estimates and the relation between measurement adjustment and the standard deviation (relative error found by data adjustment) is given in Table 2.

TABLE 2 Measured and Reconciled Values for Example 5.1 (from Ripps, 1965)

Flowrates	Measured value	Estimated value	Relative error
f_1	0.1858	0.1676	1.07
f_2	4.7936	4.8594	−1.31
f_3	1.2295	1.1730	2.35
f_4	3.88	3.854	0.13

Q-R Factorizations

Using the Q-R orthogonal factorization method described in Chapter 4, the constrained weighted least-squares estimation problem (5.4) is transformed into an unconstrained one. The following steps are required:

Step 1: Compute the general solution to the undetermined system ($\mathbf{A}_1\mathbf{x}=\mathbf{0}$). Using the procedure outlined in the previous chapter, the Q-R orthogonal factorization of $\mathbf{A}_1$ produces $\mathbf{Q}_x$, $\mathbf{R}_x$, $\mathbf{\Pi}_x$ matrices, which allows the calculation of $\mathbf{Q}_{x1}$, $\mathbf{Q}_{x2}$, $\mathbf{R}_{x1}$, $\mathbf{R}_{x2}$, $\mathbf{x}_{r_x}$, $\mathbf{x}_{g-r_x}$ such that

$$\mathbf{A}_1\mathbf{\Pi}_x = \mathbf{Q}_x\mathbf{R}_x \tag{5.12}$$

$$\mathbf{Q}_x = [\mathbf{Q}_{x1} \quad \mathbf{Q}_{x2}], \quad \mathbf{R}_x = \begin{bmatrix} \mathbf{R}_{x1} & \mathbf{R}_{x2} \\ \mathbf{0} & \mathbf{0} \end{bmatrix}, \tag{5.13}$$

$$\mathbf{\Pi}_x^T\mathbf{x} = \begin{bmatrix} \mathbf{x}_{r_x} \\ \mathbf{x}_{g-r_x} \end{bmatrix}, \tag{5.14}$$

where $r_x = \text{rank}(\mathbf{R}_{x1}) = \text{rank}(\mathbf{A}_1)$. The general solution of the problem is

$$\mathbf{x}_{r_x} = -\mathbf{R}_{x1}^{-1}\mathbf{R}_{x2}\mathbf{x}_{g-r_x}, \tag{5.15}$$

where $\mathbf{x}_{g-r_x}$ is an arbitrary vector.

Step 2: Formulation of the unconstrained problem. Applying previous results, the $(\mathbf{y}-\mathbf{x})$ vector of the objective function is modified as follows:

$$(\mathbf{y}-\mathbf{x}) = \mathbf{y} - [\mathbf{I}_{x1} \quad \mathbf{I}_{x2}]\begin{bmatrix} \mathbf{x}_{r_x} \\ \mathbf{x}_{g-r_x} \end{bmatrix} = \mathbf{y} + \mathbf{I}_{x1}\mathbf{R}_{x1}^{-1}\mathbf{R}_{x2}\mathbf{x}_{g-r_x} - \mathbf{I}_{x2}\mathbf{x}_{g-r_x}$$
$$= \mathbf{y} + \left(\mathbf{I}_{x1}\mathbf{R}_{x1}^{-1}\mathbf{R}_{x2} - \mathbf{I}_{x2}\right)\mathbf{x}_{g-r_x}, \tag{5.16}$$

where

$$\mathbf{I}\mathbf{\Pi}_x = [\mathbf{I}_{x1} \quad \mathbf{I}_{x2}], \quad \tilde{\mathbf{I}} = \mathbf{I}_{x1}\mathbf{R}_{x1}^{-1}\mathbf{R}_{x2} - \mathbf{I}_{x2}. \tag{5.17}$$

$\mathbf{I}$ represents a $(g \times g)$ identity matrix and $\tilde{\mathbf{I}}$ is a $[g \times (g - r_x)]$ matrix with independent columns.

The unconstrained minimization problem can now be stated as

$$\text{Min}\,(\mathbf{y} + \tilde{\mathbf{I}}\mathbf{x}_{g-r_x})^T\mathbf{\Psi}^{-1}(\mathbf{y} + \tilde{\mathbf{I}}\mathbf{x}_{g-r_x}). \tag{5.18}$$

Step 3: Estimation of $\mathbf{x}$. The solution of the aforementioned problem is

$$\hat{\mathbf{x}}_{g-r_x} = -(\tilde{\mathbf{I}}^T\mathbf{\Psi}^{-1}\tilde{\mathbf{I}})^{-1}\tilde{\mathbf{I}}^T\mathbf{\Psi}^{-1}\mathbf{y}. \tag{5.19}$$

Using the value of $\hat{\mathbf{x}}_{g-r_x}$, Eq. (5.15) is resolved to calculate $\hat{\mathbf{x}}_{r_x}$.

Remark. The dimension of the unconstrained optimization problem is smaller than that of the original one.

EXAMPLE 5.2

Calculate an estimate of the process variables for the system in Example 5.1 using the decomposition approach. As all variables are measured, matrices $\mathbf{A}_1$ and $\mathbf{I}$ are in this case

$$\mathbf{A}_1 = \begin{bmatrix} 0.1 & 0.6 & -0.2 & -0.7 \\ 0.8 & 0.1 & -0.2 & -0.1 \\ 0.1 & 0.3 & -0.6 & -0.2 \end{bmatrix}, \quad \mathbf{I} = \begin{bmatrix} 1 & 0 & 0 & 0 \\ 0 & 1 & 0 & 0 \\ 0 & 0 & 1 & 0 \\ 0 & 0 & 0 & 1 \end{bmatrix}.$$

Applying the Q-R decomposition to matrix $\mathbf{A}_1$, we have

$$\mathbf{Q}_\mathrm{x} = \begin{bmatrix} -0.1231 & 0.9572 & -0.2621 \\ -0.9847 & -0.1506 & -0.0874 \\ -0.1231 & 0.2474 & 0.9611 \end{bmatrix}, \quad \mathbf{R}_{\mathrm{x}1} = \begin{bmatrix} -0.8124 & 0.2093 & 0.2954 \\ 0 & -0.7044 & -0.3097 \\ 0 & 0 & -0.5067 \end{bmatrix},$$

$$\mathbf{R}_{\mathrm{x}2} = \begin{bmatrix} -0.2093 \\ 0.6334 \\ 0.1223 \end{bmatrix}.$$

Accordingly,

$$\mathbf{I}_{\mathrm{x}1} = \begin{bmatrix} 1 & 0 & 0 \\ 0 & 0 & 0 \\ 0 & 0 & 1 \\ 0 & 1 & 0 \end{bmatrix}, \quad \mathbf{I}_{\mathrm{x}_2} = \begin{bmatrix} 0 \\ 1 \\ 0 \\ 0 \end{bmatrix}, \quad \hat{\mathbf{x}}_{\mathrm{r_x}} = \begin{bmatrix} f_1 \\ f_4 \\ f_3 \end{bmatrix}, \quad \hat{\mathbf{x}}_{\mathrm{g-r_x}} = [f_2],$$

and

$$\tilde{\mathbf{I}} = \mathbf{I}_{\mathrm{x}1}\mathbf{R}_{\mathrm{x}1}^{-1}\mathbf{R}_{\mathrm{x}2} - \mathbf{I}_{\mathrm{x}2} = \begin{bmatrix} -0.0345 \\ -1.0000 \\ -0.2414 \\ -0.7931 \end{bmatrix}$$

The estimate of $\hat{\mathbf{x}}_{\mathrm{g-r_x}}$ is then given by

$$\hat{\mathbf{x}}_{\mathrm{g-r_x}} = -(\tilde{\mathbf{I}}^\mathrm{T}\boldsymbol{\Psi}^{-1}\tilde{\mathbf{I}})^{-1}\tilde{\mathbf{I}}^\mathrm{T}\boldsymbol{\Psi}^{-1}\mathbf{y} = 4.8594,$$

and finally the estimates of the remaining process variables are (from the general solution)

$$\mathbf{x}_{\mathrm{r_x}} = -\mathbf{R}_{\mathrm{x}1}^{-1}\mathbf{R}_{\mathrm{x}2}\mathbf{x}_{\mathrm{g-r_x}} = \begin{bmatrix} 0.1676 \\ 3.8540 \\ 1.1730 \end{bmatrix}$$

5.3.2. Handling Unmeasured Process Variables

The assumption that all variables are measured is usually not true, as in practice some of them are not measured and must be estimated. In the previous section the decomposition of the linear data reconciliation problem involving only measured variables was discussed, leading to a reduced least squares problem. In the following section, we will use these ideas to provide a general solution of the linear data reconciliation problem when some of the variables are considered unmeasured. The solution is based on decoupling the unmeasured variables from the measured ones using Q-R orthogonal factorizations. In this way the overall estimation problem is divided in two subproblems, as was discussed in Section 4.2.2.

Let us consider the constraint equations written as follows:

$$\mathbf{A}_1\mathbf{x} + \mathbf{A}_2\mathbf{u} = \mathbf{0}. \tag{5.20}$$

Performing a Q-R decomposition on matrix $\mathbf{A}_2$, matrices, $\mathbf{Q}_\mathrm{u}$, $\mathbf{R}_\mathrm{u}$, and $\mathbf{\Pi}_\mathrm{u}$ are obtained such that

$$\mathbf{A}_2\mathbf{\Pi}_\mathrm{u} = \mathbf{Q}_\mathrm{u}\mathbf{R}_\mathrm{u} \tag{5.21}$$

$$\mathbf{Q}_\mathrm{u} = [\mathbf{Q}_{\mathrm{u}1} \quad \mathbf{Q}_{\mathrm{u}2}], \quad \mathbf{R}_\mathrm{u} = \begin{bmatrix} \mathbf{R}_{\mathrm{u}1} & \mathbf{R}_{\mathrm{u}2} \\ \mathbf{0} & \mathbf{0} \end{bmatrix}, \tag{5.22}$$

where $r_u = \mathrm{rank}(\mathbf{A}_2) = \mathrm{rank}(\mathbf{R}_{\mathrm{u}1})$. The vector of unmeasured variables is partitioned into two subsets:

$$\mathbf{\Pi}_\mathrm{u}^\mathrm{T}\mathbf{u} = \begin{bmatrix} \mathbf{u}_{\mathrm{r_u}} \\ \mathbf{u}_{\mathrm{n-r_u}} \end{bmatrix}. \tag{5.23}$$

Now, by premultiplying the linearised constraints by $\mathbf{Q}^\mathrm{T} = \mathbf{Q}^{-1}$, we obtain

$$\begin{aligned} \mathbf{Q}_{\mathrm{u}1}^\mathrm{T}\mathbf{A}_1\mathbf{x} + \mathbf{R}_{\mathrm{u}1}\mathbf{u}_{\mathrm{r_u}} + \mathbf{R}_{\mathrm{u}2}\mathbf{u}_{\mathrm{n-r_u}} &= \mathbf{0} \\ \mathbf{Q}_{\mathrm{u}2}^\mathrm{T}\mathbf{A}_1\mathbf{x} &= \mathbf{0}. \end{aligned} \tag{5.24}$$

Performing the reconciliation on the decoupled subsystem represented by the measured variables $\mathbf{x}$ and the constraints.

$$\mathbf{Q}_{\mathrm{u}2}^\mathrm{T}\mathbf{A}_1\mathbf{x} = \mathbf{G}_\mathrm{x}\mathbf{x} = \mathbf{0}, \tag{5.25}$$

the solution of the overall system is

$$\hat{\mathbf{x}} = \mathbf{y} - \mathbf{\Psi}\mathbf{G}_\mathrm{x}^\mathrm{T}\left(\mathbf{G}_\mathrm{x}\mathbf{\Psi}\mathbf{G}_\mathrm{x}^\mathrm{T}\right)^{-1}\mathbf{G}_\mathrm{x}\mathbf{y}. \tag{5.26}$$

However, this problem can be reduced still further using the concepts developed before, for the case when all variables were considered measured, leading to the solution of a sequence of smaller subproblems.

For the unmeasured variables, we have in general

$$\mathbf{u}_{\mathrm{r_u}} = -\mathbf{R}_{\mathrm{u}1}^\mathrm{T}\mathbf{Q}_{\mathrm{u}1}^\mathrm{T}\mathbf{A}_1\hat{\mathbf{x}} - \mathbf{R}_{\mathrm{u}1}^{-1}\mathbf{R}_{\mathrm{u}2}\mathbf{u}_{\mathrm{n-r_u}}, \tag{5.27}$$

where the components $\mathbf{u}_{\mathrm{n-r_u}}$ are arbitrarily set.

We can have two cases:

1. $\mathrm{Rank}(\mathbf{R}_{\mathrm{u}1}) = n$
2. $\mathrm{Rank}(\mathbf{R}_{\mathrm{u}1}) < n$

Case (1). All unmeasured parameters are estimable and a unique solution for the unmeasured variables is possible using the rectified measured values and the balance equations.

TABLE 3 Measured Values, Variances, and Reconciled Values for Example 5.3

Flowrates	Measured values	Variances	Reconciled values
f_3	115.0663	7.8134	115.0663
f_4	111.6730	7.5556	109.6204
f_5	53.3700	1.7736	53.3700
f_8	0.8373	0.0004	0.8374
f_{11}	66.0986	2.8283	66.8670
f_{13}	95.7552	5.2739	95.7552
f_{14}	116.5318	8.4623	118.8308
f_{15}	77.8575	3.4698	76.9148

Case (2). Some unmeasured process variables are nonestimable and an infinite number of solutions are possible. Thus, the basic solution is

$$\mathbf{u}_{r_u} = -\mathbf{R}_{u1}^{-1}\mathbf{Q}_{u1}^{T}\mathbf{A}_1\hat{\mathbf{x}}; \quad \mathbf{u}_{n-r_u} = \mathbf{0}. \tag{5.28}$$

EXAMPLE 5.3

The outlined strategy has been applied to the subsystem of Example 4.4 in Chapter 4. The flow diagram, shown in Fig. 4 of Chapter 4, consists of 7 units interconnected by 15 streams. There are 8 measured flowrates and 7 unmeasured ones. The flowrate measurements with their variances are given in Table 3. In Chapter 4 we identified the subset of redundant equations. In this case it is constituted by one equation that contains the five redundant process variables. By applying the data reconciliation procedure to this reduced set of balances, we obtain the estimates of the measured variables, which are also presented in Table 3.

Regarding the unmeasured process variables, it was shown in Chapter 4 that the rank of matrix $\mathbf{R}_u$ is equal to 6; this means that at least one of the unmeasured variables is indeterminable. The remaining ones can be written in terms of it, as Eq. (5.27) indicates. In this case, from the orthogonal transformation, the subsets of $\mathbf{u}$ are defined as

$$\mathbf{u}_{r_u} = [f_1\ f_6\ f_{10}\ f_7\ f_9\ f_{12}], \quad \mathbf{u}_{n-r_u} = [f_2].$$

From an inspection of matrix $\mathbf{R}_{IU}$, we can see that only the flowrates f_6 and f_7 are independent of f_2, so they are determinable. The calculation of their values as functions of the adjusted measurements gives the following results:

$$\mathbf{u}_d = \begin{bmatrix} f_6 \\ f_7 \end{bmatrix} = \begin{bmatrix} 118.4626 \\ 172.6700 \end{bmatrix}$$

5.4. NONLINEAR DATA RECONCILIATION

The operation of a plant under steady-state conditions is commonly represented by a non-linear system of algebraic equations. It is made up of energy and mass balances and may include thermodynamic relationships and some physical behavior of the system. In this case, data reconciliation is based on the solution of a nonlinear constrained optimization problem.

The most general mathematical statement of an optimization problem is

$$
\begin{aligned}
&\underset{\mathbf{z}}{\mathrm{Min}}\ \sigma(\mathbf{z}) \\
&\text{s.t.} \\
&\qquad \varphi(\mathbf{z}) = \mathbf{0} \\
&\qquad \omega(\mathbf{z}) \leq \mathbf{0} \\
&\qquad \mathbf{z}^{\mathrm{L}} \leq \mathbf{z} \leq \mathbf{z}^{\mathrm{U}}.
\end{aligned}
\tag{5.29}
$$

The necessary conditions for an optimal solution of problem (5.29) are equivalent (Edgar and Himmelblau, 1988) to those for optimizing the Lagrange function defined as

$$L(\mathbf{z}, \lambda, \mu) = \sigma(\mathbf{z}) + \sum_{1}^{m1} \lambda_i \varphi_i(\mathbf{z}) + \sum_{1}^{m2} \mu_j \omega_j(\mathbf{z}). \tag{5.30}$$

These conditions are the well known Kuhn–Tucker (KT) conditions:

- Lineal dependency of the gradients:

$$\nabla\sigma(\mathbf{z}) + \sum_{1}^{m1} \lambda_i \nabla\varphi_i(\mathbf{z}) + \sum_{1}^{m2} \mu_j \nabla\omega_j(\mathbf{z}) = \mathbf{0}. \tag{5.31}$$

- Constraint feasibility:

$$
\begin{aligned}
\varphi_i(\mathbf{z}) &= \mathbf{0}, \quad i = 1, \ldots, m1 \\
\omega_j(\mathbf{z}) &\leq \mathbf{0}, \quad j = 1, \ldots, m2.
\end{aligned}
\tag{5.32}
$$

- Complementary conditions:

$$\mu_j \omega_j(\mathbf{z}) = 0, \quad j = 1, \ldots, m2. \tag{5.33}$$

- Nonnegativity conditions for inequality multipliers:

$$\mu_j \geq 0, \quad j = 1, \ldots, m2, \tag{5.34}$$

  where λ_i and μ_j stand for the Lagrange and Kuhn–Tucker multipliers, respectively.

The sufficient conditions for obtaining a global solution of the nonlinear programming problem are that both the objective function and the constraint set be convex. If these conditions are not satisfied, there is no guarantee that the local optima will be the global optima.

In this section we will explore the applicability of different techniques for solving the nonlinear data reconciliation problem.

5.4.1. Q-R Orthogonal Factorizations

Orthogonal factorizations may be applied to resolve problem (5.3) if the system of equations $\varphi(\mathbf{x}, \mathbf{u}) = 0$ is made up of linear mass balances and bilinear component and energy balances. After replacing the bilinear terms of the original model by the corresponding mass and energy flows, a linear data reconciliation problem results.

Using the notation of Chapter 4, it may be stated as follows:

$$\begin{aligned}&\underset{\delta,\theta}{\text{Min}}\ \left(\varepsilon_{f_m}^{T}\boldsymbol{\Psi}_{f_m}^{-1}\varepsilon_{f_m}+\varepsilon_{f_{ch}}^{T}\boldsymbol{\Psi}_{f_{ch}}^{-1}\varepsilon_{f_{ch}}+\boldsymbol{\theta}^{T}\boldsymbol{\Psi}_{\theta}^{-1}\boldsymbol{\theta}\right)\\&\text{s.t.}\\&\qquad[\mathbf{B}_{11}\ \ \mathbf{B}_{22}\ \ \mathbf{B}_{33}]\begin{bmatrix}\mathbf{a}\\\mathbf{f}_U\\\mathbf{v}\end{bmatrix}=-\begin{bmatrix}\mathbf{0}_1&\mathbf{B}_1\\\mathbf{E}_4&\mathbf{E}_1\end{bmatrix}\begin{bmatrix}\tilde{\mathbf{f}}_M\\\tilde{\mathbf{f}}_{ch}\end{bmatrix}=\mathbf{e}.\end{aligned}\tag{5.35}$$

An iterative method of solution of problem (5.35) was designed by Crowe (1986), and we have discussed, in the previous chapter, the strategy for the case when Q-R orthogonal factorizations are applied. The stages of the procedure are as follows:

- Unmeasured variable elimination through Q-R orthogonal factorizations
- Adjustment of measurements and estimation of unmeasured total flowrates
- Estimation of unmeasured component and energy flows

The procedure is not complex, and the required computation time is low. However, its use is restricted to data reconciliation problems with linear and bilinear constraints, and variable bounds cannot be handled.

5.4.2. Successive Linearizations

The method developed for linear constraints is extended to nonlinearly constrained problems. It is based on the idea that the nonlinear constraints $\varphi(\mathbf{x},\mathbf{u})=\mathbf{0}$ can be linearized in a linear Taylor series expansion around an estimation of the solution (x_i, u_i). In general, measurement values are used as initial estimations for the measured process variables. The following linear system of equations is obtained:

$$\mathbf{A}_1\mathbf{x}+\mathbf{A}_2\mathbf{u}=\mathbf{c}_1,\tag{5.36}$$

where

$$\mathbf{A}_1=\left.\frac{\partial\varphi}{\partial\mathbf{x}}\right|_{\mathrm{x_i,u_i}},\quad \mathbf{A}_2=\left.\frac{\partial\varphi}{\partial\mathbf{u}}\right|_{\mathrm{x_i,u_i}}\tag{5.37}$$

$$\mathbf{c}_1=\mathbf{A}_1\mathbf{x}_i+\mathbf{A}_2\mathbf{u}_i-\varphi(\mathbf{x}_i,\mathbf{u}_i).\tag{5.38}$$

The unmeasured variables are then eliminated using, for example, the orthogonal factorization procedure discussed before. Once the subset of equations containing only measured variables has been identified, the problem stated by Swartz (1989) is resolved:

$$\begin{aligned}&\underset{\mathrm{x}}{\text{Min}}\ (\mathbf{y}-\mathbf{x})^{T}\boldsymbol{\Psi}^{-1}(\mathbf{y}-\mathbf{x})\\&\text{s.t.}\\&\qquad\mathbf{G}_x\mathbf{x}=\mathbf{b},\end{aligned}\tag{5.39}$$

where

$$\mathbf{G}_x=\mathbf{Q}_{u2}^{T}\mathbf{A}_1,\quad \mathbf{b}=\mathbf{Q}_{u2}^{T}\mathbf{c}_1.\tag{5.40}$$

Its solution can be stated as

$$\hat{\mathbf{x}}=\mathbf{y}-\boldsymbol{\Psi}\mathbf{G}_x^{T}\left(\mathbf{G}_x\boldsymbol{\Psi}\mathbf{G}_x^{T}\right)^{-1}(\mathbf{G}_x\mathbf{y}-\mathbf{b})\tag{5.41}$$

$$\mathbf{u}_{r_u}=\mathbf{R}_{u1}^{-1}\mathbf{Q}_{u1}^{T}\mathbf{c}_1-\mathbf{R}_{u1}^{-1}\mathbf{Q}_{u1}^{T}\mathbf{A}_1\hat{\mathbf{x}}-\mathbf{R}_{u1}^{-1}\mathbf{R}_{u2}\mathbf{u}_{n-r_u}.\tag{5.42}$$

This solution is the optimal point for the linear constraints. A series of iterations are performed by linearizing the constraints about the previous iterate until a solution satisfying the nonlinear constraints is obtained.

Successive linearisation has the advantage of relative simplicity and fast calculation. In addition, it can be modified to choose a step size that minimizes a prespecified penalty function. The step size is chosen by the method of interval halving (Pai and Fisher, 1988). However, variable bounds cannot be handled; it may fail to converge to the desired minimum; and it might oscillate when multiple minima exist.

5.4.3. Nonlinear Programming Techniques

NLP algorithms allow for a general nonlinear objective function, not just a weighted least squares objective function. They can explicitly handle nonlinear constraints, inequality constraints, and variable bounds. Typical algorithms include sequential quadratic programming techniques, such as NPSOL (Gill *et al.*, 1986), and reduced gradient methods, such as GRG2 (Lasdon and Waren, 1978). The main features of both approaches are described next.

Sequential quadratic programming. A sequential quadratic programming (SQP) technique involves the resolution of a sequence of explicit quadratic programming (QP) subproblems. The solution of each subproblem produces the search direction $\mathbf{d}_k$ that has to be taken to reach the next iterate $\mathbf{z}_{k+1}$ from the current iterate $\mathbf{z}_k$. A one-dimensional search is then accomplished in the direction $\mathbf{d}_k$ to obtain the optimal step size.

To apply the procedure, the nonlinear constraints $\varphi(\mathbf{x}, \mathbf{u}) = \mathbf{0}$ are linearized using a Taylor series expansion and an optimization problem is resolved to find the solution, $\mathbf{d}_k$, that minimizes a quadratic objective function subject to linear constraints. The QP subproblem is formulated as follows:

$$\begin{aligned} &\operatorname*{Min}_{\mathrm{d}} \ \frac{1}{2}\mathbf{d}_k^{\mathrm{T}}\mathbf{HL}(\mathbf{z}_k, \boldsymbol{\lambda}_k)\mathbf{d}_k + \nabla\sigma^{\mathrm{T}}(\mathbf{z}_k)\mathbf{d}_k \\ &\text{s.t.} \\ &\qquad \nabla\varphi(\mathbf{z}_k)\mathbf{d}_k + \varphi(\mathbf{z}_k) = \mathbf{0}, \end{aligned} \tag{5.43}$$

where **HL** is the Hessian matrix of the Lagrange function formulated for problem (5.3) $\nabla\sigma$ is the objective function gradient, and $\nabla\varphi$ is the constraints gradient.

Reduced gradient method. This technique is based on the resolution of a sequence of optimization subproblems for a reduced space of variables. The process constraints are used to solve a set of variables ($\mathbf{z}_\mathrm{D}$), called basic or dependent, in terms of the others, which are known as nonbasic or independent ($\mathbf{z}_\mathrm{I}$). Using this categorization of variables, problem (5.3) is transformed into another one of fewer dimensions:

$$\begin{aligned} &\operatorname*{Min}_{\mathrm{z_I}} \ \sigma(\mathbf{z}_\mathrm{I}) \\ &\text{s.t.} \\ &\qquad \mathbf{z}_\mathrm{I}^\mathrm{L} \le \mathbf{z}_\mathrm{I} \le \mathbf{z}_\mathrm{I}^\mathrm{U}, \end{aligned} \tag{5.44}$$

where $\sigma(\mathbf{z}_\mathrm{I})$ is the reduced objective function and its reduced gradient is formulated as

$$\sigma_\mathrm{R}^\mathrm{T} = \left[\frac{\partial\sigma_{\mathrm{z_I}}}{\partial\mathbf{z}_\mathrm{I}}\right]^\mathrm{T} - \left[\frac{\partial\sigma_{\mathrm{z_I}}}{\partial\mathbf{z}_\mathrm{D}}\right]^\mathrm{T}\left[\frac{\partial\varphi}{\partial\mathbf{z}_\mathrm{D}}\right]^{-1}\left[\frac{\partial\varphi}{\partial\mathbf{z}_\mathrm{I}}\right]. \tag{5.45}$$

First the search direction, **d**, in the space of the independent variables is determined from the elements of σ_R^T, and then the search components for the dependent ones are calculated. A one-dimensional search is accomplished to obtain a solution for $\mathbf{z}_I$. The remaining dependent variables are evaluated as functions of $\mathbf{z}_I$ using the process constraints.

The disadvantage of these NLP algorithms is the large amount of computation time required relative to the successive linearisation algorithm. Nevertheless, their range of application is wider, and they are able to manage nonlinear objective functions, equality and inequality constraints, and bounds on variables.

Well-scaled problems improve the performance of NLP algorithms. Most variables are of different orders of magnitude and thus require scaling. Also, if constraints are of varying magnitudes, it is important to scale the constraints such that the constraints' residuals are of the same magnitude. In general, the accuracy of the solution returned by the NLP software is problem dependent.

EXAMPLE 5.4

Let us consider the heat exchanger network (Swartz, 1989) given in Fig. 1. In this system streams *A* receive heat from streams *B*, *C*, and *D*. The standard deviations of flow rates and temperatures are 2% and 0.75°, respectively. The measured variables are given in Table 4, where *f* and *t* indicate flowrates and temperature measurements,

TABLE 4 Measured Variables for Example 5.4

Stream	Variable	Measurement	Stream	Variable	Measurement
A1	*f*	1000.00	*A8*	*t*	614.92
A1	*t*	466.33	*B1*	*f*	253.20
A3	*f*	401.70	*B1*	*t*	618.11
A3	*t*	481.78	*C1*	*f*	308.10
A4	*t*	530.09	*C1*	*t*	694.99
A5	*t*	616.31	*D1*	*t*	667.84
A6	*f*	552.70	*D2*	*f*	680.10
A7	*t*	619.00	*D2*	*t*	558.34

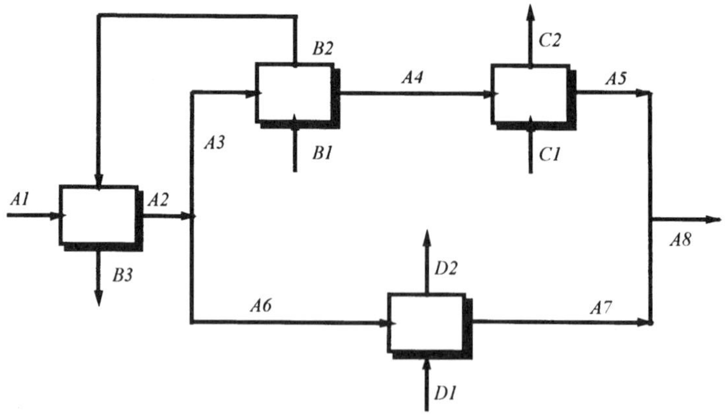

FIGURE 1 Flow diagram for Example 5.4 (from Swartz, 1989).

respectively. The model consists of 16 measured variables, 14 unmeasured ones, and 17 equations. The equations include total mass and energy balances around the heat exchangers, mixers, and dividers; they are given in Appendix A. It is assumed that the specific enthalpies for all streams can be represented by a quadratic expression in terms of temperature:

$$h_i = \eta_{1i} + \eta_{2i} t_i + \eta_{3i} t_i^2, \quad i = A, B, C, D.$$

The values of the constants in these equations are given also in Appendix A. The problem was originally solved by Swartz (1989) using successive linearizations and Q-R factorizations to eliminate the unmeasured process variables. Later, Tjoa and Biegler (1991) solved the same problem using a hybrid SQP technique, developed to exploit the advantages arising from the type of objective function in the reconciliation problem. For this example we carried out the following analysis:

1. *Effect of the a priori classification/decomposition on the solution of the reconciliation problem.*

As was discussed in Chapters 3 and 4, variable classification allows us to obtain a reduced subsystem of redundant equations that contain only measured and redundant variables. These are used in the reconciliation procedure.

A comparative analysis was performed to investigate the effect of the reduction on computing time, between the two possible approaches:

(a) Without any classification/decomposition prior to reconciliation (i.e., solving the whole set of equations containing measured and unmeasured variables)
(b) Decomposition of the global estimation problem, resolution of the reduced reconciliation problem, and then calculation of the unmeasured but determinable process variables

SQP was used for the solution, because of its good performance (Tjoa and Biegler, 1991), using the implementation available within the MATLAB environment.

From the classification it was found that, for this specific problem, there are 10 redundant and 6 nonredundant measured variables, and all the unmeasured process variables are determinable. Symbolic manipulation of the equations allowed us to obtain the three redundant equations used in the reconciliation problem:

$$f_{A1} - f_{A3} - f_{A6} = 0$$

$$f_{A3} + f_{A6} - \left(\frac{f_{A3} h(t_{A5}) + f_{A6} h(t_{A7})}{h(t_{A8})} \right) = 0$$

$$f_{A6} h(t_{A3}) + f_{D2} h(t_{D1}) - f_{D2} h(t_{D2}) - f_{A6} h(t_{A7}) = 0.$$

Expressions for the unmeasured process variables, as functions of the measured ones, were also obtained. These were solved sequentially after the reconciliation of the redundant measurements was completed:

$$\begin{aligned}
f_{A2} &= f_{A1} \\
f_{A4} &= f_{A3} \\
f_{B2} &= f_{B1} \\
f_{A5} &= f_{A3} \\
f_{C2} &= f_{C1}
\end{aligned}$$

$$f_{A7} = f_{A6}$$
$$f_{D1} = f_{D2}$$
$$t_{A2} = t_{A3}$$
$$t_{A6} = t_{A3}$$
$$f_{B3} = f_{B1}$$
$$h_{B2} = \frac{f_{B1}h_{B1} + f_{A3}h_{A3} - f_{A3}h_{A4}}{f_{B1}}$$
$$h_{C2} = \frac{f_{A3}h_{A4} + f_{C1}h_{C1} - f_{A3}h_{A5}}{f_{C1}}$$
$$f_{A8} = \frac{f_{A3}h_{A5} + f_{A6}h_{A7}}{h_{A8}}$$
$$h_{B3} = \frac{f_{A1}h_{A1} + f_{B1}h_{B1} + f_{A3}h_{A3} - f_{A3}h_{A4} - f_{A1}h_{A3}}{f_{B1}}.$$

The estimated values for the measured and unmeasured process variables for both cases are given in Table 5. As shown in Table 6, the computing time and the number of function evaluations decreased substantially when the decomposition/reconciliation approach was used in comparison with the conventional approach. This indicates the advantages of performing the decomposition before the reconciliation.

2. *Analysis of the performance of the Q-R decomposition approach for bilinear systems in correcting measured and unmeasured variables.*

Since the original problem is bilinear, a redefinition of variables was performed to transform it into a linear one, according to the strategy described in Section 4.3.

TABLE 5 Estimates of the Measured and Unmeasured Variables in Example 5.4

Stream	Flowrate	Temperature	Stream	Flowrate	Temperature
A1	963.63	466.33	*B1*	253.20	618.11
A2	963.63	481.91	*B2*	253.20	543.90
A3	407.85	481.91	*B3*	253.20	486.51
A4	407.85	530.09	*C1*	308.10	694.99
A5	407.85	615.51	*C2*	308.10	594.80
A6	555.77	481.91	*D1*	689.41	668.02
A7	555.77	617.77	*D2*	689.41	558.17
A8	963.63	616.81			

TABLE 6 Solution Performance for Example 5.4 with SQP

System of equations	Objective function evaluations	Time (sec)
Complete	2228	101
Reduced	474	18

To do this, the product of the mass flowrate and the specific enthalpy was substituted by the corresponding enthalpy flow. Results of the reconciliation procedure using the Q-R factorization are given in Table 7. Table 8 compares the residuum of the balance equations, the value of the objective function, and the computing time of the MATLAB implementation for both approaches (Q-R factorization and use of SQP with the reduced set of balance equations). These results show the improvement and the efficiency achieved using Q-R decomposition when the system can be represented as bilinear.

TABLE 7 Estimates of the Measured and Unmeasured Variables for Example 5.4

Stream	Flowrate	Temperature	Stream	Flowrate	Temperature
A1	967.43	466.33	*B1*	253.20	618.11
A2	967.43	484.11	*B2*	253.20	543.99
A3	406.58	481.82	*B3*	253.20	477.98
A4	406.58	530.09	*C1*	308.10	694.99
A5	406.58	615.51	*C2*	308.10	595.16
A6	560.85	485.76	*D1*	680.10	667.84
A7	560.85	617.87	*D2*	680.10	558.34
A8	967.43	616.87			

TABLE 8 Performance of the Different Methods in the Solution of Example 5.4

Variable	Bilinear system—SQP	Bilinear system—QR
Res. # 1	0.	0.
Res. # 2	0.	0.
Res. # 3	1×10^{-2}	1×10^{-13}
Res. # 4	0.	0.
Res. # 5	0.	0.
Res. # 6	0.	0.
Res. # 7	0.	0.
Res. # 8	1×10^{-2}	1×10^{-13}
Res. # 9	0.	0.
Res. # 10	0.	0.
Res. # 11	0.4756	−0.5071
Res. # 12	0.	−0.4720
Res. # 13	0.	0.3408
Res. # 14	−0.1268	0.3780
Res. # 15	0.6243	−0.2359
Res. # 16	−0.0446	0.0936
Res. # 17	−3.047	−1.013
Maximum norm	3.047	1.013
Objective function	14.7974	13.7962
Time (sec)	18.13	9.72

5.5. CONCLUSIONS

In this chapter we have stated the data reconciliation problem and explored some available techniques for solving it. It must be noted that the performance of reconciliation methods is strongly dependent on the particular system.

For linear models, the Q-R factorization can be successfully applied for both problem decomposition and estimation purposes. In contrast, nonlinear models require the selection of an appropriate technique that can manage the complexity of the nonlinear constraints in a reasonably low run time.

In this sense, the application of Q-R factorizations constitutes an efficient alternative for solving bilinear data reconciliation. Successive linearizations and nonlinear programming are required for more complex models. These techniques are more reliable and accurate for most problems, and thus require more computation time.

NOTATION

$\mathbf{A}_1$	submatrix corresponding to measured variables for linear models
$\mathbf{A}_2$	submatrix corresponding to unmeasured variables for linear models
$\mathbf{b}$	vector defined by Eq. (5.40)
$\mathbf{c}_1$	vector defined by Eq. (5.38)
$\mathbf{d}$	search direction
$\mathbf{f}$	vector of total flowrates
g	number of measured variables
$\mathbf{G}_\mathrm{x}$	matrix defined by Eq. (5.25)
$\mathbf{h}$	vector of stream enthalpies
$\mathbf{HL}$	Hessian matrix of the Lagrange function
$\tilde{\mathbf{I}}$	matrix defined by Eq. (5.17)
$\mathbf{I}_{\mathrm{x}_1}, \mathbf{I}_{\mathrm{x}_2}$	partition of matrix $\mathbf{I}$ considering $\mathbf{\Pi}_\mathrm{x}$
J	objective function value
k	index for the number of iterations
L	Lagrange function
m	number of process model functions
m_1	number of equality constraints
m_2	number of inequality constraints
n	number of unmeasured variables
$[\mathbf{Q}_\mathrm{x}, \mathbf{R}_\mathrm{x}, \mathbf{\Pi}_\mathrm{x}]$	$\mathrm{QR}(\mathbf{A}_1)$
$[\mathbf{Q}_\mathrm{u}, \mathbf{R}_\mathrm{u}, \mathbf{\Pi}_\mathrm{u}]$	$\mathrm{QR}(\mathbf{A}_2)$
r_x	$\mathrm{rank}(\mathbf{A}_1)$
r_u	$\mathrm{rank}(\mathbf{A}_2)$
$\mathbf{t}$	vector of stream temperatures
$\mathbf{u}$	vector of unmeasured variables ($n \times 1$)
$\mathbf{u}_{\mathrm{r_u}}, \mathbf{u}_{\mathrm{n-r_u}}$	partitions of $\mathbf{u}$
$\mathbf{x}$	vector of measured variables ($g \times 1$)
$\mathbf{x}_{\mathrm{r_x}}, \mathbf{x}_{\mathrm{g-r_x}}$	partitions of $\mathbf{x}$
$\mathbf{y}$	vector of measurements ($g \times 1$)
$\mathbf{z}$	vector of optimization variables
$\mathbf{z}_\mathrm{I}, \mathbf{z}_\mathrm{D}$	partitions of $\mathbf{z}$ in dependent and not-dependent variables

Greek

ε	vector of random errors
η	polynomial coefficients for specific enthalpies
λ	Lagrange multipliers

μ Kuhn–Tucker multipliers
Ψ covariance matrix of the measurement errors
φ nonlinear equality constraints
ω nonlinear nequality constraints
σ a general objective function

Superscripts

$\hat{}$ with reconciled values
L lower limit
U upper limit

REFERENCES

Crowe, C. M. (1986). Reconciliation of process flow rates by matrix projection. Part II: The non-linear case. *AIChE J.* **32**, 616–623.

Crowe, C. M. (1996). Data reconciliation—Progress and challenges. *J. Proc. Control*, **6**, 89–98.

Edgar, T. F., and Himmelblau, D. M. (1988). "Optimization of Chemical Processes." McGraw-Hill, New York.

Gill, P., Murray, W. A., Saunders, M. A., and Wright, M. H. (1986). "User Guide for NPSOL (Version 4.0); a FORTRAN Program for Nonlinear Programming," Tech. Rep. SOL 86-2. Standford University, Department of Operation Research, Standford, CA.

Hlavacek, V. (1977). Analysis of a complex plant—steady state and transient behaviour. I Plant data estimation and adjustment. *Comput. Chem. Eng.* **1**, 75–100.

Kuehn, D. R., and Davidson, H. (1961). Computer control. II. Mathematics of control. *Chem. Eng. Prog.* **57**, 44–47.

Lasdon, L. S., and Waren, A. D. (1978). Generalized reduced gradient software for linearly and nonlinearly constrained problems. *In* "Design and Implementation of Optimisation Software" (H. Greenberg, ed.), p. 335. Sijthoff, Holland.

Madron, F. (1992). "Process Plant Performance. Measurement and Data Processing for Optimization and Retrofits." Ellis Horwood, Chichester, England.

Mah, R. S. H. (1982). Design and analysis of performance monitoring systems. *In* "Chemical Process Control II" (D. E. Seborg and T. F. Edgat, eds.), pp. 525–540. Engineering Foundation, New York.

Mah, R. S. H. (1990). "Chemical Process Structures and Information Flows," Chem. Eng. Ser. Butterworth, Boston.

Pai, D. C., and Fisher, G. D. (1988). Application of Broyden's method to reconciliation of nonlinearly constrained data. *AIChE J.* **34**, 873–876.

Ripps, D. L. (1965). Adjustment of experimental data. *Chem. Eng. Prog., Symp. Ser.* **61**, 8–13.

Swartz, C. L. E. (1989). Data reconciliation for generalized flowsheet applications. *197th Natl. Meet., Am. Chem. Soc.* Dallas, TX

Tamhane, A. C., and Mah, R. S. H. (1985). Data reconciliation and gross error detection in chemical process networks. *Technometrics*, **27**, 409–422.

Tjoa, I., and Biegler, L. (1991), Simultaneous strategies for data reconciliation and gross error detection of nonlinear systems. *Comput. Chem. Eng.* **15**, 679.

APPENDIX A: ADDITIONAL INFORMATION FOR EXAMPLE 5.4

Model Equations

1. $f_{A1} - f_{A2} = 0$
2. $f_{B2} - f_{B3} = 0$
3. $f_{A1}h_{A1} - f_{A2}h_{A2} + f_{B2}h_{B2} - f_{B3}h_{B3} = 0$

4. $f_{A2} - f_{A3} - f_{A6} = 0$
5. $t_{A2} - t_{A3} = 0$
6. $t_{A2} - t_{A6} = 0$
7. $f_{A3} - f_{A4} = 0$
8. $f_{B1} - f_{B2} = 0$
9. $f_{A3}h_{A3} - f_{A4}h_{A4} + f_{B1}h_{B1} - f_{B2}h_{B2} = 0$
10. $f_{A4} - f_{A5} = 0$
11. $f_{C1} - f_{C2} = 0$
12. $f_{A4}h_{A4} - f_{A5}h_{A5} + f_{C1}h_{C1} - f_{C2}h_{C2} = 0$
13. $f_{A5} + f_{A7} - f_{A8} = 0$
14. $f_{A5}h_{A5} + f_{A7}h_{A7} - f_{A8}h_{A8} = 0$
15. $f_{A6} - f_{A7} = 0$
16. $f_{D1} - f_{D2} = 0$
17. $f_{A6}h_{A6} - f_{A7}h_{A7} + f_{D1}h_{D1} - f_{D2}h_{D2} = 0.$

Coefficients for the Specific Enthalpy Equations

Coefficient	A	B	C	D
η_1	−6.8909	−14.8538	−28.2807	−11.4172
η_2	0.0991	0.1333	0.1385	0.1229
η_3	1.1081×10^{-4}	0.7539×10^{-4}	0.9043×10^{-4}	0.7940×10^{-4}

6 SEQUENTIAL PROCESSING OF INFORMATION

In this chapter the mathematical formulation for the sequential processing of both constraints and measurements is presented. Some interesting features are discussed that make this approach well suited for the analysis of a set of measurements and for gross error identification, the subject of the next chapter.

6.1. INTRODUCTION

The linear/linearized data reconciliation solution deserves some special attention because it allows the formulation of alternative strategies for the processing of information. In this chapter the mathematical formulation for the sequential processing of both constraints and measurements is analyzed.

A recursive procedure is presented that exhibits some advantages over classical batch processing since it avoids the inversion of large matrices. It is shown that when only one equation is processed at a time, the inversion degenerates into the computation of the reciprocal of a scalar. Furthermore, this sequential approach can also be used to isolate systematic errors that may be present in the data set (Romagnoli and Stephanopoulos, 1981; Romagnoli, 1983).

6.2. SEQUENTIAL PROCESSING OF CONSTRAINTS

Consider in general the following set of m balance equations containing only measured variables:

$$\mathbf{A}\mathbf{x} = \mathbf{0} \tag{6.1}$$

or

$$\mathbf{A}(\mathbf{y} - \boldsymbol{\varepsilon}) = \mathbf{0}, \tag{6.2}$$

where $\mathbf{A}$ is an $(m \times g)$ matrix of known constants; $\mathbf{x}$ is the g-dimensional vector of process variables; $\boldsymbol{\varepsilon}$ is the g-dimensional vector of measurement errors whose covariance matrix is given by $\boldsymbol{\Psi}$. The residuum of the balance equations is defined as

$$\mathbf{r} = \mathbf{A}\mathbf{y}, \tag{6.3}$$

and its covariance matrix is stated as

$$\boldsymbol{\Phi} = \mathbf{A}\boldsymbol{\Psi}\mathbf{A}^{\mathrm{T}}. \tag{6.4}$$

As shown before, the general data reconciliation procedure for the overall system must solve the following constrained least squares problem:

$$\begin{aligned} &\operatorname*{Min}_{\varepsilon}\ \boldsymbol{\varepsilon}^{\mathrm{T}}\boldsymbol{\Psi}^{-1}\boldsymbol{\varepsilon} \\ &\text{s.t.} \\ &\qquad \mathbf{A}(\mathbf{y} - \boldsymbol{\varepsilon}) = \mathbf{0}. \end{aligned} \tag{6.5}$$

Now, because of the manner in which the balances arise, the total set of algebraic equations can be partitioned into two arbitrary subsystems. The first contains $(m - a)$ equations and the second contains the remaining a equations, where a is an arbitrary number $1 \leq a \leq m$. Note that the cases $a = 0$ or $a = m$ correspond to the overall reconciliation problem.

Partitioning matrix $\mathbf{A}$ along these lines, we get

$$\mathbf{A} = \begin{bmatrix} \mathbf{A}_1 \\ \mathbf{A}_2 \end{bmatrix}, \tag{6.6}$$

where $\mathbf{A}_1$ and $\mathbf{A}_2$ are compatible matrices of dimension $[(m - a) \times g]$ and $(a \times g)$, respectively. Using the partitioning just shown, the constrained least squares problem for the overall system was replaced in the work of Mikhail (1976) by the equivalent two-problem formulation:

Problem 1

$$\begin{aligned} &\operatorname*{Min}_{\hat{\varepsilon}_1}\ \boldsymbol{\varepsilon}_1^{\mathrm{T}}\boldsymbol{\Psi}_1^{-1}\boldsymbol{\varepsilon}_1 \\ &\text{s.t.} \\ &\qquad \mathbf{A}_1(\mathbf{y} - \boldsymbol{\varepsilon}_1) = \mathbf{0}. \end{aligned} \tag{6.7}$$

When the Lagrangian technique is used to solve the preceding problem, the least squares estimate of the measurement errors is given by

$$\hat{\boldsymbol{\varepsilon}}_1 = \boldsymbol{\Psi}_1\mathbf{A}_1^{\mathrm{T}}\left(\mathbf{A}_1\boldsymbol{\Psi}_1\mathbf{A}_1^{\mathrm{T}}\right)^{-1}\mathbf{A}_1\mathbf{y} \tag{6.8}$$

Also, for the covariance of the estimation error and the measurement estimates (see Appendix A) we have, respectively,

$$\boldsymbol{\Sigma}_{\hat{\varepsilon}_1} = \boldsymbol{\Psi}_1\mathbf{A}_1^{\mathrm{T}}\left(\mathbf{A}_1\boldsymbol{\Psi}_1\mathbf{A}_1^{\mathrm{T}}\right)^{-1}\mathbf{A}_1\boldsymbol{\Psi}_1 \tag{6.9}$$

and

$$\boldsymbol{\Sigma}_{\hat{\mathbf{x}}_1} = \boldsymbol{\Psi}_1 - \boldsymbol{\Sigma}_{\hat{\varepsilon}_1}. \tag{6.10}$$

Problem 2

From the solution of the first problem, the measurements are adjusted using the estimates of the measurement errors $\hat{\varepsilon}_1$. Thus, new information is available when the rest of the balances, $\mathbf{A}_2$, are processed in Problem 2.

Let us defined a matrix Ψ_2 such that

$$\mathbf{\Psi}_2 = \mathbf{\Psi}_1 - \mathbf{\Sigma}_{\hat{\varepsilon}_1}, \tag{6.11}$$

where $\mathbf{\Psi}_2$ represents the covariance matrix of the initial measurements for Problem 2.

By analogy with the resolution of the first problem, the estimate of the measurement error, $\hat{\varepsilon}_2$, associated with the subset of constraints $\mathbf{A}_2$ is given by

$$\hat{\varepsilon}_2 = \mathbf{\Psi}_2\mathbf{A}_2^{\mathrm{T}}\left(\mathbf{A}_2\mathbf{\Psi}_2\mathbf{A}_2^{\mathrm{T}}\right)^{-1}\mathbf{A}_2(\mathbf{y} - \hat{\varepsilon}_1). \tag{6.12}$$

The following relationships can also be obtained for the covariance of the estimated errors and for the measurement estimates:

$$\mathbf{\Sigma}_{\hat{\varepsilon}_2} = \mathbf{\Psi}_2\mathbf{A}_2^{\mathrm{T}}\left(\mathbf{A}_2\mathbf{\Psi}_2\mathbf{A}_2^{\mathrm{T}}\right)^{-1}\mathbf{A}_2\mathbf{\Psi}_2 \tag{6.13}$$

and

$$\mathbf{\Sigma}_{\hat{\mathbf{x}}_2} = \mathbf{\Psi}_2 - \mathbf{\Sigma}_{\hat{\varepsilon}_2}. \tag{6.14}$$

It can be shown (Mikhail, 1976), reproduced here in Appendix A, that the solution for the overall set of balance equations can be obtained from the equivalent two-stage procedure as follows:

$$\hat{\varepsilon} = \hat{\varepsilon}_1 + \hat{\varepsilon}_2 \tag{6.15}$$

and

$$J = J_1 + J_2, \tag{6.16}$$

where J is the least squares objective function.

The same kind of arguments can be extended to consider in general the partitioning of the overall system of balance equations into k smaller subsystems (defining k equivalent subproblems). At the ith stage we will have

$$\hat{\varepsilon}_i = \mathbf{\Psi}_i\mathbf{A}_i^{\mathrm{T}}\left(\mathbf{A}_i\mathbf{\Psi}_i\mathbf{A}_i^{\mathrm{T}}\right)^{-1}\mathbf{A}_i\left(\mathbf{y} - \sum_{j=1}^{i-1}\hat{\varepsilon}_j\right) \tag{6.17}$$

and

$$\mathbf{\Sigma}_{\hat{\varepsilon}_i} = \mathbf{\Psi}_i\mathbf{A}_i^{\mathrm{T}}\left(\mathbf{A}_i\mathbf{\Psi}_i\mathbf{A}_i^{\mathrm{T}}\right)^{-1}\mathbf{A}_i\mathbf{\Psi}_i \tag{6.18}$$

$$\mathbf{\Psi}_i = \mathbf{\Psi}_{i-1} - \mathbf{\Sigma}_{\hat{\varepsilon}_{i-1}}, \tag{6.19}$$

using as starting values $\mathbf{\Psi}_1 = \mathbf{\Psi}$. In general, when the total set of balances is processed in k steps, the least squares objective and the covariance of the measurement estimates are given by

$$J = \sum_{i=1}^{k} J_i, \tag{6.20}$$

where

$$J_i = \left[\mathbf{A}_i\left(\mathbf{y} - \sum_{j=1}^{i-1} \hat{\varepsilon}_j\right)\right]^{\mathrm{T}} \left(\mathbf{A}_i \boldsymbol{\Psi}_i \mathbf{A}_i^{\mathrm{T}}\right)^{-1} \left[\mathbf{A}_i\left(\mathbf{y} - \sum_{j=1}^{i-1} \hat{\varepsilon}_j\right)\right] \tag{6.21}$$

and

$$\boldsymbol{\Sigma}_{\hat{\mathbf{x}}_k} = \boldsymbol{\Psi}_k - \boldsymbol{\Sigma}_{\hat{\varepsilon}_k} \tag{6.22}$$

The recursive formulas that have been presented exhibit some advantages over classical batch processing. First, they avoid the inversion of the normal coefficient matrix, since we would usually process a few equations at a time. Obviously, when only one equation is involved each time, the inversion degenerates into computing the reciprocal of a scalar. Furthermore, these sequential relationships can also be used to isolate systematic errors that may be present in the data set, as will be shown in the next chapter.

EXAMPLE 6.1

Let us apply the sequential processing of the constraints to the system defined previously in Example 5.1. We will process one equation at a time starting from the unconstrained problem.

Step 1. According to our definition we have for the first balance

$$\mathbf{A}_1 = [0.1 \quad 0.6 \quad -0.2 \quad -0.7],$$

$$\boldsymbol{\Psi} = \boldsymbol{\Psi}_1 = \begin{bmatrix} 0.000289 & & & \\ & 0.0025 & & \\ & & 0.000576 & \\ & & & 0.04 \end{bmatrix}.$$

The scalar to be inverted is now

$$\left[\mathbf{A}_1 \boldsymbol{\Psi}_1^{-1} \mathbf{A}_1^{\mathrm{T}}\right] = 0.0205.$$

Accordingly, we have for the estimate of the variables, the measurement errors, and the error estimate covariance

$$\hat{\mathbf{x}}_1 = \begin{bmatrix} 0.1859 \\ 4.7984 \\ 1.2291 \\ 3.7883 \end{bmatrix}, \quad \hat{\varepsilon}_1 = \begin{bmatrix} -0.0001 \\ -0.0049 \\ 0.0004 \\ 0.0917 \end{bmatrix}, \quad \boldsymbol{\Sigma}_{\hat{\varepsilon}_1} = \begin{bmatrix} 0 & 0 & 0 & 0 \\ 0 & 0.0001 & 0 & -0.002 \\ 0 & 0 & 0 & 0.0002 \\ 0 & -0.002 & 0.0002 & 0.0382 \end{bmatrix}.$$

Step 2. Processing equation 2, we have

$$\mathbf{A}_2 = [0.8 \quad 0.1 \quad -0.2 \quad -0.1],$$

$$\boldsymbol{\Psi}_2 = \boldsymbol{\Psi}_1 - \boldsymbol{\Sigma}_{\hat{\varepsilon}_1} = \begin{bmatrix} 0.0003 & 0 & 0 & 0 \\ 0 & 0.0024 & 0 & 0.002 \\ 0 & 0 & 0.0006 & -0.0002 \\ 0 & 0.002 & -0.0002 & 0.0018 \end{bmatrix}$$

and

$$\left[\mathbf{A}_2 \boldsymbol{\Psi}_2 \mathbf{A}_2^{\mathrm{T}}\right] = 0.000196.$$

Therefore,

$$\hat{\mathbf{x}}_2 = \begin{bmatrix} 0.1814 \\ 0.7978 \\ 1.2311 \\ 3.7866 \end{bmatrix}, \quad \hat{\varepsilon}_2 = \begin{bmatrix} 0.0045 \\ 0.0006 \\ -0.002 \\ 0.0017 \end{bmatrix},$$

$$\boldsymbol{\Sigma}_{\hat{\varepsilon}_2} = (10^{-3}) \begin{bmatrix} 0.2633 & 0.0360 & -0.1141 & 0.1011 \\ 0.0360 & 0.0049 & -0.0156 & 0.0138 \\ -0.1141 & -0.0156 & 0.0495 & -0.0438 \\ 0.1011 & 0.0138 & -0.0438 & 0.0388 \end{bmatrix}.$$

Step 3. We process the final constraint equation. In this case

$$\mathbf{A}_3 = [0.1 \quad 0.3 \quad -0.6 \quad -0.2],$$

$$\boldsymbol{\Psi}_3 = \begin{bmatrix} 0 & 0 & 0.0001 & -0.0001 \\ 0 & 0.0024 & 0 & 0.002 \\ 0.0001 & 0 & 0.0005 & -0.0001 \\ -0.0001 & 0.002 & -0.0001 & 0.0018 \end{bmatrix}$$

and

$$\left[\mathbf{A}_3 \boldsymbol{\Psi}_3 \mathbf{A}_3^{\mathrm{T}}\right] = 0.0001816.$$

Finally, we have for the estimate of the state and the measurement error

$$\hat{\mathbf{x}}_3 = \begin{bmatrix} 0.1676 \\ 4.8594 \\ 1.1730 \\ 3.8540 \end{bmatrix}, \quad \hat{\varepsilon}_3 = \begin{bmatrix} 0.0138 \\ -0.0616 \\ 0.0581 \\ -0.0675 \end{bmatrix}$$

$$\boldsymbol{\Sigma}_{\hat{\varepsilon}_3} = (10^{-3}) \begin{bmatrix} 0.0233 & -0.1043 & 0.0983 & -0.1142 \\ -0.1043 & 0.4660 & -0.4393 & 0.5101 \\ 0.0983 & -0.4393 & 0.4140 & -0.4808 \\ -0.1142 & 0.5101 & -0.4808 & 0.5583 \end{bmatrix}.$$

6.3. SEQUENTIAL PROCESSING OF THE MEASUREMENTS

For the sequential processing of measurements data, we will consider the total set of measurements partitioned into two smaller subsets composed of $(g - c)$ and c measurements, respectively. Partitioning the matrix $\mathbf{A}$ along these lines, we have

$$\mathbf{A} = [\mathbf{A}_{\mathrm{g}} \quad \mathbf{A}_{\mathrm{c}}], \tag{6.23}$$

where $\mathbf{A}_{\mathrm{g}}$ is an $[m \times (g - c)]$ matrix whose columns are composed of the coefficients of the $(g - c)$ measurements and $\mathbf{A}_{\mathrm{c}}$ is an $(m \times c)$ matrix corresponding to the remaining c measurements. Assuming that there is no correlation between the various measurements, the covariance matrix can also be partitioned, leading to

$$\boldsymbol{\Psi} = \begin{bmatrix} \boldsymbol{\Psi}_{\mathrm{g}} & \\ & \boldsymbol{\Psi}_{\mathrm{c}} \end{bmatrix}. \tag{6.24}$$

Introducing this partitioning into the equations, we have for the covariance matrix of the residuum in the balances

$$\begin{aligned}\boldsymbol{\Phi} = \mathbf{A}\boldsymbol{\Psi}\mathbf{A}^{\mathrm{T}} &= [\mathbf{A}_{\mathrm{g}} \quad \mathbf{A}_{\mathrm{c}}]\begin{bmatrix}\boldsymbol{\Psi}_{\mathrm{g}} & 0 \\ 0 & \boldsymbol{\Psi}_{\mathrm{c}}\end{bmatrix}\begin{bmatrix}\mathbf{A}_{\mathrm{g}}^{\mathrm{T}} \\ \mathbf{A}_{\mathrm{c}}^{\mathrm{T}}\end{bmatrix} \\ &= \mathbf{A}_{\mathrm{g}}\boldsymbol{\Psi}_{\mathrm{g}}\mathbf{A}_{\mathrm{g}}^{\mathrm{T}} + \mathbf{A}_{\mathrm{c}}\boldsymbol{\Psi}_{\mathrm{c}}\mathbf{A}_{\mathrm{c}}^{\mathrm{T}}.\end{aligned} \tag{6.25}$$

The right-hand side of Eq. (6.25) represents the covariance of the residuum when both blocks of measurement equations are used. The first term in the right-hand side is the covariance resulting from considering the information provided by the first set of measurements. Hence, the second term arises as a correction to the estimate, due to the incorporation of new measurements (new information) as it becomes available:

$$\boldsymbol{\Phi}_{\mathrm{new}} = \boldsymbol{\Phi}_{\mathrm{old}} + \mathbf{A}_{\mathrm{c}}\boldsymbol{\Psi}_{\mathrm{c}}\mathbf{A}_{\mathrm{c}}^{\mathrm{T}}. \tag{6.26}$$

This procedure can be extended to the case of having an arbitrary number of blocks of information (measurements). In this case, the covariance matrix $\boldsymbol{\Phi}$, which uses the information provided by the jth block, is given by

$$\boldsymbol{\Phi}_j = \boldsymbol{\Phi}_{j-1} + \left(\mathbf{A}_j\boldsymbol{\Psi}_j\mathbf{A}_j^{\mathrm{T}}\right). \tag{6.27}$$

This formula corresponds to adding new observations to the sequential treatment. If the covariance matrix of the reduced set of measurements is to be recovered from the augmented one, then solving for $\boldsymbol{\Phi}_{j-1}$ from Eq. (6.27), we have

$$\boldsymbol{\Phi}_{j-1} = \boldsymbol{\Phi}_j - \left(\mathbf{A}_j\boldsymbol{\Psi}_j\mathbf{A}_j^{\mathrm{T}}\right). \tag{6.28}$$

Equation (6.28) yields the new covariance when measurements are deleted from the original formulation. In Eqs. (6.27) and (6.28), the subscript j denotes the jth set of measurements. If we redefine j to corresponds to the jth step of the sequential treatment of the measurements, then the covariance matrix emerging sequentially, after sets of observations are added or deleted, can be written as follows

$$\boldsymbol{\Phi}_j = \boldsymbol{\Phi}_{j-1} \pm \left(\mathbf{A}_j\boldsymbol{\Psi}_j\mathbf{A}_j^{\mathrm{T}}\right), \tag{6.29}$$

where the signs + and − correspond to addition or deletion, respectively. Matrix $\mathbf{A}_j$ in each case is the corresponding submatrix of the original matrix of coefficients.

To solve the least squares problem for the estimate of the measurement errors we need to invert the covariance matrix $\boldsymbol{\Phi}$. It is possible to relate $\boldsymbol{\Phi}_j^{-1}$ to $\boldsymbol{\Phi}_{j-1}^{-1}$ through a simple recursive formula. Let us recall the following matrix inversion lemma (Noble, 1969):

LEMMA

Let $\mathbf{Y}$ *and* $\mathbf{Z}$ *be nonsingular matrices of order* $(n \times n)$ *and* $(p \times p)$*, respectively, and* $\mathbf{U}$ *an* $(n \times p)$ *matrix. If*

$$\mathbf{X} = \mathbf{Y} \pm \mathbf{U}\mathbf{Z}\mathbf{U}^{\mathrm{T}}, \tag{6.30}$$

then

$$\mathbf{X}^{-1} = \mathbf{Y}^{-1} \mp \mathbf{Y}^{-1}\mathbf{U}[\mathbf{Z}^{-1} \pm \mathbf{U}^{\mathrm{T}}\mathbf{Y}^{-1}\mathbf{U}]^{-1}\mathbf{U}^{\mathrm{T}}\mathbf{Y}^{-1}. \tag{6.31}$$

Applying this lemma to the matrices of Eq. (6.29), we have

$$\boldsymbol{\Phi}_j^{-1} = \boldsymbol{\Phi}_{j-1}^{-1}\left\{\mathbf{I} \mp \mathbf{A}_j\left(\boldsymbol{\Psi}_j^{-1} \pm \mathbf{A}_j^{\mathrm{T}}\boldsymbol{\Phi}_{j-1}^{-1}\mathbf{A}_j\right)^{-1}\mathbf{A}_j^{\mathrm{T}}\boldsymbol{\Phi}_{j-1}^{-1}\right\}. \tag{6.32}$$

When measurements are added or deleted, this recursive formula exhibits the advantage of avoiding the inversion of large matrices. Particularly, when only one measurement is added or deleted at a time, the inversion degenerates into computing the reciprocal of a scalar.

EXAMPLE 6.2

Let us consider the same system used in the previous example to show the application of the sequential deletion of one of the sensors from the original problem formulation.

The solution of Example 6.1 is considered as the first stage of the sequential procedure. The following residuum in the balances and its corresponding covariance matrix are obtained:

$$\mathbf{r}_1 = \begin{bmatrix} -0.06722 \\ -0.00591 \\ -0.05707 \end{bmatrix}, \quad \boldsymbol{\Phi}_1^{-1} = \begin{bmatrix} 483.65 & -241.96 & -1339.82 \\ -241.96 & 5890.55 & -2071.62 \\ -1339.82 & -2071.62 & 5506.04 \end{bmatrix}.$$

In the second stage ($j = 2$), the deletion of measurement f_1 from the adjustment is desired. In this case,

$$\mathbf{A}_2 = \begin{bmatrix} 0.1 \\ 0.8 \\ 0.1 \end{bmatrix}, \quad \boldsymbol{\Psi}_2 = [2.89 \times 10^{-4}].$$

The scalar to be inverted is now given by

$$\left(\boldsymbol{\Psi}_2^{-1} + \mathbf{A}_2^{\mathrm{T}}\boldsymbol{\Phi}_1^{-1}\mathbf{A}_2\right) = \left(1/(2.89 \times 10^{-4}) + [0.1 \quad 0.8 \quad 0.1] \times \begin{bmatrix} 483.65 & -241.96 & -1339.82 \\ -241.96 & 5890.55 & -2071.62 \\ -1339.82 & -2071.62 & 5506.04 \end{bmatrix} \begin{bmatrix} 0.1 \\ 0.8 \\ 0.1 \end{bmatrix}\right),$$

and from Eq. (6.32), $\boldsymbol{\Phi}_2^{-1}$ is

$$\boldsymbol{\Phi}_2^{-1} = \begin{bmatrix} 472.34 & -60.47 & -1390.07 \\ -60.47 & 2977.47 & -1265.08 \\ -1390.07 & -1265.08 & 5282.73 \end{bmatrix},$$

with the estimate of the measurement errors and vector $\mathbf{x}$ given by

$$\hat{\varepsilon}_2 = \begin{bmatrix} -0.0425 \\ 0.0622 \\ 0.0445 \end{bmatrix}, \quad \hat{\mathbf{x}}_2 = \begin{bmatrix} 4.8360 \\ 1.1673 \\ 3.8354 \end{bmatrix}.$$

6.4. ALTERNATIVE FORMULATION FROM ESTIMATION THEORY

In this section we develop the best linear estimate of $\mathbf{x}$ for the general linear measurement model

$$\mathbf{y} = \mathbf{C}\mathbf{x} + \varepsilon, \quad \mathbf{C} \in \Re^{l \times g} \tag{6.33}$$

under a set of constraints imposed on the process variables by the balance equations. Two cases will be considered:

1. Estimate $\mathbf{x}$ given that the model (6.33) holds and, further, that the set of constraints on $\mathbf{x}$ is known to be such as indicated by Eq. (6.1):

$$\mathbf{Ax} = \mathbf{0}, \tag{6.1}$$

where

$$\operatorname{rank} \mathbf{A} = m < g. \tag{6.34}$$

2. Estimate $\mathbf{x}$ given that the previous measurement model holds and, further, that the set of linear constraints is satisfied with an additive random component $\mathbf{w}$:

$$\mathbf{Ax} + \mathbf{w} = \mathbf{0}. \tag{6.35}$$

The vector $\mathbf{w}$ is an additive error introduced to account for inaccuracies generated by approximations and reflects the expected degree of modeling errors.

6.4.1. Estimation with Nonrandom Constraints

To incorporate the constraints in our least squares problem, we consider the Lagrangian equation

$$L = (\mathbf{y} - \mathbf{Cx})^{\mathrm{T}} \boldsymbol{\Psi}^{-1} (\mathbf{y} - \mathbf{Cx}) + \boldsymbol{\lambda}^{\mathrm{T}} \mathbf{Ax}, \tag{6.36}$$

where $\boldsymbol{\lambda}$ is the vector of Lagrangian multipliers. The optimality conditions are then

$$\frac{\partial L}{\partial \mathbf{x}} = -2\mathbf{C}^{\mathrm{T}} \boldsymbol{\Psi}^{-1} (\mathbf{y} - \mathbf{Cx}) + \mathbf{A}^{\mathrm{T}} \boldsymbol{\lambda} = \mathbf{0} \tag{6.37}$$

$$\frac{\partial L}{\partial \boldsymbol{\lambda}} = \mathbf{Ax} = \mathbf{0}. \tag{6.38}$$

These give the necessary conditions for L to be minimum. The solution of the previous problem is

$$\hat{\mathbf{x}} = \hat{\mathbf{x}}_0 - (\mathbf{C}^{\mathrm{T}} \boldsymbol{\Psi}^{-1} \mathbf{C})^{-1} \mathbf{A}^{\mathrm{T}} [\mathbf{A} (\mathbf{C}^{\mathrm{T}} \boldsymbol{\Psi}^{-1} \mathbf{C})^{-1} \mathbf{A}^{\mathrm{T}}]^{-1} \mathbf{A} \hat{\mathbf{x}}_0 \tag{6.39}$$

$$\hat{\boldsymbol{\lambda}} = 2[\mathbf{A} (\mathbf{C}^{\mathrm{T}} \boldsymbol{\Psi}^{-1} \mathbf{C})^{-1} \mathbf{A}^{\mathrm{T}}]^{-1} \mathbf{A} \hat{\mathbf{x}}_0, \tag{6.40}$$

where $\hat{\mathbf{x}}_0$ is the solution of the least squares problem without constraints i.e.

$$\hat{\mathbf{x}}_0 = (\mathbf{C}^{\mathrm{T}} \boldsymbol{\Psi}^{-1} \mathbf{C})^{-1} \mathbf{C}^{\mathrm{T}} \boldsymbol{\Psi}^{-1} \mathbf{y}. \tag{6.41}$$

It is important to note that

$$\begin{aligned}
\mathbf{A}\hat{\mathbf{x}} &= \mathbf{A}[\hat{\mathbf{x}}_0 - (\mathbf{C}^{\mathrm{T}} \boldsymbol{\Psi}^{-1} \mathbf{C})^{-1} \mathbf{A}^{\mathrm{T}} [\mathbf{A} (\mathbf{C}^{\mathrm{T}} \boldsymbol{\Psi}^{-1} \mathbf{C})^{-1} \mathbf{A}^{\mathrm{T}}]^{-1} \mathbf{A} \hat{\mathbf{x}}_0] \\
&= \mathbf{A}\hat{\mathbf{x}}_0 - [\mathbf{A} (\mathbf{C}^{\mathrm{T}} \boldsymbol{\Psi}^{-1} \mathbf{C})^{-1} \mathbf{A}^{\mathrm{T}}][\mathbf{A} (\mathbf{C}^{\mathrm{T}} \boldsymbol{\Psi}^{-1} \mathbf{C})^{-1} \mathbf{A}^{\mathrm{T}}]^{-1} \mathbf{A} \hat{\mathbf{x}}_0 \\
&= \mathbf{A}\hat{\mathbf{x}}_0 - \mathbf{A}\hat{\mathbf{x}}_0 = \mathbf{0},
\end{aligned} \tag{6.42}$$

as was required. Also note that the covariance matrix of the error in the estimate is obtained as follows:

$$\boldsymbol{\Sigma}_{\hat{\mathbf{x}}} = \mathbf{B} (\mathbf{C}^{\mathrm{T}} \boldsymbol{\Psi}^{-1} \mathbf{C})^{-1} \mathbf{B}^{\mathrm{T}}, \tag{6.43}$$

where

$$\mathbf{B} = \mathbf{I} - (\mathbf{C}^{\mathrm{T}}\boldsymbol{\Psi}^{-1}\mathbf{C})^{-1}\mathbf{A}^{\mathrm{T}}[\mathbf{A}(\mathbf{C}^{\mathrm{T}}\boldsymbol{\Psi}^{-1}\mathbf{C})^{-1}\mathbf{A}^{\mathrm{T}}]^{-1}\mathbf{A}. \tag{6.44}$$

On reducing the covariance to a simpler form, we find that

$$\begin{aligned}\boldsymbol{\Sigma}_{\hat{\mathbf{x}}} &= (\mathbf{C}^{\mathrm{T}}\boldsymbol{\Psi}^{-1}\mathbf{C})^{-1} - (\mathbf{C}^{\mathrm{T}}\boldsymbol{\Psi}^{-1}\mathbf{C})^{-1}\mathbf{A}^{\mathrm{T}}[\mathbf{A}(\mathbf{C}^{\mathrm{T}}\boldsymbol{\Psi}^{-1}\mathbf{C})^{-1}\mathbf{A}^{\mathrm{T}}]^{-1}\mathbf{A}(\mathbf{C}^{\mathrm{T}}\boldsymbol{\Psi}^{-1}\mathbf{C})^{-1}\\ &= \boldsymbol{\Sigma}_{\hat{\mathbf{x}}_0} - (\mathbf{C}^{\mathrm{T}}\boldsymbol{\Psi}^{-1}\mathbf{C})^{-1}\mathbf{A}^{\mathrm{T}}[\mathbf{A}(\mathbf{C}^{\mathrm{T}}\boldsymbol{\Psi}^{-1}\mathbf{C})^{-1}\mathbf{A}^{\mathrm{T}}]^{-1}\mathbf{A}(\mathbf{C}^{\mathrm{T}}\boldsymbol{\Psi}^{-1}\mathbf{C})^{-1},\end{aligned} \tag{6.45}$$

where $\boldsymbol{\Sigma}_{\hat{\mathbf{x}}_0}$ is the original covariance without the constraints.

Notice that the addition of information in terms of the constraints leads to a reduction in the covariance of the estimate of $\mathbf{x}(\hat{\mathbf{x}})$.

6.4.2. Constraints with Additive Random Components

In this section, we first consider that, in addition to the measurement equations model (6.33), we have also the conditions defined by (6.35) imposed on the vector of process variables. Furthermore, we will assume, as a more general formulation, that a priori information enters into the estimation problem.

Case I

The objective function to be minimized is restated as follows:

$$J = \mathbf{w}^{\mathrm{T}}\mathbf{R}_1\mathbf{w} + (\mathbf{y} - \mathbf{C}\mathbf{x})^{\mathrm{T}}\mathbf{R}_2(\mathbf{y} - \mathbf{C}\mathbf{x}), \tag{6.46}$$

where $\mathbf{R}_1$ and $\mathbf{R}_2$ are symmetric positive definite weighting matrices (usually taken as the inverse of the covariance of the errors). $\mathbf{R}_1$ reflects the expected degree of satisfaction of the constraints (in our case $\mathbf{R}_1 \gg \mathbf{R}_2$). The minimization problem becomes

$$\begin{aligned}&\text{Min } \mathbf{w}^{\mathrm{T}}\mathbf{R}_1\mathbf{w} + (\mathbf{y} - \mathbf{C}\mathbf{x})^{\mathrm{T}}\mathbf{R}_2(\mathbf{y} - \mathbf{C}\mathbf{x})\\ &\text{s.t.}\\ &\qquad \mathbf{A}\mathbf{x} + \mathbf{w} = \mathbf{0}.\end{aligned} \tag{6.47}$$

Defining

$$\begin{aligned}\mathbf{s}^{\mathrm{T}} &= [\mathbf{w}^{\mathrm{T}} \quad \boldsymbol{\varepsilon}^{\mathrm{T}}]; \mathbf{s} \in \Re^{m+l}\\ \mathbf{R} &= \begin{bmatrix}\mathbf{R}_1 & \\ & \mathbf{R}_2\end{bmatrix}; \mathbf{R} \in \Re^{(m+l)\times(m+l)}\\ \mathbf{D}^{\mathrm{T}} &= [\mathbf{A}^{\mathrm{T}} \quad \mathbf{C}^{\mathrm{T}}]\\ \mathbf{z}^{\mathrm{T}} &= [\mathbf{0}^{\mathrm{T}} \quad \mathbf{y}^{\mathrm{T}}],\end{aligned} \tag{6.48}$$

the optimal solution for the estimates is given by

$$\hat{\mathbf{x}} = [\mathbf{D}^{\mathrm{T}}\mathbf{R}\mathbf{D}]^{-1}\mathbf{D}^{\mathrm{T}}\mathbf{R}\mathbf{z}. \tag{6.49}$$

Assume that $\mathbf{s}$ is a random error vector with zero mean and covariance $\mathbf{P}$, that is,

$$\begin{aligned}E(\mathbf{s}) &= \mathbf{0}\\ E(\mathbf{s}\mathbf{s}^{\mathrm{T}}) = \mathbf{P} &= \begin{bmatrix}\mathbf{P}_1 & \\ & \boldsymbol{\Psi}\end{bmatrix},\end{aligned} \tag{6.50}$$

which implies that the error vectors $\mathbf{s}_i$, $i = 1, 2$, are uncorrelated. The covariance of the estimate then becomes

$$\Sigma_{\hat{x}} = [\mathbf{D}^{\mathrm{T}}\mathbf{R}\mathbf{D}]^{-1}\mathbf{D}^{\mathrm{T}}\mathbf{R}\,\mathbf{P}\,\mathbf{R}\mathbf{D}[\mathbf{D}^{\mathrm{T}}\mathbf{R}\mathbf{D}]^{-1}. \tag{6.51}$$

Assuming that the weighting matrix $\mathbf{R}$ is taken as the inverse of the covariance matrix $\mathbf{P}$, that is, $\mathbf{R} = \mathbf{P}^{-1}$ Eq. (6.51) simplifies to

$$\Sigma_{\hat{x}} = [\mathbf{D}^{\mathrm{T}}\mathbf{P}^{-1}\mathbf{D}]^{-1} = \left[\mathbf{A}^{\mathrm{T}}\mathbf{P}_1^{-1}\mathbf{A} + \mathbf{C}^{\mathrm{T}}\Psi^{-1}\mathbf{C}\right]^{-1}. \tag{6.52}$$

Here the covariance matrix has two contributions due to model equations (constraints) and measurements, respectively.

Note: The interpretation of the matrix $\Sigma_{\hat{x}}$ as the covariance matrix of the errors in $\hat{\mathbf{x}}$ has important applications. The value of any estimate is greatly enhanced if its accuracy is known. $\Sigma_{\hat{x}}$ is also very useful in initial design and development, as it can be calculated before the estimator is implemented. $\Sigma_{\hat{x}}$ can be used to study measurement placement and what type and accuracy of information is actually needed. ♣

Now, for the sequential treatment of constraint and measurements equations, consider the system partitioned into two groups as follows:

$$\mathbf{0} = \begin{bmatrix} \mathbf{A}_1 \\ \mathbf{A}_2 \end{bmatrix} \mathbf{x} + \begin{bmatrix} \mathbf{w}_1 \\ \mathbf{w}_2 \end{bmatrix} \tag{6.53}$$

for the constraints, where $\mathbf{w}_1 \in \Re^{(m-a)}$, $\mathbf{w}_2 \in \Re^{a}$, and $\mathbf{A}_1 \in \Re^{(m-a)\times g}$, $\mathbf{A}_2 \in \Re^{a\times g}$. Also,

$$\mathbf{y} = \begin{bmatrix} \mathbf{C}_1 \\ \mathbf{C}_2 \end{bmatrix} \mathbf{x} + \begin{bmatrix} \varepsilon_1 \\ \varepsilon_2 \end{bmatrix} \tag{6.54}$$

for the measurements, where $\varepsilon_1 \in \Re^{r}$, $\varepsilon_2 \in \Re^{(l-r)}$, $\mathbf{C}_1 \in \Re^{r\times g}$, and $\mathbf{C}_2 \in \Re^{(l-r)\times g}$. Since no correlation is assumed for the corresponding noises, the weighting matrices can be partitioned as well, that is,

$$\mathbf{R}_1 = \begin{bmatrix} (\mathbf{R}_1)_1 & \\ & (\mathbf{R}_1)_2 \end{bmatrix}, \quad \mathbf{R}_2 = \begin{bmatrix} (\mathbf{R}_2)_1 & \\ & (\mathbf{R}_2)_2 \end{bmatrix}. \tag{6.55}$$

Introducing these partitions into Eq. (6.52), we have for the constraints

$$\begin{aligned} \Sigma_2^{-1} &= \mathbf{A}_1^{\mathrm{T}}(\mathbf{R}_1)_1\mathbf{A}_1 + \mathbf{C}^{\mathrm{T}}\mathbf{R}_2\mathbf{C} + \mathbf{A}_2^{\mathrm{T}}(\mathbf{R}_1)_2\mathbf{A}_2 \\ &= \Sigma_1^{-1} + \mathbf{A}_2^{\mathrm{T}}(\mathbf{R}_1)_2\mathbf{A}_2. \end{aligned} \tag{6.56}$$

A similar arrangement can be obtained for the measurements. In general,

$$\Sigma_i^{-1} = \Sigma_{i-1}^{-1} \pm \mathbf{A}_2^{\mathrm{T}}(\mathbf{R}_1)_2\mathbf{A}_2 \tag{6.57}$$

for both the addition and the deletion of constraint equations. In the same way, for the sequential addition or deletion of measurement equations, we have

$$\Sigma_i^{-1} = \Sigma_{i-1}^{-1} \pm \mathbf{C}_2^{\mathrm{T}}(\mathbf{R}_2)_2\mathbf{C}_2. \tag{6.58}$$

The application of the matrix inversion lemma leads to recursive formulas similar to those previously obtained in the conventional approach set out in Section 6.3.

Case 2

We can also consider that in addition to the model equations, a priori information enters into the estimation problem, that is, some previous knowledge about the process variables or system states is available. If the vector of a priori estimated errors $\mathbf{p}$ is defined as

$$\mathbf{p} = \bar{\mathbf{x}} - \mathbf{x}, \tag{6.59}$$

where $\bar{\mathbf{x}}$ is the vector of the a priori estimate of the process variables, then the objective function to be minimized can be expanded as follows:

$$J = \mathbf{w}^{\mathrm{T}}\mathbf{R}_1\mathbf{w} + (\mathbf{y} - \mathbf{C}\mathbf{x})^{\mathrm{T}}\mathbf{R}_2(\mathbf{y} - \mathbf{C}\mathbf{x}) + \mathbf{p}^{\mathrm{T}}\mathbf{R}_3\mathbf{p}. \tag{6.60}$$

$\mathbf{R}_3$ is a symmetric positive definite weighting matrix. The minimization problem becomes

$$\begin{aligned} &\text{Min } \mathbf{w}^{\mathrm{T}}\mathbf{R}_1\mathbf{w} + (\mathbf{y} - \mathbf{C}\mathbf{x})^{\mathrm{T}}\mathbf{R}_2(\mathbf{y} - \mathbf{C}\mathbf{x}) + \mathbf{p}^{\mathrm{T}}\mathbf{R}_3\mathbf{p},\\ &\text{s.t.}\\ &\qquad \mathbf{A}\mathbf{x} + \mathbf{w} = \mathbf{0}. \end{aligned} \tag{6.61}$$

Defining

$$\begin{aligned} \mathbf{s}^{\mathrm{T}} &= [\mathbf{w}^{\mathrm{T}} \quad \boldsymbol{\varepsilon}^{\mathrm{T}} \quad \mathbf{p}^{\mathrm{T}}];\ \mathbf{s} \in \Re^{m+l+g}\\ \mathbf{R} &= \begin{bmatrix} \mathbf{R}_1 & & \\ & \mathbf{R}_2 & \\ & & \mathbf{R}_3 \end{bmatrix};\ \mathbf{R} \in \Re^{(m+l+g)\times(m+l+g)}\\ \mathbf{D}^{\mathrm{T}} &= [\mathbf{A}^{\mathrm{T}} \quad \mathbf{C}^{\mathrm{T}} \quad \mathbf{I}]\\ \mathbf{z}^{\mathrm{T}} &= [\mathbf{0}^{\mathrm{T}} \quad \mathbf{y}^{\mathrm{T}} \quad \bar{\mathbf{x}}^{\mathrm{T}}], \end{aligned} \tag{6.62}$$

the optimal solution for the estimates is given by

$$\hat{\mathbf{x}} = [\mathbf{D}^{\mathrm{T}}\mathbf{R}\mathbf{D}]^{-1}\mathbf{D}^{\mathrm{T}}\mathbf{R}\mathbf{z}. \tag{6.63}$$

Assume that $\mathbf{s}$ is a random error vector with zero mean and covariance $\mathbf{P}$, that is,

$$\begin{aligned} E(\mathbf{s}) &= \mathbf{0}\\ E(\mathbf{s}\mathbf{s}^{\mathrm{T}}) = \mathbf{P} &= \begin{bmatrix} \mathbf{P}_1 & & \\ & \boldsymbol{\Psi} & \\ & & \mathbf{P}_3 \end{bmatrix}, \end{aligned} \tag{6.64}$$

which implies that the error vectors $\mathbf{s}_i$, $i = 1, 2, 3$, are uncorrelated. The covariance of the estimate then becomes

$$\boldsymbol{\Sigma}_{\hat{\mathbf{x}}} = [\mathbf{D}^{\mathrm{T}}\mathbf{P}^{-1}\mathbf{D}]^{-1} = \left[\mathbf{A}^{\mathrm{T}}\mathbf{P}_1^{-1}\mathbf{A} + \mathbf{C}^{\mathrm{T}}\boldsymbol{\Psi}^{-1}\mathbf{C} + \mathbf{P}_3^{-1}\right]^{-1}. \tag{6.65}$$

EXAMPLE 6.3

Let us now take the same system used in Example 4.1 but represented under the general formulation. We have

$$\mathbf{A} = \begin{bmatrix} 0.1 & 0.6 & -0.2 & -0.7\\ 0.8 & 0.1 & -0.2 & -0.1\\ 0.1 & 0.3 & -0.6 & -0.2 \end{bmatrix}, \quad \mathbf{C} = \begin{bmatrix} 1 & 0 & 0 & 0\\ 0 & 1 & 0 & 0\\ 0 & 0 & 1 & 0\\ 0 & 0 & 0 & 1 \end{bmatrix},$$

from which the augmented matrix is

$$\mathbf{D} = \begin{bmatrix} 0.1 & 0.6 & -0.2 & -0.7 \\ 0.8 & 0.1 & -0.2 & -0.1 \\ 0.1 & 0.3 & -0.6 & -0.2 \\ 1 & 0 & 0 & 0 \\ 0 & 1 & 0 & 0 \\ 0 & 0 & 1 & 0 \\ 0 & 0 & 0 & 1 \end{bmatrix}.$$

We will assume that there is no a priori information and that we also need to assign the weights corresponding to the balance equations. They represent the expected degree of satisfaction of the constraints. Assuming as a first approximation $\mathbf{R}_1 = \mathbf{I}$, the estimate of the process variables is given by the formula

$$\hat{\mathbf{x}} = [\mathbf{D}^{\mathrm{T}}\mathbf{R}\mathbf{D}]^{-1}\mathbf{D}^{\mathrm{T}}\mathbf{R}\mathbf{z},$$

that is,

$$\hat{\mathbf{x}} = \begin{bmatrix} 0.1858 \\ 4.7936 \\ 1.2295 \\ 3.8777 \end{bmatrix},$$

which are very close to the measured values. This is consistent because we are giving a large amount of confidence to the measurements. Next, we consider the case of giving larger weights to the satisfaction of the constraints. Table 1 shows the results of the reconciliation for different values of the weighting matrix $\mathbf{R}_1$. As expected, larger corrections are performed to the measured values in order to satisfy the constraints. In the limit, the results of the general approach and the data reconciliation results from Section 6.2 coincide.

It is clear that using this general procedure we have additional degrees of freedom in terms of assigning weights to the satisfaction of constraints. Furthermore, this general formulation clarifies the contributions to the error in the final estimate from the processing of both the constraints and the measurements. It is now possible to study separately aspects related to modeling problems and aspects related to measurement problems.

TABLE 1 Results for Example 6.3

	$\hat{\mathbf{x}}$ $\mathbf{R}_1 = \mathbf{I}$	$\hat{\mathbf{x}}$ $\mathbf{R}_1 = 1000\mathbf{I}$	$\hat{\mathbf{x}}$ $\mathbf{R}_1 = \mathbf{I} \times 10^6\,\mathbf{I}$
f_1	0.1858	0.1852	0.1678
f_2	4.7936	4.8076	4.8591
f_3	1.2295	1.2204	1.1733
f_4	3.877	3.7920	3.8536

6.5. CONCLUSIONS

This chapter discussed the idea of exploiting the sequential processing of information (both constraints and measurements), to allow computations to be done in a recursive way without solving the full-scale reconciliation problem.

The sequential procedure can be implemented on-line, in real time, for any processing plant without much computational effort. Furthermore, by sequentially deleting one measurement at a time, it is possible to quantify the effect of that measurement on the reconciliation procedure, making this approach very suitable for gross error detection/identification, as discussed in the next chapter.

NOTATION

a	number of rows of $\mathbf{A}_2$
$\mathbf{A}$	matrix of linear constraints
$\mathbf{A}_1$	$[(m-a) \times g]$ submatrix of $\mathbf{A}$
$\mathbf{A}_2$	$(a \times g)$ submatrix of $\mathbf{A}$
$\mathbf{A}_g$	$[m \times (g-c)]$ submatrix of $\mathbf{A}$
$\mathbf{A}_c$	$(m \times c)$ submatrix of $\mathbf{A}$
$\mathbf{B}$	$(g \times g)$ matrix defined by Eq. (6.44)
$\mathbf{C}$	$(l \times g)$ matrix of the general measurement model
c	number of columns of $\mathbf{A}_c$
$\mathbf{D}$	matrix defined by Eqs. (6.48) and (6.62)
g	number of measured variables
i	index for the row sequential processing stage
j	index for the measurement sequential processing stage
J	least square objective function
k	number of subsystems for the row sequential processing of $\mathbf{A}$
m	number of process constraints
n	dimension of matrix $\mathbf{Y}$
p	dimension of matrix $\mathbf{Z}$
$\mathbf{P}$	covariance matrix of the vector $\mathbf{s}$
R	number of rows of $\mathbf{C}_1$
$\mathbf{r}$	residuum of process constraints
$\mathbf{R}$	weighting matrix
$\mathbf{s}$	vector defined by Eqs. (6.48) and (6.62)
$\mathbf{U}$	$(p \times n)$ general matrix
$\mathbf{w}$	vector of expected degree of modeling errors
$\mathbf{x}$	vector of true value of measured variables
$\mathbf{X}$	$(n \times n)$ general matrix
$\mathbf{y}$	vector of measurements
$\mathbf{Y}$	$(n \times n)$ general matrix
$\mathbf{z}$	vector defined by Eqs. (6.48) and (6.62)
$\mathbf{Z}$	$(p \times p)$ general matrix

Greek Symbols

ε	measurement random errors
$\mathbf{\Psi}$	covariance matrix of measurement errors
$\mathbf{\Phi}$	covariance matrix of residuum
$\mathbf{\Sigma}_{\hat{\varepsilon}}$	covariance matrix of the error estimates
$\mathbf{\Sigma}_{\hat{\mathbf{x}}}$	covariance matrix of variable estimates
λ	Lagrangian multipliers

Superscripts

$\hat{}$ least square estimation

REFERENCES

Mikhail, E. (1976). "Observations and Least Squares," IEP Ser. Harper & Row, New York.

Noble, E. (1969). "Applied Linear Algebra." Prentice-Hall, Englewood Cliffs, NJ.

Romagnoli, J. (1983). On data reconciliation: constraints processing and treatment of bias. *Chem. Eng. Sci.* **38**, 1107–1117.

Romagnoli, J., and Stephanopoulos, G. (1981). Rectification of process measurement data in the presence of gross errors. *Chem. Eng. Sci.* **36**, 1849–1863.

APPENDIX A

Covariance Matrix of the Estimated Measurement Error

When the Lagrangian technique is used to solve problem (6.5), the following expression for the estimate of the measurement error is obtained:

$$\hat{\varepsilon} = \mathbf{\Psi A}^{\mathrm{T}}(\mathbf{A\Psi A}^{\mathrm{T}})^{-1}\mathbf{Ay}. \tag{A6.1}$$

Its covariance matrix $\mathbf{\Sigma}_{\hat{\varepsilon}}$ is defined as

$$\mathbf{\Sigma}_{\hat{\varepsilon}} = \mathbf{E}(\hat{\varepsilon}\hat{\varepsilon}^{\mathrm{T}}) = \mathbf{E}(\mathbf{\Psi A}^{\mathrm{T}}(\mathbf{A\Psi A}^{\mathrm{T}})^{-1}\mathbf{Ayy}^{\mathrm{T}}\mathbf{A}^{\mathrm{T}}(\mathbf{A\Psi A}^{\mathrm{T}})^{-1}\mathbf{A\Psi}) \tag{A6.2}$$

$$\mathbf{\Sigma}_{\hat{\varepsilon}} = \mathbf{E}(\hat{\varepsilon}\hat{\varepsilon}^{\mathrm{T}}) = \mathbf{\Psi A}^{\mathrm{T}}(\mathbf{A\Psi A}^{\mathrm{T}})^{-1}\mathbf{AE}(\mathbf{yy}^{\mathrm{T}})\mathbf{A}^{\mathrm{T}}(\mathbf{A\Psi A}^{\mathrm{T}})^{-1}\mathbf{A\Psi}. \tag{A6.3}$$

As $\mathbf{\Psi} = \mathbf{E}(\mathbf{yy}^{\mathrm{T}})$, the covariance matrix of the estimated measurement error is

$$\mathbf{\Sigma}_{\hat{\varepsilon}} = \mathbf{E}(\hat{\varepsilon}\hat{\varepsilon}^{\mathrm{T}}) = \mathbf{\Psi A}^{\mathrm{T}}(\mathbf{A\Psi A}^{\mathrm{T}})^{-1}\mathbf{A\Psi}. \tag{A6.4}$$

Covariance Matrix of the Adjusted Measurements

The adjusted measurements can be stated as

$$\hat{\mathbf{x}} = \mathbf{y} - \hat{\varepsilon} = \mathbf{y} - \mathbf{\Psi A}^{\mathrm{T}}(\mathbf{A\Psi A}^{\mathrm{T}})^{-1}\mathbf{Ay}. \tag{A6.5}$$

By applying the propagation rules of the covariance, the following expression is obtained for the covariance of $\hat{\mathbf{x}}$:

$$\mathbf{\Sigma}_{\hat{\mathbf{x}}} = \mathrm{cov}(\mathbf{y}) - \mathrm{cov}(\hat{\varepsilon}) = \mathbf{\Psi} - \mathbf{\Sigma}_{\hat{\varepsilon}}. \tag{A6.6}$$

Equivalence between the Solution of the Overall Set of Balance Equations and the Two-Stage Procedure

1. Proof that $\hat{\varepsilon} = \hat{\varepsilon}_1 + \hat{\varepsilon}_2$ (from Mikhail, 1976):

   The total system of equations

$$\mathbf{A\hat{x}} = \mathbf{0} \tag{A6.7}$$

can be partitioned to

$$\begin{bmatrix} \mathbf{A}_1 \\ \mathbf{A}_2 \end{bmatrix} \hat{\varepsilon} = \begin{bmatrix} \mathbf{A}_1 \\ \mathbf{A}_2 \end{bmatrix} \mathbf{y} \tag{A6.8}$$

or

$$\mathbf{A}_1 \hat{\varepsilon} = \mathbf{A}_1 \mathbf{y} \tag{A6.9}$$

$$\mathbf{A}_2 \hat{\varepsilon} = \mathbf{A}_2 \mathbf{y}. \tag{A6.10}$$

Now, if the vectors of measurement errors resulting from the resolution of each stage are added and multiplied by matrix $\mathbf{A}_2$, we obtain

$$\mathbf{A}_2(\hat{\varepsilon}_1 + \hat{\varepsilon}_2) = \mathbf{A}_2\hat{\varepsilon}_1 + \mathbf{A}_2\mathbf{\Psi}_2\mathbf{A}_2^{\mathrm{T}}\left(\mathbf{A}_2\mathbf{\Psi}_2\mathbf{A}_2^{\mathrm{T}}\right)^{-1}\mathbf{A}_2(\mathbf{y} - \hat{\varepsilon}_1) = \mathbf{A}_2\mathbf{y}. \tag{A6.11}$$

So from Eqs. (A6.10) and (A6.11) it follows that

$$\hat{\varepsilon} = \hat{\varepsilon}_1 + \hat{\varepsilon}_2. \tag{A6.12}$$

2. Proof that $J = J_1 + J_2$ (from Mikhail, 1976):

The computation of J can be accomplished by the expression

$$J = \mathbf{r}^{\mathrm{T}}(\mathbf{A}\mathbf{\Psi}\mathbf{A}^{\mathrm{T}})^{-1}\mathbf{r}, \tag{A6.13}$$

which is equivalent to

$$J = \varepsilon^{\mathrm{T}}\mathbf{\Psi}^{-1}\varepsilon \tag{A6.14}$$

because of

$$\begin{aligned} J = \varepsilon^{\mathrm{T}}\mathbf{\Psi}^{-1}\varepsilon &= [\mathbf{\Psi}\mathbf{A}^{\mathrm{T}}(\mathbf{A}\mathbf{\Psi}\mathbf{A}^{\mathrm{T}})^{-1}\mathbf{r}]^{\mathrm{T}}\mathbf{\Psi}^{-1}[\mathbf{\Psi}\mathbf{A}^{\mathrm{T}}(\mathbf{A}\mathbf{\Psi}\mathbf{A}^{\mathrm{T}})^{-1}\mathbf{r}] \\ &= [\mathbf{A}^{\mathrm{T}}(\mathbf{A}\mathbf{\Psi}\mathbf{A}^{\mathrm{T}})^{-1}\mathbf{r}]^{\mathrm{T}}\mathbf{\Psi}[\mathbf{A}^{\mathrm{T}}(\mathbf{A}\mathbf{\Psi}\mathbf{A}^{\mathrm{T}})^{-1}\mathbf{r}] \\ &= \mathbf{r}^{\mathrm{T}}(\mathbf{A}\mathbf{\Psi}\mathbf{A}^{\mathrm{T}})^{-1}\mathbf{A}\mathbf{\Psi}\mathbf{A}^{\mathrm{T}}(\mathbf{A}\mathbf{\Psi}\mathbf{A}^{\mathrm{T}})^{-1}\mathbf{r} = \mathbf{r}^{\mathrm{T}}(\mathbf{A}\mathbf{\Psi}\mathbf{A}^{\mathrm{T}})^{-1}\mathbf{r}. \end{aligned}$$

If the total system of equations is partitioned into two subsets, the computation of J by Eq. (A6.13) can be accomplished as follows:

$$J = \mathbf{r}^{\mathrm{T}}\left\{\begin{bmatrix} \mathbf{A}_1 \\ \mathbf{A}_2 \end{bmatrix}\mathbf{\Psi}\begin{bmatrix} \mathbf{A}_1^{\mathrm{T}} & \mathbf{A}_2^{\mathrm{T}} \end{bmatrix}\right\}^{-1}\mathbf{r} = \mathbf{r}^{\mathrm{T}}\begin{bmatrix} \mathbf{A}_1\mathbf{\Psi}\mathbf{A}_1^{\mathrm{T}} & \mathbf{A}_1\mathbf{\Psi}\mathbf{A}_2^{\mathrm{T}} \\ \mathbf{A}_2\mathbf{\Psi}\mathbf{A}_1^{\mathrm{T}} & \mathbf{A}_2\mathbf{\Psi}\mathbf{A}_2^{\mathrm{T}} \end{bmatrix}^{-1}\mathbf{r} \tag{A6.15}$$

$$J = \begin{bmatrix} \mathbf{r}_1^{\mathrm{T}} & \mathbf{r}_2^{\mathrm{T}} \end{bmatrix}\begin{bmatrix} \dot{\mathbf{M}} & \bar{\mathbf{M}} \\ \tilde{\mathbf{M}}^{\mathrm{T}} & \ddot{\mathbf{M}} \end{bmatrix}^{-1}\begin{bmatrix} \mathbf{r}_1^{\mathrm{T}} \\ \mathbf{r}_2^{\mathrm{T}} \end{bmatrix} = \begin{bmatrix} \mathbf{r}_1^{\mathrm{T}} & \mathbf{r}_2^{\mathrm{T}} \end{bmatrix}\begin{bmatrix} \mathbf{E} & \mathbf{G} \\ \mathbf{G}^{\mathrm{T}} & \mathbf{H} \end{bmatrix}\begin{bmatrix} \mathbf{r}_1 \\ \mathbf{r}_2 \end{bmatrix} \tag{A6.16}$$

$$J = \left(\mathbf{r}_1^{\mathrm{T}}\dot{\mathbf{M}}^{-1}\mathbf{r}_1 - \mathbf{r}_1^{\mathrm{T}}\dot{\mathbf{M}}^{-1}\bar{\mathbf{M}}\mathbf{G}^{\mathrm{T}}\mathbf{r}_1\right) + 2\mathbf{r}_1^{\mathrm{T}}\mathbf{G}\mathbf{r}_2 + \mathbf{r}_2^{\mathrm{T}}\mathbf{H}\mathbf{r}_2, \tag{A6.17}$$

where

$$\mathbf{H} = (\ddot{\mathbf{M}} - \bar{\mathbf{M}}^{\mathrm{T}}\dot{\mathbf{M}}^{-1}\bar{\mathbf{M}})^{-1} \tag{A6.18}$$

$$\mathbf{G} = -\dot{\mathbf{M}}^{-1}\bar{\mathbf{M}}\mathbf{H} \tag{A6.19}$$

$$\mathbf{E} = \dot{\mathbf{M}}^{-1} - \dot{\mathbf{M}}^{-1}\bar{\mathbf{M}}\mathbf{G}^{\mathrm{T}}. \tag{A6.20}$$

Now with regard to the adjustment in steps, from the first step,

$$J_1 = \mathbf{r}_1^{\mathrm{T}}\left(\mathbf{A}_1\mathbf{\Psi}\mathbf{A}_1^{\mathrm{T}}\right)^{-1}\mathbf{r}_1 = \mathbf{r}_1^{\mathrm{T}}\dot{\mathbf{M}}^{-1}\mathbf{r}_1 \tag{A6.21}$$

$$\hat{\varepsilon}_1 = \mathbf{\Psi}\mathbf{A}_1^{\mathrm{T}}\left(\mathbf{A}_1\mathbf{\Psi}\mathbf{A}_1^{\mathrm{T}}\right)^{-1}\mathbf{r}_1 = \mathbf{\Psi}\mathbf{A}_1^{\mathrm{T}}\dot{\mathbf{M}}^{-1}\mathbf{r}_1. \tag{A6.22}$$

From the second step,

$$\mathbf{r}_2' = \mathbf{A}_2\hat{\varepsilon}_2 = \mathbf{r}_2 - \mathbf{A}_2\hat{\varepsilon}_1 = \mathbf{A}_2(\mathbf{y} - \hat{\varepsilon}_1) \tag{A6.23}$$

and

$$J_2 = \left(\mathbf{r}_2'\right)^{\mathrm{T}}\left(\mathbf{A}_2\boldsymbol{\Psi}_2\mathbf{A}_2^{\mathrm{T}}\right)^{-1}\mathbf{r}_2'. \tag{A6.24}$$

Since

$$\begin{aligned}\left(\mathbf{A}_2\boldsymbol{\Psi}_2\mathbf{A}_2^{\mathrm{T}}\right)^{-1} &= \left[\mathbf{A}_2\left(\boldsymbol{\Psi} - \boldsymbol{\Psi}\mathbf{A}_1^{\mathrm{T}}\dot{\mathbf{M}}^{-1}\mathbf{A}_1\boldsymbol{\Psi}\right)\mathbf{A}_2^{\mathrm{T}}\right]^{-1} \qquad \text{(A6.25)}\\ &= \left[\left(\mathbf{A}_2\boldsymbol{\Psi}\mathbf{A}_2^{\mathrm{T}}\right) - \left(\mathbf{A}_2\boldsymbol{\Psi}\mathbf{A}_1^{\mathrm{T}}\right)\dot{\mathbf{M}}^{-1}(\mathbf{A}_1\boldsymbol{\Psi})\mathbf{A}_2^{\mathrm{T}})\right]^{-1}\\ &= (\ddot{\mathbf{M}} - \bar{\mathbf{M}}^{\mathrm{T}}\dot{\mathbf{M}}^{-1}\bar{\mathbf{M}})^{-1} = \mathbf{H},\end{aligned}$$

Eq. (A6.24) can be expressed as

$$J_2 = \left(\mathbf{r}_2'\right)^{\mathrm{T}}\mathbf{H}\mathbf{r}_2' \tag{A6.26}$$

$$\begin{aligned}J_2 &= \left(\mathbf{r}_2^{\mathrm{T}} - \hat{\varepsilon}_1^{\mathrm{T}}\mathbf{A}_2^{\mathrm{T}}\right)\mathbf{H}(\mathbf{r}_2 - \mathbf{A}_2\hat{\varepsilon}_1)\\ &= \mathbf{r}_2^{\mathrm{T}}\mathbf{H}\mathbf{r}_2 - 2\hat{\varepsilon}_1^{\mathrm{T}}\mathbf{A}_2^{\mathrm{T}}\mathbf{H}\mathbf{r}_2 + \hat{\varepsilon}_1^{\mathrm{T}}\mathbf{A}_2^{\mathrm{T}}\mathbf{H}\mathbf{A}_2\hat{\varepsilon}_1 \qquad \text{(A6.27)}\end{aligned}$$

$$J_2 = \mathbf{r}_2^{\mathrm{T}}\mathbf{H}\mathbf{r}_2 - 2\mathbf{r}_1^{\mathrm{T}}\dot{\mathbf{M}}^{-1}\bar{\mathbf{M}}\mathbf{H}\mathbf{r}_2 + \mathbf{r}_1^{\mathrm{T}}\dot{\mathbf{M}}^{-1}\bar{\mathbf{M}}(\mathbf{H}\bar{\mathbf{M}}^{\mathrm{T}}\dot{\mathbf{M}}^{-1})\mathbf{r}_1 \tag{A6.28}$$

$$J_2 = -\mathbf{r}_1^{\mathrm{T}}\dot{\mathbf{M}}^{-1}\bar{\mathbf{M}}\mathbf{G}^{\mathrm{T}}\mathbf{r}_1 + 2\mathbf{r}_1^{\mathrm{T}}\mathbf{G}\mathbf{r}_2 + \mathbf{r}_2^{\mathrm{T}}\mathbf{H}\mathbf{r}_2. \tag{A6.29}$$

It is obvious from Eqs. (A6.21) and (A6.29) that $J = J_1 + J_2$.

7 TREATMENT OF GROSS ERRORS

In this chapter we start by defining the data reconciliation problem in the presence of gross errors and by describing the different stages to follow for the treatment of biased data. Different strategies for testing a set of data are then described and a serial elimination strategy discussed for identifying sources of gross errors. Then, a method for estimating the amount of gross error (measurement bias and/or leaks) is presented that will allow the continuing use of faulty sensors. Finally, a strategy for the simultaneous estimation and identification of gross errors will be discussed.

7.1. INTRODUCTION

In the previous development it was assumed that only random, normally distributed measurement errors, with zero mean and known covariance, are present in the data. In practice, process data may also contain other types of errors, which are caused by nonrandom events. For instance, instruments may not be adequately compensated, measuring devices may malfunction, or process leaks may be present. These biases are usually referred as gross errors. The presence of gross errors invalidates the statistical basis of data reconciliation procedures. It is also impossible, for example, to prepare an adequate process model on the basis of erroneous measurements or to assess production accounting correctly. In order to avoid these shortcomings we need to check for the presence of gross systematic errors in the measurement data.

There are various ways to identify a systematic large error: (1) with a theoretical analysis of all the effects leading to a gross error, (2) with measurements of a given process variable by two methods with different precision, or (3) by checking the satisfaction of the balance equations. This third alternative is particularly attractive because it is relatively simple and is based on relations of absolute validity, namely on the conservation of mass and energy.

Several works in the literature have attempted to deal with the problem of the location of gross errors. A simple heuristic criterion has been proposed by Vaclaveck and Vosolsobe (1975) to test each of the balances. This approach evaluates a ratio of the balance residuum to the measured average flow of balanced variables through the node, testing after all the double nodes to localize the position of the gross error. Almasy and Sztano (1975) suggest a different procedure based on the statistical properties of the measurements. Their method of search for the source of large errors is limited to systems containing a single element with systematic error, and those cases where the ratio of the extreme error to the dispersion of the regular error is not too small. Mah *et al.* (1976) extensively studied the problem of the identification of the source of gross errors and developed a series of rules (based on graph-theoretical results) that enhance the effectiveness of the algorithmic search. Exploiting the topological character of the process, and using available statistical information, a test function for each node in the flow graph is developed, which is used in an identification scheme by searching along the internal streams.

A serial elimination algorithm was first proposed by Ripps (1965) and extended later by Nogita (1972). This approach eliminates one measuring element at a time from the set of measurements and each time checks the value of a test function, subsequently choosing the consistent set of data with the minimum variance. In this case, after a new measurement has been deleted, the test function and the variance for the resulting system have to be recomputed; when the number of suspect measurements is increased, this may become a laborious solution.

A more systematic approach was developed by Romagnoli and Stephanopoulos (1981) and Romagnoli (1983) to analyze a set of measurement data in the presence of gross errors. The method is based on the idea of exploiting the sequential processing of the information (constraints and measurements), thus allowing the computations to be done in a recursive way without solving the full-scale reconciliation problem.

Of the various available techniques, the most widely used are based on the Measurement Test (Mah and Tamhane, 1982). These are the Modified Iterative Measurement Test (MIMT) developed by Serth and Heenan (1986) and the Generalized Likelihood Ratio (GLR) method presented by Narasimhan and Mah (1987). The MIMT method uses a serial elimination strategy to detect and identify only biases in measuring instruments. The GLR method allows us to identify multiple gross errors of any type. It uses a serial compensation strategy.

The Unbiased Estimation Technique (UBET) was developed by Rollins and Davis (1992). This approach simultaneously provides unbiased estimates and confidence intervals of process variables when biased measurements and process leaks exist.

Gross error detection in steady-state systems has received great attention in the past 10 years. Good surveys of the available methodologies can be found in the monograph by Mah (1990) and the paper of Crowe (1996).

In this chapter we start by defining the data reconciliation problem in the presence of gross errors and by describing the stages to follow for the treatment of gross biased data. Different strategies for testing a set of data are then described and a serial elimination strategy discussed for identifying sources of gross errors. Then, methods for estimating the amount of gross error (measurement bias and/or leaks) are discussed, which will allow the continuing use of faulty sensors. Finally, a strategy for the simultaneous estimation and identification of gross errors (SEGE) is briefly described and compared to GLR and MIMT methodologies.

7.2. GROSS ERROR DETECTION

As was shown before, the measurement vector in the absence of gross errors can be written as

$$\mathbf{y} = \mathbf{x} + \varepsilon, \quad \mathbf{y} \in \Re^g, \tag{7.1}$$

with the following assumptions:

1. The expected value of ε, i.e., $E(\varepsilon) = \mathbf{0}$
2. The covariance matrix of ε is known and given by $E(\varepsilon_i \varepsilon_i^{\mathrm{T}}) = \mathbf{\Psi}$

Furthermore, the balance (constraint) equations for the linear or linearized case are

$$\mathbf{A}\begin{bmatrix}\mathbf{x}\\ \mathbf{u}\end{bmatrix} = \mathbf{0}. \tag{7.2}$$

In Chapters 3 and 4 we have shown that the vector of process variables can be partitioned into four different subsets: (1) overmeasured, (2) just-measured, (3) determinable, and (4) indeterminable. It is clear from the previous developments that only the overmeasured (or overdetermined) process variables provide a spatial redundancy that can be exploited for the correction of their values. It was also shown that the general data reconciliation problem for the whole plant can be replaced by an equivalent two-problem formulation. This partitioning allows a significant reduction in the size of the constrained least squares problem. Accordingly, in order to identify the presence of gross (bias) errors in the measurements and to locate their sources, we need only to concentrate on the largely reduced set of balances

$$\mathbf{A}_1 \mathbf{x}_1 = \mathbf{0}, \tag{7.3}$$

where $\mathbf{A}_1$ corresponds to the reduced subset of redundant equations and $\mathbf{x}_1$, in this case, corresponds to the redundant measured variables (overmeasured). In the following we will make the following nomenclature changes: $\mathbf{A} \equiv \mathbf{A}_1$, $\mathbf{x} \equiv \mathbf{x}_1$, with the balances taking the form

$$\mathbf{A}\mathbf{x} = \mathbf{0}. \tag{7.4}$$

Introducing the measurements within the balances, we obtain the vector of the residua in the balances, $\mathbf{r}$:

$$\mathbf{r} = \mathbf{A}(\mathbf{x} + \varepsilon) = \mathbf{A}\mathbf{x} + \mathbf{A}\varepsilon = \mathbf{A}\varepsilon. \tag{7.5}$$

The most common techniques for detecting the presence of gross errors are based on so-called *statistical hypothesis testing*. This is based on the idea of testing the data set against alternative hypotheses: (1) the *null hypothesis*, H_0, that no gross error is present, and (2) the *alternative hypothesis*, H_1, that gross errors are present.

Under the H_0 hypothesis, if we postulate that the measurement errors are normally distributed, then Eq. (7.5) implies that the residua in the balances are also normally distributed. As it was assumed that $E(\varepsilon) = \mathbf{0}$, we will have

$$E[\mathbf{r}] = E[\mathbf{A}\varepsilon] = \mathbf{A}E[\varepsilon] = \mathbf{0}. \tag{7.6}$$

Furthermore, the propagation of variances and covariance yields

$$\mathbf{\Phi} = E[\mathbf{r}\mathbf{r}^{\mathrm{T}}] = \mathbf{A}E[\varepsilon\varepsilon^{\mathrm{T}}]\mathbf{A}^{\mathrm{T}} = \mathbf{A}\mathbf{\Psi}\mathbf{A}^{\mathrm{T}}. \tag{7.7}$$

At this point, it is important to make a distinction between the random errors and the systematic errors. When the same variable is measured repeatedly, we usually get a series of measurement. Because of a number of uncontrollable factors of small importance, varying at random, the measurement errors are normally distributed, in this case about zero.

Contrary to random errors, systematic (gross) errors are usually due to one or more isolated factors that cause the displacement of the measurement in one direction. The observations are distributed around a certain value y, different from the true value of the variable x. The systematic error is consequently equal to $(y - x) = a$.

In our case the measured quantity is $\mathbf{r}$ and the theoretical value is assumed to be known and equal to zero. The mean of the residua, that is, $E[\mathbf{r}]$, is calculated, and the question now is to decide whether this mean differs significantly from zero, such that the null hypothesis of zero mean is not satisfied.

The following two test functions can be formulated:

1. Global test.

$$\tau = \mathbf{r}^{\mathrm{T}}\mathbf{\Phi}^{-1}\mathbf{r}, \tag{7.8}$$

where $\mathbf{r}$ is the residua in the unsatisfied balances and $\mathbf{\Phi}$ is the covariance matrix of the residua. If matrix $\mathbf{A}$ has full row rank m, τ will have chi-square distribution with m degrees of freedom. Thus, at a specified level of significance, α, such that

$$\text{Prob}\{\tau \geq \chi^2_{1-\alpha}(m)\} = \alpha, \tag{7.9}$$

a test for the inconsistency of a set of measurements in the presence of gross errors, at a preassigned probability, is available. We need only preassign an allowable error probability, which gives us a critical value of τ. The choice of the error probability depends on the process characteristics. For example, a value of 0.10 is acceptable in many cases.

The advantages of the global test are that it is simple, and that the test statistic applies to the whole set of data a priori, that is, before any reconciliation of the measurements is attempted. However, it provides only a global test, and no indication of the origin of the failure is provided. Once the presence of gross errors is detected; a separate procedure is required to identify their source.

2. Individual constraints test. The test statistic,

$$h_i = \frac{|r_i|}{\sqrt{\mathbf{\Phi}_{ii}}}, \tag{7.10}$$

is used for each constraint in the problem, where h_i follows a standard normal distribution, that is, $h_i \sim N(0, 1)$. In this case, in place of a single test, we have a different number of tests for each constraint in the problem. As with the global test, a separate analysis is required to identify the source of the gross error. However, in this case the presence of a gross error can be related to one equation, or to a subset of equations, within the global problem.

Now we consider the estimate of the measurement error, $\hat{\varepsilon}$,

$$\hat{\varepsilon} = \mathbf{y} - \hat{\mathbf{x}} = (\mathbf{I} - \mathbf{M})\mathbf{y}. \tag{7.11}$$

In the absence of gross errors, the vector $\hat{\varepsilon}$ has the following properties:

$$E(\hat{\varepsilon}) = \mathbf{0} \tag{7.12}$$

and

$$\text{Cov}(\hat{\varepsilon}) = (\mathbf{I} - \mathbf{M})\boldsymbol{\Psi}(\mathbf{I} - \mathbf{M})^{\text{T}} = \mathbf{V}, \tag{7.13}$$

where

$$\mathbf{M} = \mathbf{I} - \boldsymbol{\Psi}\mathbf{A}^{\text{T}}(\mathbf{A}\boldsymbol{\Psi}\mathbf{A}^{\text{T}})^{-1}\mathbf{A}. \tag{7.14}$$

Vector $\hat{\varepsilon}$ has a multivariate normal distribution. Mah and Tamhane (1982) proposed the use of the *test on the estimates*,

$$|z_i| = \frac{|d_i|}{\sqrt{W_{ii}}}, \tag{7.15}$$

where

$$\mathbf{d} = \boldsymbol{\Psi}^{-1}(\mathbf{y} - \hat{\mathbf{x}}) \tag{7.16}$$

$$\text{Cov}(\mathbf{d}) = \mathbf{A}^{\text{T}}(\mathbf{A}\boldsymbol{\Psi}\mathbf{A}^{\text{T}})^{-1}\mathbf{A} = \mathbf{W}, \tag{7.17}$$

which follows a standard normal distribution. Unlike the previous test statistics, the test on the estimates of the measurement errors is applied after the reconciliation procedure. On the other hand, again in contrast to the previous tests, it provides a test associated with each measurement in such a way that no additional identification procedure is required.

In practice, linear balances are only encountered for total mass balances. The equations for general component and energy balances are nonlinear. Consequently, Eq. (7.5), relating the balance residuals to the measurement errors, requires linearization around the approximate values $\varepsilon^{(0)}$. For nonlinear balances, Eq. (7.7) may be further generalized as

$$\boldsymbol{\Phi} = \mathbf{G}_{\text{r}\varepsilon}\boldsymbol{\Psi}\mathbf{G}_{\text{r}\varepsilon}^{\text{T}} \tag{7.18}$$

with

$$G_{\text{r}\varepsilon} = \left.\frac{\partial \mathbf{r}}{\partial \varepsilon}\right|_{\varepsilon=\varepsilon^{(0)}}. \tag{7.19}$$

Although in practical applications linearized functions are used regularly for the propagation of variances and covariance, it should be pointed out (Brandt, 1970; Mikhail, 1976) that this is appropriate only if the range of dispersion in ε is small for linear approximations compared to the curvature of the function in the neighborhood of $\varepsilon^{(0)}$. In other words, the functions would be well approximated by their tangents within the region of interest, that is, the region of dispersion of the random variable. From a formal point of view it should be noted that in linearization the properties of random variables change from the variables themselves (in the nonlinear form) to the increments

$$\varepsilon = \varepsilon^{(0)} + \Delta\varepsilon \quad \text{and} \quad \mathbf{r} = \mathbf{r}^{(0)} + \Delta\mathbf{r}. \tag{7.20}$$

Thus, the error properties (i.e., the characteristics of the probability distribution) are now associated with $\Delta\varepsilon$ and $\Delta\mathbf{r}$ instead of ε and $\mathbf{r}$, respectively.

EXAMPLE 7.1

To illustrate the usage of test functions for the detection of gross errors, we will consider the problem presented by Ripps (1965), which has been presented in Example 5.1. The residua in the balances and their corresponding variances are given by

$$\mathbf{r} = \begin{bmatrix} -0.0672 \\ -0.0059 \\ -0.0571 \end{bmatrix}, \quad \mathbf{\Phi}^{-1} = \begin{bmatrix} 483.7 & -242.0 & -1339.8 \\ -242.0 & 5890.6 & -2071.6 \\ -1339.8 & -2071.6 & 5506.0 \end{bmatrix}.$$

By direct application of the global test defined in Eq. (7.8), we have

$$\tau = \mathbf{r}^{\mathrm{T}}\mathbf{\Phi}^{-1}\mathbf{r} = 8.5347.$$

If we consider an error probability of 0.10, the critical value for τ with three degrees of freedom is $\tau_c = 6.251$. Since in this case $\tau \gg \tau_c$, we can say that the inconsistency is significant at an error probability level of 0.10, and gross errors are present in the set of data.

7.3. IDENTIFICATION OF THE MEASUREMENTS WITH GROSS ERROR

If an extreme error is found by the test function, we need to identify which measurement has the gross error in order to guide instrument repair and to correct the corresponding measurement or to delete it from the set of data during the reconciliation procedure. In typical processes, small errors are usual and gross errors are unusual. A gross error is likely to occur only occasionally in any of the instruments; therefore, in any given set of measurements the number of such large errors is generally small compared with the total number of measurements taken.

In the following a serial elimination procedure (Romagnoli and Stephanopoulos, 1981; Romagnoli, 1983) is described. This scheme isolates the sources of gross errors by a systematic treatment of the measurements.

7.3.1. A Serial Elimination Procedure

Let us consider the system of g overmeasured (redundant) variables in m balance equations. Assuming that all of the errors are normally distributed with zero mean and variance $\mathbf{\Psi}$, it has been shown that the least squares estimate of the measurement errors is given by the solution of the following problem:

$$\begin{aligned} &\min_{\varepsilon} \ \varepsilon^{\mathrm{T}}\mathbf{\Psi}^{-1}\varepsilon \\ &\text{s.t.} \\ &\qquad \mathbf{A}\varepsilon = \mathbf{r}, \end{aligned} \tag{7.21}$$

with the solution given by

$$\hat{\varepsilon} = \boldsymbol{\Psi}\mathbf{A}^{\mathrm{T}}(\mathbf{A}\boldsymbol{\Psi}\mathbf{A}^{\mathrm{T}})^{-1}\mathbf{A}\mathbf{y} = \boldsymbol{\Psi}\mathbf{A}^{\mathrm{T}}\boldsymbol{\Phi}^{-1}\mathbf{r}. \tag{7.22}$$

If certain measurements have gross biases, this solution is not valid.

Assume now that c measurements have gross errors while the rest $(g - c)$ have only random errors with zero mean. Partitioning the matrix $\mathbf{A}$ along these lines, we have

$$\mathbf{A} = [\mathbf{A}_{\mathrm{g}} \quad \mathbf{A}_{\mathrm{c}}], \tag{7.23}$$

where $\mathbf{A}_{\mathrm{g}}$ is an $[m \times (g - c)]$ matrix whose columns are composed of the coefficients of the $(g - c)$ measurements, and $\mathbf{A}_{\mathrm{c}}$ is an $(m \times c)$ matrix corresponding to the c measurements with gross errors. Assuming that there is no correlation between the various measurements, the covariance matrix can be also partitioned, leading to

$$\boldsymbol{\Psi}_n = \begin{bmatrix} \boldsymbol{\Psi}_{\mathrm{g}} & \mathbf{0} \\ \mathbf{0} & \boldsymbol{\Psi}_{\mathrm{c}} + \Delta\boldsymbol{\Psi} \end{bmatrix}, \tag{7.24}$$

where $\Delta\Psi$ is the increase in the variance of the c measurements with gross errors. Accordingly, the variance for the residuals in the balances can be expressed as

$$\begin{aligned} \boldsymbol{\Phi}_n &= \mathbf{A}\boldsymbol{\Psi}_n\mathbf{A}^{\mathrm{T}} = [\mathbf{A}_{\mathrm{g}} \quad \mathbf{A}_{\mathrm{c}}]\begin{bmatrix} \boldsymbol{\Psi}_{\mathrm{g}} & \mathbf{0} \\ \mathbf{0} & \boldsymbol{\Psi}_{\mathrm{c}} + \Delta\boldsymbol{\Psi} \end{bmatrix}\begin{bmatrix} \mathbf{A}_{\mathrm{g}}^{\mathrm{T}} \\ \mathbf{A}_{\mathrm{c}}^{\mathrm{T}} \end{bmatrix} \\ &= \mathbf{A}_{\mathrm{g}}\boldsymbol{\Psi}_{\mathrm{g}}\mathbf{A}_{\mathrm{g}}^{\mathrm{T}} + \mathbf{A}_{\mathrm{c}}(\boldsymbol{\Psi}_{\mathrm{c}} + \Delta\boldsymbol{\Psi})\mathbf{A}_{\mathrm{c}}^{\mathrm{T}} \\ &= \boldsymbol{\Phi} + \mathbf{A}_{\mathrm{c}}(\Delta\boldsymbol{\Psi})\mathbf{A}_{\mathrm{c}}^{\mathrm{T}}, \end{aligned} \tag{7.25}$$

where

$$\boldsymbol{\Psi} = \begin{bmatrix} \boldsymbol{\Psi}_{\mathrm{g}} & \mathbf{0} \\ \mathbf{0} & \boldsymbol{\Psi}_{\mathrm{c}} \end{bmatrix} \tag{7.26}$$

and

$$\begin{aligned} \boldsymbol{\Phi} &= \mathbf{A}\boldsymbol{\Psi}\mathbf{A}^{\mathrm{T}} = [\mathbf{A}_{\mathrm{g}} \quad \mathbf{A}_{\mathrm{c}}]\begin{bmatrix} \boldsymbol{\Psi}_{\mathrm{g}} & \mathbf{0} \\ \mathbf{0} & \boldsymbol{\Psi}_{\mathrm{c}} \end{bmatrix}\begin{bmatrix} \mathbf{A}_{\mathrm{g}}^{\mathrm{T}} \\ \mathbf{A}_{\mathrm{c}}^{\mathrm{T}} \end{bmatrix} \\ &= \mathbf{A}_{\mathrm{g}}\boldsymbol{\Psi}_{\mathrm{g}}\mathbf{A}_{\mathrm{g}}^{\mathrm{T}} + \mathbf{A}_{\mathrm{c}}\boldsymbol{\Psi}_{\mathrm{c}}\mathbf{A}_{\mathrm{c}}^{\mathrm{T}} \end{aligned} \tag{7.27}$$

are the variances of the measurements and the balance residua if all the measurements are assumed to possess only random errors.

In order to estimate the vector $\hat{\varepsilon}$ in the presence of gross errors, we need to invert the covariance matrix, $\boldsymbol{\Phi}_n$, as Eq. (7.22) indicates. It is possible, though, to relate $\boldsymbol{\Phi}_n^{-1}$ to $\boldsymbol{\Phi}^{-1}$ (the inverse of the balance residuals in the absence of gross errors) through the simple recursive formula (6.32), which was presented in the previous chapter. In this case we obtain the following relation:

$$\boldsymbol{\Phi}_n^{-1} = \boldsymbol{\Phi}^{-1}\left\{\mathbf{I} - \mathbf{A}_{\mathrm{c}}\left[(\Delta\boldsymbol{\Psi})^{-1} + \mathbf{A}_{\mathrm{c}}^{\mathrm{T}}\boldsymbol{\Phi}^{-1}\mathbf{A}_{\mathrm{c}}\right]^{-1}\mathbf{A}_{\mathrm{c}}^{\mathrm{T}}\boldsymbol{\Phi}^{-1}\right\}. \tag{7.28}$$

Equation (7.28) can also be used for a different situation. Consider that initially c_i specified measurements are suspected to possess gross errors, and let $\boldsymbol{\Phi}_i$ be the corresponding covariance matrix of the residuals in the balances. If a different set c_{i+1} of suspect measurements is obtained by adding measurements to the set c_i, the covariance matrix of the residual for the new case, denoted as $\boldsymbol{\Phi}_{i+1}$, can be calculated as follows:

$$\boldsymbol{\Phi}_{i+1}^{-1} = \boldsymbol{\Phi}_i^{-1}\{\mathbf{I} - \mathbf{A}_{c(i+1)}\left[(\Delta\boldsymbol{\Psi}_{i+1})^{-1} + \mathbf{A}_{c(i+1)}^{\mathrm{T}}\boldsymbol{\Phi}_i^{-1}\mathbf{A}_{c(i+1)}\right]^{-1}\mathbf{A}_{c(i+1)}^{\mathrm{T}}\boldsymbol{\Phi}_i^{-1}\}. \tag{7.29}$$

It is interesting to note that in Eq. (7.29) we need to invert the matrix

$$\left[(\Delta\boldsymbol{\Psi})_{i+1}^{-1} + \mathbf{A}_{c(i+1)}^{\mathrm{T}}\boldsymbol{\Phi}_i^{-1}\mathbf{A}_{c(i+1)}\right], \tag{7.30}$$

which is a square matrix of order equal to the number of new suspect measurements c_{i+1}. Given the fact that in usual practical situations the number of suspect measurements is small, it is clear that Eq. (7.30) provides a simple recursive formula for computing the variance of the residua in the balance equations when the effect of different sets of suspect measurements is analyzed.

For each set of suspect measurements, the estimate of the measurement error is given by

$$\hat{\varepsilon}_i = \boldsymbol{\Psi}_i\mathbf{A}^{\mathrm{T}}\boldsymbol{\Phi}_i^{-1}\mathbf{r}, \tag{7.31}$$

where

$$\boldsymbol{\Psi}_i = \begin{bmatrix} \boldsymbol{\Psi}_{g(i)} & \mathbf{0} \\ \mathbf{0} & \boldsymbol{\Psi}_{c(i)} + (\Delta\boldsymbol{\Psi})_i \end{bmatrix}. \tag{7.32}$$

Also, the value of the least squares objective is given by

$$J_i = \hat{\varepsilon}_i^{\mathrm{T}}\boldsymbol{\Psi}_i^{-1}\hat{\varepsilon}_i = \left(\boldsymbol{\Psi}_i\mathbf{A}^{\mathrm{T}}\boldsymbol{\Phi}_i^{-1}\mathbf{r}\right)^{\mathrm{T}}\boldsymbol{\Psi}_i^{-1}\left(\boldsymbol{\Psi}_i\mathbf{A}^{\mathrm{T}}\boldsymbol{\Phi}_i^{-1}\mathbf{r}\right) = \mathbf{r}^{\mathrm{T}}\boldsymbol{\Phi}_i^{-1}\mathbf{r}. \tag{7.33}$$

Therefore, in order to compute the value of the least squares objective resulting from a given set of suspect measurements, we can use the recursive formula of Eq. (7.29) for $\boldsymbol{\Phi}_i^{-1}$, exploiting the information already available from previous calculations, as is codified by the value $\boldsymbol{\Phi}_{i-1}^{-1}$.

It is now clear how we can establish a recursive strategy to identify the measurements with gross errors:

1. Consider the ith set of c_i suspect measurements and assign to them a variance

$$\boldsymbol{\Psi}_{c(i)} + (\Delta\boldsymbol{\Psi})_i, \tag{7.34}$$

with

$$\Delta\boldsymbol{\Psi}_i \to \infty \tag{7.35}$$

while the remaining $(g - c_i)$ measurements have a variance $\boldsymbol{\Psi}_{g(i)}$.

2. Compute the inverse of the variance of the residua in the balances using the recursive formula

$$\boldsymbol{\Phi}_i^{-1} = \boldsymbol{\Phi}_{i-1}^{-1}\{\mathbf{I} - \mathbf{A}_{c(i)}\left[\mathbf{A}_{c(i)}^{\mathrm{T}}\boldsymbol{\Phi}_{i-1}^{-1}\mathbf{A}_{c(i)}\right]^{-1}\mathbf{A}_{c(i)}^{\mathrm{T}}\boldsymbol{\Phi}_{i-1}^{-1}\} \tag{7.36}$$

and by inverting the matrix

$$\left[\mathbf{A}_{c(i)}^{\mathrm{T}}\boldsymbol{\Phi}_{i-1}^{-1}\mathbf{A}_{c(i)}\right] \tag{7.37}$$

of low order, equal to the number c_i of suspect measurements.

3. The estimate of the measurement errors after the deletion of the suspect measurements is given by

$$\hat{\varepsilon}_i = \mathbf{\Psi}_{g(i)} \mathbf{A}_{g(i)}^{T} \mathbf{\Phi}_i^{-1} \mathbf{r} \tag{7.38}$$

and the objective function is given by Eq. (7.33).

4. If $J_i \ll J_{i-1}$, then we can conclude that some of the c_i that were deleted as suspect measurements possess gross errors.

Note: If one measurement is deleted at a time, the matrix $[\mathbf{A}_{c(i)}^{T} \mathbf{\Phi}_{i-1}^{-1} \mathbf{A}_{c(i)}]$ is a scalar quantity and then the inverse $\mathbf{\Phi}_i^{-1}$ can be easily calculated using a previously computed $\mathbf{\Phi}_{i-1}^{-1}$. ♣

The foregoing procedure can be implemented on-line, in real time, for any processing plant without much computational effort, by sequentially deleting one measurement at a time to quantify the effect of that measurement on the reconciliation procedure.

EXAMPLE 7.2

Let us consider again the system defined in Example 5.1. From the application of the global statistical test, gross errors were detected among the data set as indicated in Example 7.1. Now the serial elimination strategy will be applied to isolate the source of gross error, that is to identify which set of measurements contains gross error.

Step 1. Deletion of measurement f_1. The new system matrices, according to our definition, are

$$\mathbf{A}_u = \begin{bmatrix} 0.6 & -0.2 & -0.7 \\ 0.1 & -0.1 & -0.1 \\ 0.3 & -0.6 & -0.2 \end{bmatrix}, \quad \mathbf{A}_c = \begin{bmatrix} 0.1 \\ 0.8 \\ 0.1 \end{bmatrix}, \quad \mathbf{\Psi}_u = \begin{bmatrix} 0.0025 & & \\ & 0.000576 & \\ & & 0.04 \end{bmatrix}.$$

The scalar to be inverted is given by

$$\mathbf{A}_c^{T} \mathbf{\Phi}^{-1} \mathbf{A}_c = [0.1 \quad 0.8 \quad 0.1] \begin{bmatrix} 483.7 & -242.0 & -1339.8 \\ -242.0 & 5890.6 & -2071.6 \\ -1339.8 & -2071.6 & 5506.0 \end{bmatrix} \begin{bmatrix} 0.1 \\ 0.8 \\ 0.1 \end{bmatrix} = 3432.88,$$

and $\mathbf{\Phi}_n^{-1}$ from Eq. (7.28) is

$$\mathbf{\Phi}_n^{-1} = \begin{bmatrix} 460.9 & 122.5 & -1440.7 \\ 122.5 & 41.2 & -452.1 \\ -1440.7 & -452.1 & 5057.6 \end{bmatrix}.$$

From here the least squares objective is given directly by

$$J_1 = \mathbf{r}^{T} \mathbf{\Phi}_n^{-1} \mathbf{r} = 7.2953.$$

Also, the estimates of the measurement error $\hat{\varepsilon}$ and vector $\hat{\mathbf{f}}$ are given by

$$\hat{\varepsilon}_n = \begin{bmatrix} -0.0617 \\ 0.0575 \\ 0.0293 \end{bmatrix}, \quad \hat{\mathbf{f}} \begin{bmatrix} 4.8552 \\ 1.1720 \\ 3.8507 \end{bmatrix}.$$

Step 2. Deleting measurement f_2. For this case the system matrices are now

$$\mathbf{A}_g = \begin{bmatrix} 0.1 & -0.2 & -0.7 \\ 0.8 & -0.2 & -0.1 \\ 0.1 & -0.6 & -0.2 \end{bmatrix}, \quad \mathbf{A}_c = \begin{bmatrix} 0.6 \\ 0.1 \\ 0.3 \end{bmatrix}, \quad \boldsymbol{\Psi}_g = \begin{bmatrix} 0.00029 & & \\ & 0.000576 & \\ & & 0.04 \end{bmatrix}.$$

Consequently, we have

$$\mathbf{A}_c^T \boldsymbol{\Phi}^{-1} \mathbf{A}_c = [0.6 \quad 0.1 \quad 0.3] \begin{bmatrix} 483.7 & -242.0 & -1339.8 \\ -242.0 & 5890.6 & -2071.6 \\ -1339.8 & -2071.6 & 5506.0 \end{bmatrix} \begin{bmatrix} 0.6 \\ 0.1 \\ 0.3 \end{bmatrix} = 92.8938,$$

and from Eq. (6.33), $\boldsymbol{\Phi}_n^{-1}$ is

$$\boldsymbol{\Phi}_n^{-1} = \begin{bmatrix} 284.7 & -501.9 & -402.1 \\ -501.9 & 5551.0 & -846.5 \\ -402.1 & -846.5 & 1086.3 \end{bmatrix}.$$

The new value of the least squares objective is

$$J_2 = \mathbf{r}^T \boldsymbol{\Phi}_n^{-1} \mathbf{r} = 0.9636,$$

with estimates of the measurement error $\hat{\varepsilon}$ and vector $\hat{\mathbf{f}}$ given by

$$\hat{\varepsilon}_n = \begin{bmatrix} 0.0107 \\ 0.0039 \\ -0.1470 \end{bmatrix}, \quad \hat{\mathbf{f}}_n = \begin{bmatrix} 0.1751 \\ 1.2256 \\ 4.0270 \end{bmatrix}.$$

Following the same procedure, we have considered the deletion of f_3 and f_4. The corresponding values of the least squares objectives were $J_3 = 1.5702$ and $J_4 = 8.4374$, respectively. The results indicate that measurement f_2 contains a gross error and has to be deleted, or some corrective action has to be undertaken with the corresponding measurement device.

7.3.2. A Combined Procedure

The previous approach for solving the reconciliation problem allows the calculation, in a systematic recursive way, of the residual covariance matrix after a measurement is added or deleted from the original adjustment. A combined procedure can be devised by using the sequential treatment of measurements together with the sequential processing of the constraints.

Consider, in general, the overall problem consisting of m balances and divide it into m smaller subproblems, that is, we will be processing one equation at a time. Then, after the ith balance has been processed, a new value of the least squares objective (test function) can be computed. Let J_i denote the value of the objective evaluated after the ith equation has been considered. The approach for the detection of a gross error in this balance is based on the fact that J_i is a random variable whose probability distribution can be calculated.

When a normal distribution of the errors can be assumed, the least squares fit can be combined with a χ^2-test after each step of the sequential processing. We will

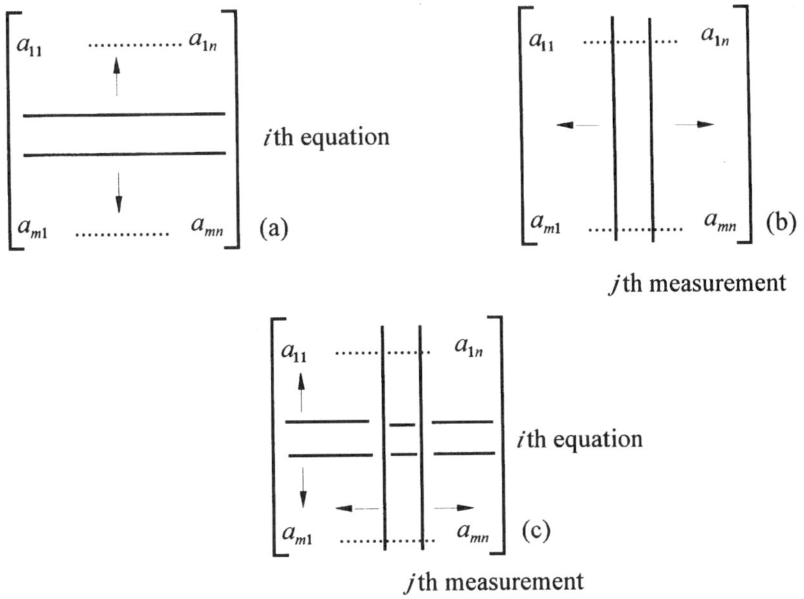

FIGURE 1 Steps in the identification procedure: (a) searching along equations; (b) searching along measurements; (c) combined procedure (from Romagnoli, 1983).

reject the result of the adjustment if

$$J_i > \chi^2_{1-\alpha}(i), \tag{7.39}$$

that is, if J_i exceeds the critical value of χ^2 belonging to a level of significance, α. Thus, J_i gives us an individualized criterion as each balance is processed.

Now if the test fails at the ith step, a systematic error has been detected. The source of the gross error will be located in one or more of the new measurements incorporated by the ith balance, and so they constitute a subset of suspect measurements. In practical applications we will be faced with two possible alternatives:

1. Only one measurement is suspected to have gross error; then it can be deleted and a new reliable estimate obtained
2. More than one measurement is suspected; then the serial elimination of the measurements is initiated until a new reliable estimate is obtained

Figure 1 illustrates, schematically, the use of this recursive scheme for constraints/measurements processing in any identification procedure.

EXAMPLE 7.3

In this example we will show the combined procedure. Let us take the simple serial system in Fig. 2 for which the available data is given in Table 1. Only total mass balances are considered and the covariance matrix is the identity matrix, that is, $\mathbf{\Psi} = \mathbf{I}$.

Now, applying the sequential approach we have the following:

Step 1. Processing Equation 1, we have

$$\mathbf{A}_1 = [1 \quad -1 \quad 1 \quad 0 \quad 0 \quad 0], \quad \mathbf{\Psi} = \mathbf{\Psi}_1 = \mathbf{I}, \quad \left[\mathbf{A}_1\mathbf{\Psi}_1\mathbf{A}_1^{\mathrm{T}}\right] = 3, \quad \mathbf{r}_1 = 1.5,$$

TABLE 1 Data for Example 7.3 (from Romagnoli and Stephanopoulos, 1981)

	Measured values	True values
f_1	10.5	10
f_2	14.5	15
f_3	5.5	5
f_4	14*	10
f_5	19.5	20
f_6	20.5	20

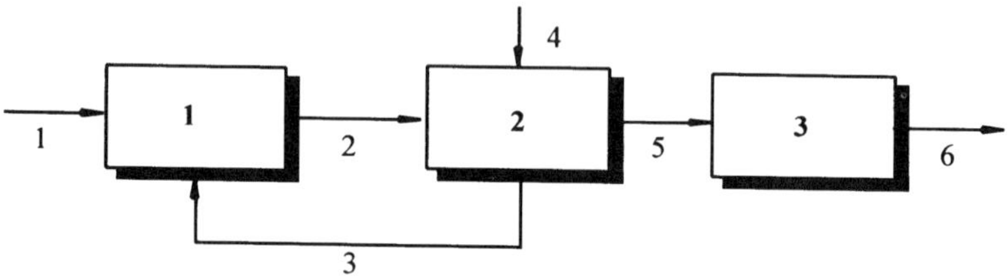

FIGURE 2 Simple serial system for Example 7.3 (from Romagnoli and Stephanopoulos, 1981).

with the corresponding estimates given by

$$\hat{\varepsilon}_1 = \begin{bmatrix} 0.5 \\ 0.5 \\ 0.5 \\ 0 \\ 0 \\ 0 \end{bmatrix}, \quad \boldsymbol{\Sigma}_{\hat{\varepsilon}_1} = \begin{bmatrix} 1/3 & -1/3 & 1/3 & 0 & 0 & 0 \\ -1/3 & 1/3 & -1/3 & 0 & 0 & 0 \\ 1/3 & -1/3 & 1/3 & 0 & 0 & 0 \\ 0 & 0 & 0 & 0 & 0 & 0 \\ 0 & 0 & 0 & 0 & 0 & 0 \\ 0 & 0 & 0 & 0 & 0 & 0 \end{bmatrix}.$$

The value of the objective is now

$$J_1 = 0.75 < \chi^2_{0.1}(1) = 2.75 \qquad \text{(acceptable)}.$$

Step 2. Processing Equation 2, we have

$$\mathbf{A}_2 = [0 \quad 1 \quad -1 \quad 1 \quad -1 \quad 0], \quad \boldsymbol{\Psi}_2 = \boldsymbol{\Psi}_1 - \boldsymbol{\Sigma}_{\hat{\varepsilon}_1},$$

and

$$\left[\mathbf{A}_2 \boldsymbol{\Psi}_2 \mathbf{A}_2^{\mathrm{T}}\right] = 2.6667, \quad \mathbf{r}_2 = 4.5,$$

with the estimates given by

$$\hat{\varepsilon}_2 = \begin{bmatrix} 1.1250 \\ 0.5625 \\ -0.5625 \\ 1.6875 \\ -1.6875 \end{bmatrix}$$

and

$$J_2 = 13.67 > \chi^2_{0.1}(2) = 4.61.$$

This estimate is not acceptable. Since measurements f_1, f_2, and f_3 were used in step 1 and they gave a good estimate, the gross error is in either measurement f_4 or f_5 or both. Now, applying the combined approach, we delete sequentially measurements f_4 and f_5 from the adjustment.

Step 3.

(a) Deleting measurement f_4, we have

$$\mathbf{A}_g = \begin{bmatrix} 1 & -1 & 1 & 0 & 0 \\ 0 & 1 & -1 & -1 & 0 \\ 0 & 0 & 0 & 1 & -1 \end{bmatrix}, \quad \mathbf{A}_c = \begin{bmatrix} 0 \\ 1 \\ 0 \end{bmatrix}, \quad \mathbf{\Phi} = \begin{bmatrix} 3 & -2 & 0 \\ -2 & 4 & -1 \\ 0 & -1 & 2 \end{bmatrix},$$

with

$$\left[\mathbf{A}_c^T \Phi \mathbf{A}_c\right] = 0.4615, \quad \mathbf{\Phi}_n^{-1} = \begin{bmatrix} 0.337 & 0 & 0 \\ 0 & 0 & 0 \\ 0 & 0 & 0.5 \end{bmatrix}.$$

From here, the value of the least squares objective becomes

$$J_4 = \mathbf{r}^T \mathbf{\Phi}_n^{-1} \mathbf{r} = 1.25.$$

(b) Deleting measurement f_5,

$$J_5 = \mathbf{r}^T \mathbf{\Phi}_n^{-1} \mathbf{r} = 5.34.$$

Thus,

$$J_4 \ll J_5.$$

It is obvious from these results that measurement f_4 is grossly faulty and has to be deleted during the rectification process. Finally, estimates of the measurement error and the process variables when measurement f_4 is deleted are given by

$$\hat{\varepsilon} = \begin{bmatrix} 0.5 \\ -0.5 \\ 0.5 \\ -0.5 \\ 0.5 \end{bmatrix}, \quad \hat{\mathbf{f}} = \begin{bmatrix} 10 \\ 15 \\ 5 \\ 20 \\ 20 \end{bmatrix}.$$

Solving for f_4 from the balances using the estimates of $\hat{\mathbf{f}}$,

$$\hat{f}_4 = -15 + 5 + 20 = 10.$$

7.4. ESTIMATION OF THE MAGNITUDE OF BIAS AND LEAKS

Once the existence of systematic errors is ascertained, their effect is modeled functionally. In the following, three cases of gross error estimation are discussed:

1. Bias in the measurements
2. Leaks in units
3. Combinations of leaks and measurement biases

7.4.1. Bias in the Measurements

In the following discussion, one or several sensor failures are assumed, so a constant bias of magnitude $\mathbf{m}_b$ is added to the measurement vector $\mathbf{y}$. In the presence of a failure in the sensors, the measurement equation takes the form

$$\mathbf{y} = \mathbf{x} + \varepsilon + \mathbf{B}_{rm}\mathbf{m}_b, \tag{7.40}$$

where s is the number of gross errors in the data set (which is known from the gross error identification phase) and $\mathbf{B}_{rm}$ is a matrix of order $(g \times s)$ whose elements are all zero or one depending on whether the corresponding measurement is faulty or not. The vector of constants $\mathbf{m}_b$ represents the sign and magnitude (bias) of the failure. We proceed, in consequence, to minimize the error criterion:

$$\begin{aligned} &\underset{\varepsilon}{\text{Min}}\ \varepsilon^T \Psi^{-1} \varepsilon \\ &\text{s.t.} \\ &\qquad \mathbf{A}(\mathbf{y} - \varepsilon - \mathbf{B}_{rm}\mathbf{m}_b) = \mathbf{0}. \end{aligned} \tag{7.41}$$

Using the method of Lagrange multipliers, the solution of the problem is given by

$$\hat{\varepsilon} = \Psi\mathbf{A}^T(\mathbf{A}\Psi\mathbf{A}^T)^{-1}[\mathbf{A}\mathbf{y} - \mathbf{P}_b\hat{\mathbf{m}}_b], \tag{7.42}$$

where

$$\hat{\mathbf{m}}_b = \left[\mathbf{P}_b^T(\mathbf{A}\Psi\mathbf{A}^T)^{-1}\mathbf{P}_b\right]^{-1}\mathbf{P}_b^T(\mathbf{A}\Psi\mathbf{A}^T)^{-1}\mathbf{A}\mathbf{y} \tag{7.43}$$

and

$$\mathbf{P}_b = \mathbf{A}\mathbf{B}_{rm}. \tag{7.44}$$

This can be rewritten more conveniently as

$$\hat{\varepsilon} = \Psi\mathbf{A}^T(\mathbf{A}\Psi\mathbf{A}^T)^{-1}\mathbf{A}\mathbf{y} - \Psi\mathbf{A}^T(\mathbf{A}\Psi\mathbf{A}^T)^{-1}\mathbf{P}_b\hat{\mathbf{m}}_b \tag{7.45}$$

or in compact notation as

$$\hat{\varepsilon} = \hat{\varepsilon}_c + \mathbf{x}_b, \tag{7.46}$$

where $\hat{\varepsilon}_c$ is the bias-free estimate of the measurement error and $\mathbf{x}_b$ is a correction term arising because of the biases in some measurements.

This means that the computation of the optimum estimate of $\mathbf{x}$ is effectively decoupled from the estimate of the bias. Moreover, it can be computed in terms of the residuals of the bias-free estimate. In terms of $\mathbf{x}$, this can be expressed as follows:

$$\hat{\mathbf{x}} = \mathbf{y} - \Psi\mathbf{A}^T(\mathbf{A}\Psi\mathbf{A}^T)^{-1}\mathbf{A}\mathbf{y} - \Psi\mathbf{A}^T(\mathbf{A}\Psi\mathbf{A}^T)^{-1}\mathbf{P}_b\hat{\mathbf{m}}_b, \tag{7.47}$$

or more generally, this can be rewritten as

$$\hat{\mathbf{x}} = \mathbf{x}_0 + \mathbf{x}_c + \mathbf{x}_b. \tag{7.48}$$

That is, the least squares estimate can be finally expressed as the contribution of three terms. The first one arises from the solution of the original problem, without constraints (for data reconciliation $\mathbf{x}_0 = \mathbf{y}$); the next is a correction term due to the presence of constraints; and the last one takes into account failures in the model (systematic errors).

A quite important part of the adjustment is the determination of the precision in the estimation problem. Such precision is in the form of the covariance matrix. Following a similar procedure we arrive at

$$\begin{aligned}\boldsymbol{\Sigma} &= \mathbf{E}\{\mathbf{x}_0\mathbf{x}_0^{\mathrm{T}}\} + \mathbf{E}\{\mathbf{x}_c\mathbf{x}_c^{\mathrm{T}}\} + \mathbf{E}\{\mathbf{x}_b\mathbf{x}_b^{\mathrm{T}}\} \\ &= \boldsymbol{\Psi} - \boldsymbol{\Psi}\mathbf{A}^{\mathrm{T}}[\mathbf{A}\boldsymbol{\Psi}\mathbf{A}^{\mathrm{T}}]^{-1}\mathbf{A}\boldsymbol{\Psi} - \boldsymbol{\Psi}\mathbf{A}^{\mathrm{T}}[\mathbf{A}\boldsymbol{\Psi}\mathbf{A}^{\mathrm{T}}]^{-1}\mathbf{P}_b\boldsymbol{\Sigma}_{m_b}\mathbf{P}_b^{\mathrm{T}}[\mathbf{A}\boldsymbol{\Psi}\mathbf{A}^{\mathrm{T}}]^{-1}\mathbf{A}\boldsymbol{\Psi} \\ &= \boldsymbol{\Sigma}_0 + \boldsymbol{\Sigma}_c + \boldsymbol{\Sigma}_b,\end{aligned} \tag{7.49}$$

where

$$\boldsymbol{\Sigma}_{m_b} = \left[\mathbf{P}_b^{\mathrm{T}}(\mathbf{A}\boldsymbol{\Psi}\mathbf{A}^{\mathrm{T}})^{-1}\mathbf{P}_b\right]^{-1}. \tag{7.50}$$

The covariance of the residuals in the estimate can be expressed as the contribution of two terms, the first corresponding to the original adjustment and the second to a correction term. Furthermore, the quadratic objective can be expressed as

$$J = \varepsilon^{\mathrm{T}}\boldsymbol{\Psi}^{-1}\varepsilon = \mathbf{r}^{\mathrm{T}}(\mathbf{A}\boldsymbol{\Psi}\mathbf{A}^{\mathrm{T}})^{-1}\mathbf{r} - \hat{\mathbf{m}}_b^{\mathrm{T}}\mathbf{P}_b^{\mathrm{T}}(\mathbf{A}\boldsymbol{\Psi}\mathbf{A}^{\mathrm{T}})^{-1}\mathbf{r} = J_c - J_b, \tag{7.51}$$

which can be used as a checking procedure after the correction for bias has been performed.

EXAMPLE 7.4

The same example proposed for the sequential treatment of the measurements (Example 7.2) serves as a test case to illustrate the application of the scheme developed for estimation of the magnitude of the bias term. The data for the new problem is given in Table 2.

Case 1. A bias present in the measurement of f_2 was identified by the sequential processing of the measurements (see Example 7.2). We augment, in consequence, the vector of parameters of the original problem by adding an additional component to represent the uncertain parameter (bias term).

Accordingly,

$$\mathbf{B}_{rm} = [0 \quad 1 \quad 0 \quad 0]^{\mathrm{T}}, \qquad \mathbf{P}_b = [0.6 \quad 0.1 \quad 0.3]^{\mathrm{T}}.$$

Using this value for $\mathbf{P}_b$, we have

$$\left[\mathbf{P}_b^{\mathrm{T}}(\mathbf{A}\boldsymbol{\Psi}\mathbf{A}^{\mathrm{T}})^{-1}\mathbf{P}_b\right]^{-1} = 0.010733,$$

from which the bias is calculated as

$$\hat{\mathbf{m}}_b = \left[\mathbf{P}_b^{\mathrm{T}}(\mathbf{A}\boldsymbol{\Psi}\mathbf{A}^{\mathrm{T}})^{-1}\mathbf{P}_b\right]^{-1}\mathbf{P}_b^{\mathrm{T}}(\mathbf{A}\boldsymbol{\Psi}\mathbf{A}^{\mathrm{T}})^{-1}\mathbf{A}\mathbf{y} = -0.2840.$$

TABLE 2 Data for Example 7.4 (from Romagnoli, 1983)

	Measurement, Case 1	Measurement, Case 2	True value
f_1	0.1858	0.2058	0.1739
f_2	4.7935	4.7935	5.0435
f_3	1.2295	1.2295	1.2175
f_4	3.88	3.88	4.00

TABLE 3 Results for Example 7.4 (Case 2)

	Measured value	Estimate with bias	Bias-free estimate
f_1	0.2058	0.1677	0.1747
f_2	4.7935	4.8640	5.0667
f_3	1.2295	1.1741	1.2230
f_4	3.88	3.8577	4.0183

With this numerical value, the correction vector of the measurements due to the presence of systematic errors and the estimates of the process variables may be computed:

$$\mathbf{x}_b = -\mathbf{\Psi A}^T(\mathbf{A\Psi A}^T)^{-1}\mathbf{P}_b\hat{\mathbf{m}}_b = [-0.0075 \quad -0.0659 \quad -0.0526 \quad -0.1729]^T$$

and

$$\hat{\mathbf{x}} = [0.1751 \quad 5.0775 \quad 1.2256 \quad 4.0270].$$

Case 2. In this case an additional bias present in measurement of f_1 was considered. In accordance with the previous development,

$$\mathbf{B}_{rm} = \begin{bmatrix} 1 & 0 & 0 & 0 \\ 0 & 1 & 0 & 0 \end{bmatrix}^T, \quad \mathbf{P}_b = \begin{bmatrix} 0.1 & 0.8 & 0.1 \\ 0.6 & 0.1 & 0.3 \end{bmatrix}^T$$

$$\hat{\mathbf{m}}_b = \left[\mathbf{P}_b^T(\mathbf{A\Psi A}^T)^{-1}\mathbf{P}_b\right]^{-1}\mathbf{P}_b^T(\mathbf{A\Psi A}^T)^{-1}\mathbf{Ay} = \begin{bmatrix} 0.0311 \\ -0.2730 \end{bmatrix}.$$

With these numerical values, the correction vector for the measurement and the estimate of the process variables may be computed. They are given in Table 3.

7.4.2. Estimation of Leaks

If one or more leaks are considered, the constraint model for the process must be modified to take them into account. Now, the least squares formulation of the problem, when measurement bias are absent, can be stated as

$$\begin{aligned} &\underset{\varepsilon}{\text{Min}}\ \varepsilon^T\mathbf{\Psi}^{-1}\varepsilon \\ &\text{s.t.} \\ &\qquad \mathbf{A}(\mathbf{y}-\varepsilon) - \mathbf{B}_{rp}\mathbf{m}_p = \mathbf{0}, \end{aligned} \tag{7.52}$$

where

$\mathbf{B}_{rp}$ is an $(m \times p)$ matrix with $\mathbf{e}_i$ column vectors indicating the positions of the leaks

$\mathbf{m}_p$ is the p-dimensional vector of the magnitudes of the leaks

Again, using the Lagrangian approach, the estimates of $\mathbf{m}_p$, ε, and $\mathbf{x}$ are given by

$$\hat{\mathbf{m}}_p = \left[\mathbf{B}_{rp}^T(\mathbf{A\Psi A}^T)^{-1}\mathbf{B}_{rp}\right]^{-1}\mathbf{B}_{rp}^T(\mathbf{A\Psi A}^T)^{-1}\mathbf{Ay} \tag{7.53}$$

$$\hat{\varepsilon} = \mathbf{\Psi A}^T(\mathbf{A\Psi A}^T)^{-1}[\mathbf{Ay} - \mathbf{B}_{rp}\hat{\mathbf{m}}_p] \tag{7.54}$$

$$\hat{\mathbf{x}} = \mathbf{y} - \mathbf{\Psi A}^T(\mathbf{A\Psi A}^T)^{-1}[\mathbf{Ay} - \mathbf{B}_{rp}\hat{\mathbf{m}}_p]. \tag{7.55}$$

7.4.3. Estimation of Leaks and Biases

If combinations of leaks and measurement biases are considered, both the measurement model and the process constraints equations need to be modified. The formulation for the least squares problem is now

$$\begin{aligned} &\operatorname*{Min}_{x} \varepsilon^{T}\Psi^{-1}\varepsilon \\ &\text{s.t.} \\ &\qquad \mathbf{A}(\mathbf{y}-\varepsilon-\mathbf{B}_{\mathrm{rm}}\mathbf{m}_{\mathrm{b}})-\mathbf{B}_{\mathrm{rp}}\mathbf{m}_{\mathrm{p}}=\mathbf{0}. \end{aligned} \tag{7.56}$$

Again, we can use the Lagrangian approach to obtain estimates for $\mathbf{m}_p$, $\mathbf{m}_b$, ε, and $\mathbf{x}$. The estimates for $\mathbf{m}_p$ and $\mathbf{m}_b$ are obtained by solving the following system:

$$\begin{bmatrix} \mathbf{A}_b\mathbf{P}_b & \mathbf{A}_b\mathbf{B}_{rp} \\ \mathbf{C}_p\mathbf{P}_b & \mathbf{C}_p\mathbf{B}_{rp} \end{bmatrix} \begin{bmatrix} \mathbf{m}_b \\ \mathbf{m}_p \end{bmatrix} = \begin{bmatrix} \mathbf{A}_b\mathbf{A}\mathbf{y} \\ \mathbf{C}_b\mathbf{A}\mathbf{y} \end{bmatrix}, \tag{7.57}$$

where $\mathbf{A}_b = \mathbf{P}_b^T(\mathbf{A}\Psi\mathbf{A}^T)^{-1}$ and $\mathbf{C}_p = \mathbf{B}_{rp}^T(\mathbf{A}\Psi\mathbf{A}^T)^{-1}$.

Finally, for the estimates of $\mathbf{x}$ and ε we have

$$\hat{\varepsilon} = \Psi\mathbf{A}^T(\mathbf{A}\Psi\mathbf{A}^T)^{-1}[\mathbf{A}\mathbf{y} - \mathbf{P}_b\hat{\mathbf{m}}_b - \mathbf{B}_{rp}\hat{\mathbf{m}}_p] \tag{7.58}$$

$$\hat{\mathbf{x}} = \mathbf{y} - \Psi\mathbf{A}^T(\mathbf{A}\Psi\mathbf{A}^T)^{-1}[\mathbf{A}\mathbf{y} - \mathbf{P}_b\hat{\mathbf{m}}_b - \mathbf{B}_{rp}\hat{\mathbf{m}}_p] - \mathbf{B}_{rm}\hat{\mathbf{m}}_b. \tag{7.59}$$

7.5. A RECURSIVE SCHEME FOR GROSS ERROR IDENTIFICATION AND ESTIMATION

This section briefly discusses an approach that combines statistical tests with simultaneous gross error identification and estimation. The strategy is called SEGE (Simultaneous Estimation of Gross Error Method). It was proposed by Sánchez and Romagnoli (1994).

Recall that, in the absence of gross errors, the measurement and linear constraint models are given by Eqs. (7.1) and (7.4), respectively. Furthermore, the solution of the least square estimation problem of $\mathbf{x}$ variables is

$$\hat{\mathbf{x}} = \mathbf{y} - \Psi\mathbf{A}^T(\mathbf{A}\Psi\mathbf{A}^T)^{-1}\mathbf{A}\mathbf{y}. \tag{7.60}$$

By a direct application of propagation rules, the covariance matrix of $\hat{\mathbf{x}}$ is

$$\mathbf{Cov}(\hat{\mathbf{x}}) = \Sigma = \Psi - \Psi\mathbf{A}^T(\mathbf{A}\Psi\mathbf{A}^T)^{-1}\mathbf{A}\Psi, \tag{7.61}$$

so $\hat{\mathbf{x}}$ can be expressed in terms of its covariance matrix using Eq. (7.61) as follows:

$$\hat{\mathbf{x}} = \Sigma\Psi^{-1}\mathbf{y} = (\Psi - \Psi\mathbf{A}^T(\mathbf{A}\Psi\mathbf{A}^T)^{-1}\mathbf{A}\Psi)\Psi^{-1}\mathbf{y}, \tag{7.62}$$

where it should be noted that the set of process constraints is only involved in the calculation of Σ.

As was indicated in Section 7.2, the vector of measurement adjustments, $\hat{\varepsilon}$, has a multivariate normal distribution with zero mean and covariance matrix $\mathbf{V}$. Thus, the objective function value of the least square estimation problem (7.21), $ofv = \hat{\varepsilon}^T\Psi^{-1}\hat{\varepsilon}$, has a central chi-square distribution with a number of degrees of freedom equal to the rank of $\mathbf{A}$.

In order to detect the presence of gross errors in the proposed measurement and constraint models, the strategy SEGE applies a collective hypothesis statistical test based on the vector of measurement adjustments, $\hat{\varepsilon}$. The null and alternative hypotheses are stated as follows:

$$\mathrm{H}_0 = E(\hat{\varepsilon}) = 0$$
$$\mathrm{H}_\mathrm{a} = E(\hat{\varepsilon}) \neq 0.$$

The global test statistic $\hat{\varepsilon}^{\mathrm{T}}\boldsymbol{\Psi}^{-1}\hat{\varepsilon}$ is used to compare both alternatives. Thus, an appropriate α-level test is to reject H_0 in favor of H_a, iff $\hat{\varepsilon}^{\mathrm{T}}\boldsymbol{\Psi}^{-1}\hat{\varepsilon} \geq \chi^2_{m,\alpha}$, where $\chi^2_{m,\alpha}$ is the upper (100α)th percentile of the χ^2_m with $m = \mathrm{rank}(\mathbf{A})$.

If H_0 is rejected, a two-stage procedure is initiated. First, a list of candidate biases and leaks is constructed by means of the recursive search scheme outlined by Romagnoli (1983). All possible combinations of gross errors (measurement biases and/or process leaks) from this subset are analyzed in the second stage. Gross error magnitudes are estimated simultaneously for each combination and chi-square test statistic calculations are performed to identify the suspicious combinations. We will now explain the stages of the procedure.

Stage I

The recursive search is undertaken to identify a set of constraints that do not satisfy the proposed models for measurements and constraints.

Let us suppose that an initial data reconciliation problem has been resolved using a set of process constraints and the covariance matrix for the estimated variables ($\boldsymbol{\Sigma}^{\mathrm{old}}$) is available. If a set of constraints $\mathbf{B}_\mathrm{i}$ is incorporated into the least square estimation problem, the covariance matrix $\boldsymbol{\Sigma}$ for the new case ($\boldsymbol{\Sigma}^{\mathrm{new}}$) can be estimated using the previous one by means of the formula

$$\boldsymbol{\Sigma}^{\mathrm{new}} = \boldsymbol{\Sigma}_\mathrm{c}^{\mathrm{old}} - \boldsymbol{\Sigma}^{\mathrm{old}}\mathbf{B}_\mathrm{i}^{\mathrm{T}}\left(\mathbf{B}_\mathrm{i}\boldsymbol{\Sigma}^{\mathrm{old}}\mathbf{B}_\mathrm{i}^{\mathrm{T}}\right)^{-1}\mathbf{B}_\mathrm{i}\boldsymbol{\Sigma}^{\mathrm{old}}. \tag{7.63}$$

The corresponding demonstrations are included in Appendix A.

Also, the vector $\hat{\mathbf{x}}$ for the new case is calculated as

$$\hat{\mathbf{x}}^{\mathrm{new}} = \boldsymbol{\Sigma}_\mathrm{c}^{\mathrm{new}}\boldsymbol{\Psi}^{-1}\mathbf{y}. \tag{7.64}$$

It should be noted that if one constraint is added at a time, the vector $\hat{\mathbf{x}}$ is easily estimated as function of $\boldsymbol{\Sigma}^{\mathrm{new}}$ by the inversion of a scalar.

The preceding results are applied to develop a strategy that allows us to isolate the source of gross errors from a set of constraints and measurements. Different least squares estimation problems are resolved by adding one equation at a time to the set of process constraints. After each incorporation, the least square objective function value is calculated and compared with the critical value.

Two different situations can arise when an equation is incorporated:

- If $ofv > \tau_c$, gross error is detected. The equation is eliminated from the set of constraints. The procedure is repeated with the following equation.
- If $ofv < \tau_c$, no gross error is detected, so the following equation is analyzed.

By this procedure, all of the measurements involved in the deleted constraints and the leaks from the corresponding units are included in the list of suspected gross errors.

Stage 2

In this stage, all possible combinations of suspicious leaks and/or bias are analyzed. For each combination, the following tasks are accomplished:

- Modification of measurement/constraint models to take into account the presence of gross errors
- Least square estimation of $\hat{\mathbf{x}}$ and gross error magnitudes
- Test statistic (ofv) calculation

Then the values of the test statistic for all combinations are compared with the critical value. The presence of gross errors correspond to the combinations with the low objective function value (ofv). Detailed algorithms for Stages 1 and 2 are included in Appendix B.

7.5.1. Performance Evaluation

A simulation procedure can be applied to study the performance of different methods in detecting gross errors and in estimating their magnitude.

As it is common practice (Iordache *et al.*, 1985), each result is based on 10,000 simulation trials for given magnitudes of gross errors. In each simulation trial, a different set of measurements is randomly generated. Three performance measures can be used:

1. The overall power (OP)
2. Average number of type I errors (AVTI)
3. Expected fraction of correct identification (OPF)

They are defined as follows:

$$\mathrm{OP} = \frac{\text{No. of gross errors correctly identified}}{\text{No. of gross errors simulated}}$$

$$\mathrm{AVTI} = \frac{\text{No. of gross errors wrongly identified}}{\text{No. of simulations trials}}$$

$$\mathrm{OPF} = \frac{\text{No. of trials with perfect identification}}{\text{No. of simulations trials}}.$$

The first two measures were proposed by Narasimhan and Mah (1987) and the last one was presented by Rollins and Davis (1992).

EXAMPLE 7.5

Consider the well-known example (Rosemberg *et al.*, 1987) consisting of a recycle system with four units and seven streams (Fig. 3). In this example, true flowrate values are considered to be $\hat{\mathbf{x}} = [5\ \ 15\ \ 15\ \ 5\ \ 10\ \ 5\ \ 5]$ and the standard deviations of the flowrates are taken as 2.5% of the true flowrate values. As in previous publications, all possible combinations of two measurement biases are simulated. Fixed gross error magnitudes of 7 and 4 standard deviations are considered for the corresponding flowrates. The measurement value for each simulation trial is taken as the average of 10 randomly generated values.

Three methods were applied to detect bias measurements in this process network where only total mass balances are considered. These are the Generalized Ratio

TABLE 4 Performance Results for MIMT, GLR, and SEGE for Example 7.5

	MIMT			GLR			SEGE		
Str.	AVTI	OP	OPF	AVTI	OP	OPF	AVTI	OP	OPF
1–2	0.152	0.969	0.910	0.167	0.971	0.904	0.008	0.996	0.992
1–3	0.153	0.969	0.908	0.168	0.972	0.903	0.000	1.000	1.000
1–4	0.070	0.974	0.929	0.063	0.973	0.934	0.049	0.974	0.948
1–5	2.088	0.035	0.028	1.950	0.145	0.021	0.615	0.704	0.692
1–6	1.048	0.500	0.000	0.143	0.993	0.868	1.000	0.997	1*6*7*
1–7	0.403	0.821	0.614	1.074	0.504	0.000	1.000	0.997	1*6*7*
2–3	1.045	0.500	0.000	1.041	0.997	0.005	1.000	0.999	2*3*4*
2–4	0.963	0.501	0.000	0.961	0.500	0.000	0.948	0.958	2*3*4*
2–5	0.043	1.000	0.956	0.089	0.999	0.911	0.000	0.999	0.999
2–6	0.050	0.989	0.951	0.130	0.966	0.889	0.027	0.987	0.973
2–7	0.046	0.999	0.954	0.057	0.999	0.944	0.002	0.999	0.998
3–4	0.965	0.501	0.000	0.964	0.500	0.000	0.951	0.960	2*3*4*
3–5	0.044	0.999	0.956	0.086	1.000	0.913	0.000	1.000	1.000
3–6	0.051	0.988	0.949	0.125	0.967	0.893	0.027	0.987	0.973
3–7	0.049	0.999	0.951	0.054	0.999	0.947	0.001	0.999	0.999
4–5	1.037	0.500	0.000	0.077	0.999	0.925	1.000	0.999	4*5*6*
4–6	0.189	0.923	0.814	1.063	0.577	0.002	0.999	0.998	4*5*6*
4–7	0.035	0.999	0.965	0.067	0.998	0.936	0.004	0.998	0.996
5–6	0.062	0.978	0.933	1.026	0.500	0.000	0.996	0.977	4*5*6*
5–7	0.051	0.996	0.949	0.144	0.996	0.864	0.006	0.997	0.994
6–7	0.045	1.000	0.955	0.820	0.886	0.324	1.000	1.000	1*6*7*

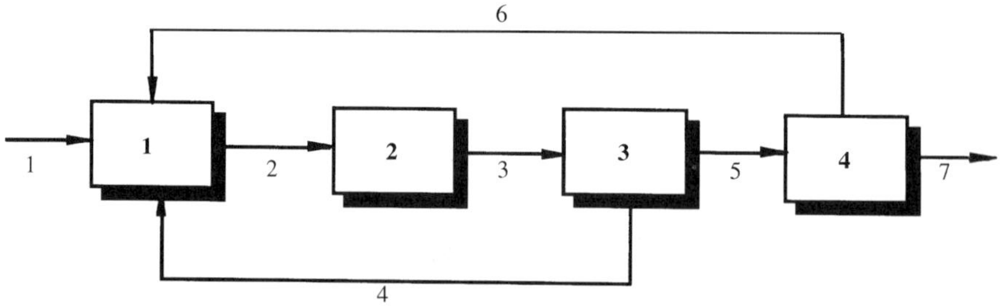

FIGURE 3 Recycle process network (from Rosenberg *et al.*, 1987).

Test Method (Narasimhan and Mah, 1987), the Modified Iterative Measurement Test (MIMT), and the Simultaneous Estimation of Gross Error Method (SEGE). In order to compare results on the same basis, the level of significance of each method is chosen such that it gives an AVTI, under null hypothesis, equal to 0.1.

In Table 4 the results are presented in terms of identification performance measures for the three methods. Two different situations arise from this table:

1. For 12 of the 21 cases, a numerical value can be calculated for the overall performance of the SEGE strategy. For these combinations of gross errors, SEGE exhibits a larger fraction of perfect identification runs (OPF) compared to the other two methods.

2. For the other nine cases, a numerical value for the SEGE's overall performance is not available. Nevertheless, the OP for this methodology is high and, in many cases,

TABLE 5 Bias Estimation Results for SEGE and GLR for Example 7.5

		SEGE		GLR	
Stream	SIM.M	EST.M	ST.DEV	EST.M	ST.DEV
1–2	0.875 1.5	0.876 1.503	0.053 0.127	0.805 1.474	0.052 0.126
1–3	0.875 1.5	0.876 1.501	0.053 0.129	0.805 1.475	0.052 0.126
1–4	0.875 0.5	0.876 0.505	0.054 0.099	0.925 0.487	0.053 0.095
2–5	2.625 1.0	2.625 0.999	0.130 0.091	2.404 0.973	0.128 0.089
2–6	2.625 0.5	2.627 0.503	0.137 0.082	2.330 0.439	0.128 0.071
2–7	2.625 0.5	2.625 0.499	0.129 0.053	2.478 0.491	0.128 0.052
3–5	2.625 1.0	2.626 0.998	0.131 0.091	2.405 0.973	0.129 0.089
3–6	2.625 0.5	2.628 0.503	0.138 0.082	2.331 0.440	0.129 0.071
3–7	2.625 0.5	2.626 0.499	0.130 0.053	2.478 0.492	0.129 0.053
4–7	0.875 0.5	0.875 0.499	0.103 0.054	0.895 0.554	0.168 0.087
5–7	1.750 0.5	1.748 0.499	0.092 0.054	1.552 0.470	0.090 0.051

is greater than the others. This indicates that the strategy identifies the simulated gross errors, but includes an extra flowrate in the suspicious set. It should be noted that GLR and MIMT perform poorly for seven and six of these nine cases, respectively.

SEGE is unable to give a perfect identification with the current process knowledge in the following situations:

1. The objective function value is equal for different combinations of gross errors
2. The matrices $[\mathbf{P}_b^T(\mathbf{A}\Psi\mathbf{A}^T)^{-1}\mathbf{P}_b]$ in (7.43) or $[\mathbf{B}_{rp}^T(\mathbf{A}\Psi\mathbf{A}^T)^{-1}\mathbf{B}_{rp}]$ in (7.53) or the system of equations in (7.57) is singular

In these cases it gives a suspicious subset of gross errors that includes the simulated ones but is larger than the real set. The user is advised that a unique solution is not possible. To sort out these difficulties, additional information on the process, for example, component or energy balances using fixed values of composition or temperature, may be included in Stage 2 of the procedure.

In Table 5, bias estimation results are given for GLR and SEGE. This table contains the cases from Table 4 with overall performance values greater than 0.9. The values in the SIM.M column are the simulated gross error magnitudes. EST.M and ST.DEV stand for the bias magnitude and its standard deviation estimate, respectively. They are obtained from the simulation trials with perfect identification. The SEGE method shows superior performance in estimating the size of gross errors; it gives unbiased and minimum variance estimators.

7.6. CONCLUSIONS

In this chapter we first presented a number of different, simple strategies for gross error identification. The serial elimination of measurements, the search along equations, and a combined procedure have been demonstrated to be simple and efficient ways for identifying gross errors. The estimation of gross errors due to both bias and leaks was also considered. These estimates have been evaluated using the least square technique.

Furthermore, we discussed a new strategy for the simultaneous identification and estimation of gross errors that can be applied for both bias in the measurements and leaks. This method is especially suited to large-scale, automated plants, where a low amount of gross error is present. The first stage of the procedure is the quick identification of suspicious systematic errors, which are then investigated in the second stage in order to identify the faulty sensors and leaks. This method significantly reduces the number of combinations for the second stage of the analysis.

The simultaneous estimation of gross errors enhances identification performance and the accuracy of the estimation. This is a key characteristic when instruments cannot be repaired until the units are out of service. In these situations the corrected measurement data are used for control and optimization purposes.

NOTATION

a	magnitude of a gross error
$\mathbf{A}$	matrix of linear constraints
$\mathbf{A}_g$	$[(m \times (g-c)]$ submatrix of $\mathbf{A}$
$\mathbf{A}_c$	$(m \times c)$ submatrix of $\mathbf{A}$
$\mathbf{A}_{ux}$	auxiliary matrix for constraints
AVTI	average number of type I errors
$\mathbf{B}_i$	row of $\mathbf{A}$
$\mathbf{B}_{rm}$	$(g \times s)$ matrix with $\mathbf{e}_i$ column vectors indicating bias positions
$\mathbf{B}_{rp}$	$(m \times p)$ matrix with $\mathbf{e}_i$ column vectors indicating leaks positions
c	number of measurements with gross error
$\mathbf{d}$	vector defined by Eq. (7.16)
ec	number of incorporated equations
EG	set of gross errors
$\mathbf{f}$	vector of total flowrates
g	number of measured variables
$\mathbf{G}$	Jacobian matrix of nonlinear constraints
$\mathbf{H}_0$	null hyphotesis
$\mathbf{H}_1$	alternative hyphotesis
h	nodal test statistic
$\mathbf{I}$	identity matrix
J	least square objective function
m	number of process constraints
$\mathbf{m}_b$	vector of bias magnitudes
$\mathbf{m}_p$	vector of leaks magnitudes
mnh	maximum number of hypothetical gross errors
mfo	minimum objective function value
$\mathbf{M}$	matrix defined by Eq. (7.14)
nh	number of hypothetical gross errors
ofv	vector of least square objective function values
OP	overall performance
OPF	expected fraction of perfect identification
$\mathbf{P}_b$	matrix defined by Eq. (7.44)
$\mathbf{r}$	residuum of process constraints
s	number of gross errors in the data set
SE	set of suspicious equations
SM	set of suspicious measurements
$\mathbf{u}$	vector of unmeasured variables

$\mathbf{V}$	covariance matrix of measurement error estimates
$\mathbf{W}$	covariance matrix of $\mathbf{d}$
$\mathbf{y}$	vector of measurements
$\mathbf{x}$	vector of true value of measured variables
$\mathbf{x}_1$	vector of redundant measured variables
z	measurement test statistic

Greek Symbols

ε	measurement random errors
$\boldsymbol{\Psi}$	covariance matrix of measurement errors
$\boldsymbol{\Phi}$	covariance matrix of residuum
τ	global test statistic
α	level of significance
χ_m^2	chi-square distribution with m degrees of freedom
$\boldsymbol{\Sigma}$	covariance matrix of $\hat{\mathbf{x}}$
$\boldsymbol{\Sigma}_{\mathrm{m_b}}$	matrix defined by Eq. (7.50)

Superscripts

$\frown$	least square estimation
0	linearization point
new	new case
old	old case

Subscripts

i	index for the sequential processing stage
0	without constraints
c	with process constraints
b	due to the presence of bias

REFERENCES

Almasy, G., and Sztano, T. (1975). Checking and correction of measurements on the basis of linear system model. *Probl. Control Inf. Theory* **4**, 57.

Brandt, S. (1970). "Statistical and Computational Methods in Data Analysis." North-Holland Publ., Amsterdam.

Crowe, C. M. (1996). Data reconciliation—Progress and challenges. *J. Proc. Control* **6**, 89–98.

Iordache, C., Mah, R., and Tamhane, A. (1985). Performance studies of the measurement test for detection of gross errors in process data. *AIChE J.* **31**, 1187–1201.

Mah, R. S. H. (1990). "Chemical Process Structures and Information Flows," Chem. Eng. Ser. Butterworth, Boston.

Mah, R. S. H., and Tamhane, A. C. (1982). Generalised likelihood ratio method for gross error identification. *AIChE J.* **28**, 828–830.

Mah, R. S., Stanley, G. M., and Downing, D. M. (1976). Reconciliation and rectification of process flow and inventory data. *Ind. Eng. Chem. Process Des. Dev.* **15**, 175.

Mikhail, E. (1976). "Observations and Least Squares," IEP Ser. Harper & Row, New York.

Narasimhan, S., and Mah, R. S. H. (1987). Generalised likelihood ratio method for gross error identification. *AIChE J.* **33**, 1514–1521.

Nogita, S. (1972). Statistical test and adjustment of process data. *Ind. Eng. Chem. Process Des. Dev.* **11** (2), 197–200.

Ripps, D. L. (1965). Adjustment of experimental data. *Chem. Eng. Prog. Symp. Ser.* **61** (55), 8–13.

Rollins, D., and Davis, J. (1992). Unbiased estimation of gross errors in process measurements. *AIChE J.* **38**, 563–572.

Romagnoli, J. (1983). On data reconciliation constraints processing and treatment of bias. *Chem. Eng. Sci.* **38**, 1107–1117.

Romagnoli, J., and Stephanopoulos, G. (1981). Rectification of process measurement data in the presence of gross errors. *Chem. Eng. Sci.* **36**, 1849–1863.

Rosemberg J., Mah, R. S. H., and Iordache, C. (1987). Evaluation of schemes for detecting and identifying gross errors in process data. *Ind. Eng. Chem. Res.* **26**, 555–564.

Sánchez, M., and Romagnoli, J. (1994). Comparative analysis of identification/bias estimation techniques. *AADECA' 94—14th Nat. Symp. Autom. Control.*

Serth, R., and Heenan, W. (1986). Gross error detection and data reconciliation in steam metering systems. *AIChE J.* **32**, 733–742.

Vaclaveck, V., and Vosolsobe, J. (1975). "The Design and Erection of Chemical Plants." Karlovy Vary, Czechoslovakia.

APPENDIX A

If an equation, or a set of equations, $\mathbf{B}_i$, is incorporated into a system of constraints defined by a matrix $\mathbf{A}$, the new process model can be stated as

$$\begin{bmatrix} \mathbf{B}_i \\ \mathbf{A} \end{bmatrix} \mathbf{x} = \mathbf{B}\mathbf{x} = \mathbf{0}. \tag{7A.1}$$

The covariance matrix of $\hat{\mathbf{x}}$ for the new case is by definition

$$\boldsymbol{\Sigma}^{\text{new}} = \boldsymbol{\Psi} - \boldsymbol{\Psi}\mathbf{B}^{\mathrm{T}}(\mathbf{B}\boldsymbol{\Psi}\mathbf{B}^{\mathrm{T}})^{-1}\mathbf{B}\boldsymbol{\Psi}. \tag{7A.2}$$

$\boldsymbol{\Sigma}^{\text{new}}$ can be calculated as function of $\mathbf{B}_i$ and the covariance matrix for the old case, $\boldsymbol{\Sigma}^{\text{old}}$, by the following expression:

$$\boldsymbol{\Sigma}^{\text{new}} = \boldsymbol{\Sigma}^{\text{old}} - \boldsymbol{\Sigma}^{\text{old}}\mathbf{B}_i^{\mathrm{T}}(\mathbf{B}_i\boldsymbol{\Sigma}^{\text{old}}\mathbf{B}_i^{\mathrm{T}})^{-1}\mathbf{B}_i\boldsymbol{\Sigma}^{\text{old}} \tag{7A.3}$$

Demonstration

The difference between the rigorous definitions of the covariance matrix of $\hat{\mathbf{x}}$ for the new and old cases is

$$\boldsymbol{\Sigma}^{\text{new}} - \boldsymbol{\Sigma}^{\text{old}} = \boldsymbol{\Psi}\mathbf{A}^{\mathrm{T}}(\mathbf{A}\boldsymbol{\Psi}\mathbf{A}^{\mathrm{T}})^{-1}\mathbf{A}\boldsymbol{\Psi} - \boldsymbol{\Psi}\mathbf{B}^{\mathrm{T}}(\mathbf{B}\boldsymbol{\Psi}\mathbf{B}^{\mathrm{T}})^{-1}\mathbf{B}\boldsymbol{\Psi}, \tag{7A.4}$$

so

$$\boldsymbol{\Sigma}^{\text{new}} = \boldsymbol{\Sigma}^{\text{old}} + \boldsymbol{\Psi}\mathbf{A}^{\mathrm{T}}(\mathbf{A}\boldsymbol{\Psi}\mathbf{A}^{\mathrm{T}})^{-1}\mathbf{A}\boldsymbol{\Psi} - \boldsymbol{\Psi}\mathbf{B}^{\mathrm{T}}(\mathbf{B}\boldsymbol{\Psi}\mathbf{B}^{\mathrm{T}})^{-1}\mathbf{B}\boldsymbol{\Psi}. \tag{7A.5}$$

$\mathbf{B}^{\mathrm{T}}(\mathbf{B}\boldsymbol{\Psi}\mathbf{B}^{\mathrm{T}})^{-1}\mathbf{B}$ can be written in terms of $\mathbf{A}$ and $\mathbf{B}_i$ matrices as follows:

$$\begin{aligned}
\mathbf{B}^{\mathrm{T}}(\mathbf{B}\boldsymbol{\Psi}\mathbf{B}^{\mathrm{T}})^{-1}\mathbf{B} &= \begin{bmatrix} \mathbf{B}_i^{\mathrm{T}} & \mathbf{A}^{\mathrm{T}} \end{bmatrix} \begin{bmatrix} \mathbf{B}_i\boldsymbol{\Psi}\mathbf{B}_i^{\mathrm{T}} & \mathbf{B}_i\boldsymbol{\Psi}\mathbf{A}^{\mathrm{T}} \\ \mathbf{A}\boldsymbol{\Psi}\mathbf{B}_i^{\mathrm{T}} & \mathbf{A}\boldsymbol{\Psi}\mathbf{A}^{\mathrm{T}} \end{bmatrix}^{-1} \begin{bmatrix} \mathbf{B}_i \\ \mathbf{A} \end{bmatrix} \\
&= \begin{bmatrix} \mathbf{B}_i^{\mathrm{T}} & \mathbf{A}^{\mathrm{T}} \end{bmatrix} \begin{bmatrix} \mathbf{C}_{11} & \mathbf{C}_{12} \\ \mathbf{C}_{21} & \mathbf{C}_{22} \end{bmatrix} \begin{bmatrix} \mathbf{B}_i \\ \mathbf{A} \end{bmatrix} \\
&= \begin{bmatrix} (\mathbf{B}_i^{\mathrm{T}}\mathbf{C}_{11} + \mathbf{A}^{\mathrm{T}}\mathbf{C}_{21}) & (\mathbf{B}_i^{\mathrm{T}}\mathbf{C}_{12} + \mathbf{A}^{\mathrm{T}}\mathbf{C}_{22}) \end{bmatrix} \begin{bmatrix} \mathbf{B}_i \\ \mathbf{A} \end{bmatrix} \\
&= \mathbf{B}_i^{\mathrm{T}}\mathbf{C}_{11}\mathbf{B}_i + \mathbf{A}^{\mathrm{T}}\mathbf{C}_{21}\mathbf{B}_i + \mathbf{B}_i^{\mathrm{T}}\mathbf{C}_{12}\mathbf{A} + \mathbf{A}^{\mathrm{T}}\mathbf{C}_{22}\mathbf{A},
\end{aligned} \tag{7A.6}$$

where the elements $\mathbf{C}_{ij}$ come from the development of the inverse of a partitioned matrix (Noble, 1969):

$$\begin{aligned}\mathbf{C}_{11} &= \left[\mathbf{B}_i\boldsymbol{\Psi}\mathbf{B}_i^{\mathrm{T}} - \mathbf{B}_i\boldsymbol{\Psi}\mathbf{A}^{\mathrm{T}}(\mathbf{A}\boldsymbol{\Psi}\mathbf{A}^{\mathrm{T}})^{-1}\mathbf{A}\boldsymbol{\Psi}\mathbf{B}_i^{\mathrm{T}}\right]^{-1}\\ &= \left[\mathbf{B}_i(\boldsymbol{\Psi} - \boldsymbol{\Psi}\mathbf{A}^{\mathrm{T}}(\mathbf{A}\boldsymbol{\Psi}\mathbf{A}^{\mathrm{T}})^{-1}\mathbf{A}\boldsymbol{\Psi})\mathbf{B}_i^{\mathrm{T}}\right]^{-1}\\ &= \left[\mathbf{B}_i\boldsymbol{\Sigma}_{\mathrm{c}}^{\mathrm{old}}\mathbf{B}_i^{\mathrm{T}}\right]^{-1}\end{aligned}$$

$$\mathbf{C}_{12} = -\,\mathbf{C}_{11}\mathbf{B}_i\boldsymbol{\Psi}\mathbf{A}^{\mathrm{T}}(\mathbf{A}\boldsymbol{\Psi}\mathbf{A}^{\mathrm{T}})^{-1}$$

$$\mathbf{C}_{21} = -\,(\mathbf{A}\boldsymbol{\Psi}\mathbf{A}^{\mathrm{T}})^{-1}\mathbf{A}\boldsymbol{\Psi}\mathbf{B}_i^{\mathrm{T}}\mathbf{C}_{11}$$

$$\mathbf{C}_{22} = (\mathbf{A}\boldsymbol{\Psi}\mathbf{A}^{\mathrm{T}})^{-1} + (\mathbf{A}\boldsymbol{\Psi}\mathbf{A}^{\mathrm{T}})^{-1}\mathbf{A}\boldsymbol{\Psi}\mathbf{B}_i^{\mathrm{T}}\mathbf{C}_{11}\mathbf{B}_i\boldsymbol{\Psi}\mathbf{A}^{\mathrm{T}}(\mathbf{A}\boldsymbol{\Psi}\mathbf{A}^{\mathrm{T}})^{-1}.$$

Replacing these formulas in (7A.5), we obtain

$$\begin{aligned}\boldsymbol{\Sigma}^{\mathrm{new}} = \boldsymbol{\Sigma}^{\mathrm{old}} &+ \left[-\boldsymbol{\Psi}\mathbf{B}_i^{\mathrm{T}}\left[\mathbf{B}_i\boldsymbol{\Sigma}^{\mathrm{old}}\mathbf{B}_i^{\mathrm{T}}\right]^{-1}\mathbf{B}_i\boldsymbol{\Psi}\right]\\ &+ \boldsymbol{\Psi}\mathbf{A}^{\mathrm{T}}(\mathbf{A}\boldsymbol{\Psi}\mathbf{A}^{\mathrm{T}})^{-1}\mathbf{A}\boldsymbol{\Psi}\mathbf{B}_i^{\mathrm{T}}\left[\mathbf{B}_i\boldsymbol{\Sigma}^{\mathrm{old}}\mathbf{B}_i^{\mathrm{T}}\right]^{-1}\mathbf{B}_i\boldsymbol{\Psi}\\ &+ \boldsymbol{\Psi}\mathbf{B}_i^{\mathrm{T}}\left[\mathbf{B}_i\boldsymbol{\Sigma}^{\mathrm{old}}\mathbf{B}_i^{\mathrm{T}}\right]^{-1}\mathbf{B}_i\boldsymbol{\Psi}\mathbf{A}^{\mathrm{T}}(\mathbf{A}\boldsymbol{\Psi}\mathbf{A}^{\mathrm{T}})^{-1}\mathbf{A}\boldsymbol{\Psi}\\ &- \boldsymbol{\Psi}\mathbf{A}^{\mathrm{T}}(\mathbf{A}\boldsymbol{\Psi}\mathbf{A}^{\mathrm{T}})^{-1}\mathbf{A}\boldsymbol{\Psi}\mathbf{B}_i^{\mathrm{T}}\left[\mathbf{B}_i\boldsymbol{\Sigma}^{\mathrm{old}}\mathbf{B}_i^{\mathrm{T}}\right]^{-1}\mathbf{B}_i\boldsymbol{\Psi}\mathbf{A}^{\mathrm{T}}(\mathbf{A}\boldsymbol{\Psi}\mathbf{A}^{\mathrm{T}})^{-1}\mathbf{A}\boldsymbol{\Psi}\end{aligned}$$

$$\begin{aligned}\boldsymbol{\Sigma}^{\mathrm{new}} = \boldsymbol{\Sigma}^{\mathrm{old}} &- \boldsymbol{\Psi}\mathbf{B}_i^{\mathrm{T}}\left[\mathbf{B}_i\boldsymbol{\Sigma}^{\mathrm{old}}\mathbf{B}_i^{\mathrm{T}}\right]^{-1}\mathbf{B}_i[\boldsymbol{\Psi} - \boldsymbol{\Psi}\mathbf{A}^{\mathrm{T}}(\mathbf{A}\boldsymbol{\Psi}\mathbf{A}^{\mathrm{T}})^{-1}\mathbf{A}\boldsymbol{\Psi}]\\ &+ \boldsymbol{\Psi}\mathbf{A}^{\mathrm{T}}(\mathbf{A}\boldsymbol{\Psi}\mathbf{A}^{\mathrm{T}})^{-1}\mathbf{A}\boldsymbol{\Psi}\mathbf{B}_i^{\mathrm{T}}\left[\mathbf{B}_i\boldsymbol{\Sigma}^{\mathrm{old}}\mathbf{B}_i^{\mathrm{T}}\right]^{-1}\mathbf{B}_i[\boldsymbol{\Psi} - \boldsymbol{\Psi}\mathbf{A}^{\mathrm{T}}(\mathbf{A}\boldsymbol{\Psi}\mathbf{A}^{\mathrm{T}})^{-1}\mathbf{A}\boldsymbol{\Psi}]\end{aligned}$$

$$\begin{aligned}\boldsymbol{\Sigma}^{\mathrm{new}} = \boldsymbol{\Sigma}^{\mathrm{old}} &- \boldsymbol{\Psi}\mathbf{B}_i^{\mathrm{T}}\left[\mathbf{B}_i\boldsymbol{\Sigma}^{\mathrm{old}}\mathbf{B}_i^{\mathrm{T}}\right]^{-1}\mathbf{B}_i\boldsymbol{\Sigma}^{\mathrm{old}}\\ &+ \boldsymbol{\Psi}\mathbf{A}^{\mathrm{T}}(\mathbf{A}\boldsymbol{\Psi}\mathbf{A}^{\mathrm{T}})^{-1}\mathbf{A}\boldsymbol{\Psi}\mathbf{B}_i^{\mathrm{T}}\left[\mathbf{B}_i\boldsymbol{\Sigma}^{\mathrm{old}}\mathbf{B}_i^{\mathrm{T}}\right]^{-1}\mathbf{B}_i\boldsymbol{\Sigma}^{\mathrm{old}}\end{aligned}$$

$$\boldsymbol{\Sigma}^{\mathrm{new}} = \boldsymbol{\Sigma}^{\mathrm{old}} - \left[\boldsymbol{\Psi} - \boldsymbol{\Psi}\mathbf{A}^{\mathrm{T}}(\mathbf{A}\boldsymbol{\Psi}\mathbf{A}^{\mathrm{T}})^{-1}\mathbf{A}\boldsymbol{\Psi}\right]\mathbf{B}_i^{\mathrm{T}}\left[\mathbf{B}_i\boldsymbol{\Sigma}^{\mathrm{old}}\mathbf{B}_i^{\mathrm{T}}\right]^{-1}\mathbf{B}_i\boldsymbol{\Sigma}^{\mathrm{old}}$$

$$\boldsymbol{\Sigma}^{\mathrm{new}} = \boldsymbol{\Sigma}^{\mathrm{old}} - \boldsymbol{\Sigma}^{\mathrm{old}}\mathbf{B}_i^{\mathrm{T}}\left[\mathbf{B}_i\boldsymbol{\Sigma}^{\mathrm{old}}\mathbf{B}_i^{\mathrm{T}}\right]^{-1}\mathbf{B}_i\boldsymbol{\Sigma}^{\mathrm{old}}.$$

APPENDIX B

Stage I

The recursive search algorithm is as follows:

(a) Variable initialization:

- α = level of significance
- $ec = 0$; number of incorporated equations
- $\mathbf{A}_{\mathrm{ux}} = \mathbf{0}$; auxiliary matrix for constraints
- $\boldsymbol{\Sigma}^{\mathrm{old}} = \boldsymbol{\Psi}$
- $\mathbf{SE} = \varnothing$; set of suspicious equations
- $\mathbf{SM} = \varnothing$; set of suspicious measurements

(b) For $i = 1$: number of equations:

- $\mathbf{A}_{\mathrm{ux}} = \left[\begin{smallmatrix}\mathbf{B}_i\\ \mathbf{A}_{\mathrm{ux}}\end{smallmatrix}\right]$; the row corresponding to the ith equation is incorporated into the auxiliary matrix

- $ec = ec + 1$
- Calculate
  - $\boldsymbol{\Sigma}^{\text{new}}$ using $\boldsymbol{\Sigma}^{\text{old}}$ [Eq. (7.63)]
  - $\hat{\mathbf{x}}$ using $\boldsymbol{\Sigma}^{\text{new}}$ [Equation (7.64)]
  - ofv as function of $\hat{\mathbf{x}}$
  - $\tau_c = \chi^2_{1-\alpha}(ec)$
- If $ofv > \tau_c$,
  - Equation i is included in **SE**
  - Variables corresponding to equation i are included in **SM**, avoiding repetitions
  - $ec = ec - 1$
  - The last incorporated row of $\mathbf{A}_{\text{ux}}$ is eliminated

  Else

$$\boldsymbol{\Sigma}^{\text{old}} = \boldsymbol{\Sigma}^{\text{new}}.$$

(c) Go to Stage 2.

Stage 2

Gross error identification is accomplished by the following procedure:

(a) Variable initialization:

- mnh = maximum number of hypothetical gross errors
- $nh = 1$; number of hypothetical gross errors
- τ_c = critical value for the Chi-square distribution with $(m - 1)$ degrees of freedom
- $\mathbf{ofv} = \varnothing$
- $\mathbf{EG} = \varnothing$

(b) For each hypothetical gross error i, that is, for each leak corresponding to a unit whose balance equation is included in **SE**, or for each bias of a measurement included in **SM**, calculate

- Bias or leak magnitude
- $ofv(i)$; objective function value of the least square technique

(c) Selection of the minimum value of the objective functions, min $(\mathbf{ofv}) = mfo$.

(d) For each gross error in (b),

- If $ofv(i) < \tau_c$ and $ofv(i) = mof$, include i in the set of gross errors **EG**

(e) If $\mathbf{EG} \neq \varnothing$, go to (1).

(f) Variable initialization:

- $nh = nh + 1$
- $\mathbf{ofv} = \varnothing$
- τ_c = critical value for the chi-square distribution for $(m - nh)$ degrees of freedom

(g) Formulation of all nh possible combinations of leaks, biases, and leaks with bias.

(h) For each combination of gross error, i, formulated in (g), calculate:

- Gross error magnitude
- $ofv(i)$

(i) Selection of the minimum value of the objective functions; min (**ofv**) $= mfo$.

(j) For each combination of gross error, i, formulated in (g),

- If $ofv(i) < \tau_c$ and $ofv(i) = mof$, include i in the set of gross errors **EG**

(k) If **EG** $\neq \varnothing$, go to (1).
Else

If $nh < mnh$, go to (f)
Else **EG** $=$ {combination $i/ofv(i) = mfo$}

(l) End.

8

RECTIFICATION OF PROCESS MEASUREMENT DATA IN DYNAMIC SITUATIONS

In this chapter, the data reconciliation problem for dynamic/quasi-steady-state evolving processes is considered. The problem of measurement bias is extended to consider dynamic situations. Finally in this chapter, an alternative approach for nonlinear dynamic data reconciliation using nonlinear programming techniques will be discussed.

8.1. INTRODUCTION

In the previous chapters the data reconciliation problem was analyzed for systems that could be assumed to be operating at steady state. Consequently, only one set of data was available. In some practical situations, the occurrence of various disturbances generates a dynamic or quasi-steady-state response of the process, thus nullifying this steady-state assumption. In this chapter, the notions previously developed are extended to cover these cases.

Under dynamic or quasi-steady-state conditions, a continuously monitored process will reveal changes in the operating conditions. When the process is sampled regularly, at discrete periods of time, then along with the *spatial redundancy* previously defined, we will have *temporal redundancy*. If the estimation methods presented in the previous chapters were used, the estimates of the desired process variables calculated for two different times, t_1 and t_2, are obtained independently, that is, no previous information is used in the generation of estimates for other times. In other words, temporal redundancy is ignored and past information is discarded.

In this chapter, the data reconciliation problem for dynamically evolving processes is considered. Thus, temporal redundancy is taken into account by using filtering techniques. A scheme for sequential processing information within a dynamic environment is also developed, allowing us to check for the presence and source of gross errors.

The second problem to be tackled is data reconciliation for applications in which the dominant time constant of the dynamic response of the system is much smaller than the period in which disturbances enter the system. Under this assumption the system displays quasi-steady-state behavior. Thus, we are concerned with a process that is essentially at steady state, except for slow drifts or occasional sudden transitions between steady states. In such cases, the estimates should be consistent, that is, they should satisfy the mass and energy balances.

Finally in this chapter, an alternative approach for nonlinear dynamic data reconciliation, using nonlinear programming techniques, is discussed. This formulation involves the optimization of an objective function through the adjustment of estimate functions constrained by differential and algebraic equalities and inequalities and thus requires efficient and novel solution techniques.

8.2. DYNAMIC DATA RECONCILIATION: A FILTERING APPROACH

8.2.1. Problem Statement

Let us now consider a system modeled by the following system of equations:

$$\dot{\mathbf{x}} = \mathbf{f}(\mathbf{x}) + \mathbf{w}(t) \tag{8.1}$$

$$\mathbf{y} = \phi(\mathbf{x}) + \varepsilon(t) \tag{8.2}$$

$$\mathbf{x}(0) = \mathbf{x}_0, \tag{8.3}$$

where modeling and observation errors are taken to be Gaussian, white noise processes, that is,

$$\begin{aligned} \mathbf{w}(t) &\approx N[\mathbf{0}; \mathbf{Q}(t)], \quad & \mathbf{Q}(t) = \mathbf{Q}^{\mathrm{T}}(t) > \mathbf{0} \\ \varepsilon(t) &\approx N[\mathbf{0}; \mathbf{R}(t)] \quad & \mathbf{R}(t) = \mathbf{R}^{\mathrm{T}}(t) > \mathbf{0}. \end{aligned} \tag{8.4}$$

The first term in the bracket stands for the mean, and the second for the spectral density matrix. For the continuous formulation, the covariances for the model and observation errors are given as

$$\begin{aligned} \mathbf{E}[\mathbf{w}(t)\mathbf{w}^{\mathrm{T}}(t)] &= \mathbf{Q}(t)\delta(t-\xi) \\ \mathbf{E}[\varepsilon(t)\varepsilon(t)^{\mathrm{T}}] &= \mathbf{R}(t)\delta(t-\xi), \end{aligned} \tag{8.5}$$

where the operator δ is the Dirac delta function.

The distinctive feature of the dynamic case is the time evolution of the estimate and its error covariance matrix. Their time dependence is given by

$$\dot{\hat{\mathbf{x}}} = \hat{\mathbf{f}}(\mathbf{x}, t) \tag{8.6}$$

$$\dot{\Sigma}(t) = \mathrm{E}(\mathbf{x}\mathbf{f}^{\mathrm{T}}) - \hat{\mathbf{x}}\hat{\mathbf{f}}^{\mathrm{T}} + \mathrm{E}(\mathbf{f}\mathbf{x}^{\mathrm{T}}) - \hat{\mathbf{f}}\hat{\mathbf{x}}^{\mathrm{T}} + \mathbf{Q}(t) \tag{8.7}$$

$$\mathbf{x}(0) = \mathbf{x}_0, \qquad \Sigma(0) = \Sigma_0, \tag{8.8}$$

where the caret (^) implies the expectation operator.

These differential equations depend on the entire probability density function $p(\mathbf{x}, t)$ for $\mathbf{x}(t)$. The evolution with time of the probability density function can, in principle, be solved with Kolmogorov's forward equation (Jazwinski, 1970), although this equation has been solved only in a few simple cases (Bancha-Reid, 1960). The implementation of practical algorithms for the computation of the estimate and its error covariance requires methods that do not depend on knowing $p(\mathbf{x}, t)$.

An often-used method consists of expanding $\mathbf{f}$ in Eq. (8.1) as a Taylor series about a certain vector that is close to $\mathbf{x}(t)$. In particular, if a first-order expansion is carried out on the current estimate of the state vector, we obtain

$$\mathbf{f}(\mathbf{x}, t) = \mathbf{f}(\hat{\mathbf{x}}, t) + \mathbf{A}(\hat{\mathbf{x}}, t)(\mathbf{x} - \hat{\mathbf{x}}) + \dots, \tag{8.9}$$

where

$$\mathbf{A}_{ij}(\hat{\mathbf{x}}, t) = \left.\frac{\partial f_i(\mathbf{x}, t)}{\partial x_j}\right|_{x(t)=\hat{x}(t)}. \tag{8.10}$$

Application of the expectation operation on both sides of equation (8.9) and substitution into Eq. (8.6) yields

$$\dot{\hat{\mathbf{x}}}(t) = \mathbf{f}(\hat{\mathbf{x}}, t); \quad \mathbf{x}(0) = \mathbf{x}_0. \tag{8.11}$$

Substitution of Eq. (8.11) into Eq. (8.7) allows us to obtain an expression for the differential equation of the estimation error covariance matrix:

$$\begin{aligned} \dot{\boldsymbol{\Sigma}}(t) &= \mathbf{A}(\hat{\mathbf{x}}, t)\boldsymbol{\Sigma}(t) + \boldsymbol{\Sigma}(t)\mathbf{A}^{\mathrm{T}}(\hat{\mathbf{x}}, t) + \mathbf{Q}(t) \\ \boldsymbol{\Sigma}(\mathbf{0}) &= \boldsymbol{\Sigma}_0; \quad \boldsymbol{\Sigma}_0 = \boldsymbol{\Sigma}_0^{\mathrm{T}} > \mathbf{0}. \end{aligned} \tag{8.12}$$

Equations (8.11) and (8.12) are approximate expressions for propagating the estimate and the error covariance, and in the literature they are referred to as the extended Kalman filter (EKF) propagation equations (Jaswinski, 1970). Other methods for dealing with the same problem are discussed in Gelb (1974) and Anderson and Moore (1979).

When implementing the solution, some discretization in time has to be done; therefore, it may be convenient to divide the system into time intervals, and approximate (8.11) and (8.12) with the difference equations, that is,

$$\mathbf{x}_{k+1} = \mathbf{F}_k\mathbf{x}_k + (\mathbf{w})_k, \tag{8.13}$$

where Δ is the sampling interval, $\Delta' = t_{k+1} - t_k$, $i = \Delta/\Delta'$, that is, Δ' is contained i times in Δ; $(\mathbf{w})_k$ is now a stochastic sequence $\mathbf{w} \sim N[\mathbf{0}, \mathbf{Q}_k]$; and $\mathbf{Q}_k = \mathbf{Q}(t)\Delta'$.

The discrete version of the equation for the estimate error covariance propagation is now

$$\boldsymbol{\Sigma}_{k+1} = \mathbf{F}_k\boldsymbol{\Sigma}_k\mathbf{F}_k^{\mathrm{T}} + \mathbf{Q}_k, \tag{8.14}$$

where $\mathbf{F}_k$ represents the transition matrix for the system equation (8.11).

Summarizing, the statistical characterisation of the random process (mean and covariance) can be projected through the interval $t_k \le t \le t_{k+1}$, and in this process there is an input noise that will increase the error, damaging the quality of the estimate.

Suppose that at a time t_{k-1}, the updated values of the mean and the estimate error covariance ($\hat{\mathbf{x}}(t_{k-1}|t_{k-1})$ and $\boldsymbol{\Sigma}(t_{k-1}|t_{k-1})$) are already available where the argument means "at time t_{k-1}, given information up to time t_{k-1}." These values are then used as initial values for the propagation in time, that is, for $t_{k-1} \leq t < t_k$, of the mean and covariance via the model equations.

If the predicted values $\hat{\mathbf{x}}_{k/k-1}$ and $\boldsymbol{\Sigma}_{k/k-1}$ are already computed, the minimum variance estimates of the states are obtained as the solution of the minimization problem

$$\text{Min } J = \mathbf{a}_k^{\mathrm{T}} \boldsymbol{\Sigma}_{k/k-1}^{-1} \mathbf{a}_k + \varepsilon_k^{\mathrm{T}} \mathbf{R}_k^{-1} \varepsilon_k, \tag{8.15}$$

where

$$\begin{aligned} \mathbf{a}_k &= \mathbf{x}_k - \hat{\mathbf{x}}_{k/k-1} \\ \varepsilon_k &= \mathbf{y}_k - \mathbf{C}\mathbf{x}_k, \end{aligned} \tag{8.16}$$

where the first term in the objective function accounts for the modeling prediction errors, and the second term for the observation errors. The solution to this minimization problem is given by the formula

$$\hat{\mathbf{x}}_k = \boldsymbol{\Sigma}_k \left[\boldsymbol{\Sigma}_{k/k-1}^{-1} \hat{\mathbf{x}}_{k/k-1} + \mathbf{C}^{\mathrm{T}} \mathbf{R}_k^{-1} \mathbf{y}_k \right], \tag{8.17}$$

where

$$\boldsymbol{\Sigma}_k = \left[\boldsymbol{\Sigma}_{k/k-1}^{-1} + \mathbf{C}^{\mathrm{T}} \mathbf{R}_k^{-1} \mathbf{C} \right]^{-1}. \tag{8.18}$$

Since $\boldsymbol{\Sigma}$ is a positive definite symmetric matrix, its trace can be taken as a measure of the estimate error covariance.

As in the static analysis, the processing of the information (provided by the addition of the new measurements) can be done systematically by means of a recursion formula. As a result, the computational effort is reduced considerably. The procedure is initialized with the determination of the error covariance for a single measurement, say $\mathbf{c}_1$, where $\mathbf{c}_1$ is the first row vector of the measurement matrix $\mathbf{C}$:

$$\boldsymbol{\Sigma}_k^{(1)} = \left[\boldsymbol{\Sigma}_{k/k-1}^{-1} + \mathbf{c}_1^{\mathrm{T}} \mathbf{R}_1^{-1} \mathbf{c}_1 \right]^{-1}. \tag{8.19}$$

This formula requires the inversion of a $(g \times g)$ matrix. The remaining $(l-1)$ pieces of sensor information are added one at a time. As in the static case, this is accomplished with the recursion formula

$$\boldsymbol{\Sigma}_k^{(i)} = \boldsymbol{\Sigma}_k^{(i-1)} \left\{ \mathbf{I} \mp \mathbf{c}_i^{\mathrm{T}} \left[\mathbf{R}_i \pm \mathbf{c}_i \boldsymbol{\Sigma}_k^{(i-1)} \mathbf{c}_i^{\mathrm{T}} \right]^{-1} \mathbf{c}_i \boldsymbol{\Sigma}_k^{(i-1)} \right\}, \tag{8.20}$$

where the upper and lower signs correspond to addition and deletion, respectively.

It should be noted that the solution of the minimization problem simplifies to the updating step of a Kalman filter. In fact, if instead of applying the matrix inversion lemma to Eq. (8.19) to produce Eq. (8.20), the inversion is performed on the estimate equation (8.18), the well-known form of the Kaman filter equations is obtained.

8.2.2. Analysis of Systems under Quasi-Steady-State Operation

In this section, the analysis of the data reconciliation problem is restricted to quasi-steady-state process operations. That is, those processes where the dominant time constant of the dynamic response of the system is much smaller than the period with which disturbances enter the system. Under this assumption the system displays quasi-steady-state behavior. The disturbances that cause the change in the operating conditions may be due to a slow variation in the heat transfer coefficients, catalytic activity in reactors, etc., or the source of a disturbance may be a sudden but lasting change. This is a process that is virtually at steady state but that exhibits slow, or occasionally sharp, transitions between steady states. The model for these kinds of processes consists of the following (Stanley and Mah, 1977).

1. A set of g transition equations

$$\mathbf{x}_{k+1} = \mathbf{x}_k + \mathbf{n}_k, \quad k = 0, 1, \ldots, \tag{8.21}$$

where $\mathbf{n}_k$ is the process noise.

2. A set of l measurement equations at time k,

$$\mathbf{y}_k = \mathbf{C}\mathbf{x}_k + \varepsilon_k, \quad k = 0, 1, \ldots. \tag{8.22}$$

3. A set of m steady-state algebraic balance equations at time k,

$$\mathbf{A}\mathbf{x}_k + \mathbf{w}_k = \mathbf{0}, \tag{8.23}$$

where $\mathbf{w}_k$ accounts for the modeling errors.

In this model, the noises $\mathbf{n}$, ε, and $\mathbf{w}$ are assumed to be uncorrelated, with zero mean and known covariance, that is,

$$\begin{aligned} \mathbf{n}_k &\approx N[\mathbf{0}, \mathbf{S}] \\ \varepsilon_k &\approx N[\mathbf{0}, \mathbf{R}] \\ \mathbf{w}_k &\approx N[\mathbf{0}, \mathbf{Q}], \end{aligned} \tag{8.24}$$

and the state vector $\mathbf{x}$ is also assumed to be a random variable with $\mathbf{x}_0 \approx N[\bar{\mathbf{x}}_0, \boldsymbol{\Sigma}_0]$.

Now consider the situation at time t_k. Suppose the predicted values are already available. As in the completely dynamic situation, the minimum variance estimates of the states are obtained as the solution of the following minimization problem:

$$\text{Min } J(\mathbf{x}) = [\mathbf{x}_k - \hat{\mathbf{x}}_{k/k-1}]^{\mathrm{T}} \boldsymbol{\Sigma}_{k/k-1}^{-1} [\mathbf{x}_k - \hat{\mathbf{x}}_{k/k-1}] + \varepsilon_k^T \mathbf{R}_k^{-1} \varepsilon_k + \mathbf{w}_k^{\mathrm{T}} \mathbf{Q}_k^{-1} \mathbf{w}_k. \tag{8.25}$$

Now, the solution of the data reconciliation problem at time k is given by

$$\hat{\mathbf{x}}_k = \boldsymbol{\Sigma}_k \left[\boldsymbol{\Sigma}_{k/k-1}^{-1} \hat{\mathbf{x}}_{k/k-1} + \mathbf{C}^{\mathrm{T}} \mathbf{R}_k^{-1} \mathbf{y}_k\right], \tag{8.26}$$

where

$$\boldsymbol{\Sigma}_k = \left[\boldsymbol{\Sigma}_{k/k-1}^{-1} + \mathbf{C}^{\mathrm{T}} \mathbf{R}_k^{-1} \mathbf{C} + \mathbf{A}^{\mathrm{T}} \mathbf{Q}_k^{-1} \mathbf{A}\right]^{-1} \tag{8.27}$$

is the estimate error covariance. These values of $\hat{\mathbf{x}}_k$ and $\boldsymbol{\Sigma}_k$ are used as initial values for the propagation in time, via the model equation for $t_k < t < t_{k+1}$, that is,

$$\hat{\mathbf{x}}_{k+1/k} = \hat{\mathbf{x}}_k \tag{8.28}$$

and

$$\boldsymbol{\Sigma}_{k+1/k} = \boldsymbol{\Sigma}_k + \mathbf{Q}_k. \tag{8.29}$$

The initial values for the recursive calculations are assumed to be available and given by

$$\mathbf{x}(\mathbf{0}) = \bar{\mathbf{x}}_0 \tag{8.30}$$

and

$$\Sigma(\mathbf{0}) = \Sigma_0. \tag{8.31}$$

The solution of the minimization problem again simplifies to updating steps of a static Kalman filter. For the linear case, matrices **A** and **C** do not depend on **x** and the covariance matrix of error can be calculated in advance, without having actual measurements. When the problem is nonlinear, these matrices depend on the last available estimate of the state vector, and we have the extended Kalman filter.

8.2.3. Fault Detection–Identification in Dynamic Processes

Problem Formulation

In the following we will be mostly concerned with the analysis of linear, stochastic models in the standard state space form:
System equations:

$$\mathbf{x}_{k+1} = \mathbf{A}\mathbf{x}_k + \mathbf{B}\mathbf{u}_k + \mathbf{w}_k. \tag{8.32}$$

Sensor equations:

$$\mathbf{y}_k = \mathbf{C}\mathbf{x}_k + \varepsilon_k, \tag{8.33}$$

where $\mathbf{u}_k$ is a known input, and **w** and ε are assumed to be zero mean and independent white sequences with covariances defined by

$$\begin{aligned} E\left[\mathbf{w}_k\,\mathbf{w}_j^{\mathrm{T}}\right] &= \mathbf{Q}\delta_{kj} \\ \mathbf{E}\left[\varepsilon_k\,\varepsilon_j^T\right] &= \mathbf{R}\delta_{kj}, \end{aligned} \tag{8.34}$$

where δ_{kj} is Kronecker delta function.

Equations (8.32) and (8.33) describe what we call the "normal" or "no failure" operation of the system of interest. The problem of failure detection is concerned with the detection of abrupt changes in a system, as modeled in Eqs. (8.32) and (8.33). Changes in (8.33) will be referred to as sensor failures. The main task of failure detection and compensation design is to modify the normal mode configuration to add the capability of detecting abrupt changes and compensating for them. In order to do that, we need to formulate what is called the "failure" model system:

$$\mathbf{x}_{k+1} = \mathbf{A}\mathbf{x}_k + \mathbf{B}\mathbf{u}_k + \mathbf{E}\mathbf{q}_k + \mathbf{w}_k + \mathbf{P}\mathbf{b}_k \tag{8.35}$$

$$\mathbf{y}_k = \mathbf{C}\mathbf{x}_k + \mathbf{G}\mathbf{g}_k + \varepsilon_k + \mathbf{D}\mathbf{b}_k, \tag{8.36}$$

where $\mathbf{b}_{k_0} \in \Re^{sb}$, $\mathbf{g}_{k_0} \in \Re^{sg}$, and $\mathbf{q}_{k_0} \in \Re^{sq}$ are the jumps of unknown magnitudes at $k = k_0$ of the inputs, outputs, and states, respectively.

Note: The previous formulations for both normal and abnormal situations are very general and include inputs to the process as well as different types of perturbations (jumps) in normal process behavior. Later on in this chapter we will consider a reduced version of this formulation, since we will be mainly interested in the measurement bias detection and identification problem. ♣

Now the failure detection problem consists of three tasks: alarm, isolation, and estimation. The alarm task consists of making a binary decision: either something has gone wrong or everything is fine. The problem of isolation is that of identifying the source of the failure, that is, which sensor has failed. The estimation problem involves the determination of the extent of the failure. For example, a sensor may become completely inoperable or it may suffer degradation in the form of a bias. In the latter case, estimation of the bias may allow the continued use of this sensor.

An important part of the failure detection problem involves the analysis of the necessary and sufficient conditions for detectability of jumps in the system. These are related to properties of matrix $\mathbf{A}$ and are discussed in detail by Caglayan (1980). The detectability of jumps in the outputs will be dictated by the observability of the discrete system defined by the pair $(\mathbf{A}_0, \mathbf{C}_0)$ (Caglayan, 1980).

Fault Detection by Statistical Tests

Failure detection methods are based on successive monitoring of the innovation sequence and statistical tests. Basically, the standard filter calculations are performed until some form of aberrant behavior is detected. A test was suggested first by Wilsky and Jones (1976) and is based on the following.

Under normal behavior of the filter, the innovation sequence is as follows:

$$\gamma_k = \mathbf{y}_k - \hat{\mathbf{y}}_{k/k-1} = \mathbf{y}_k - \mathbf{C}\hat{\mathbf{x}}_{k/k-1}, \tag{8.37}$$

where γ_k is a Gaussian white noise sequence with $E\{\gamma_k\} = \mathbf{0}$ and covariance matrix

$$\mathbf{M} = E\{\gamma_j \gamma_k^{\mathrm{T}}\} = \begin{cases} \mathbf{0}, & j \neq k \\ [\mathbf{C}\boldsymbol{\Sigma}_{k/k-1}\mathbf{C}^{\mathrm{T}} + \mathbf{R}_k], & j = k \end{cases}. \tag{8.38}$$

This will be termed the null hypothesis H_0. If γ_k belongs to a distribution with some other mean, the alternative hypothesis H_1 is satisfied, so the fault is declared. That is,

$$\begin{aligned} &\mathrm{H}_0 : E[\gamma_k] = \mathbf{0} \quad \text{(measurements are good)} \\ &\mathrm{H}_1 : \mathrm{E}[\gamma_k] \neq \mathbf{0} \quad \text{(fault is declared).} \end{aligned} \tag{8.39}$$

The following test statistic is applied in order to compare the hypotheses:

$$\mathbf{T} = \gamma_k^{\mathrm{T}}\mathbf{M}^{-1}\gamma_k \tag{8.40}$$

This statistic has chi-square distribution with l degrees of freedom under the null hypothesis, where l is the number of elements of γ_k. If $\mathbf{T} > \chi^2_{\alpha,l}$, H_0 is rejected, otherwise H_0 is accepted. α is the significance level of the test.

As in the steady-state case, the implementation of the chi-square test is quite simple, but has its limitations. One can use a more sophisticated technique such as the Generalized Likelihood Ratio (GLR) test. An alternative formulation of the chi-square test is to consider the components of γ_k separately (this may be useful for failure isolation information). In this case we compute the innovation of the ith measurement as

$$\gamma_{k_i} = y_{k_i} - \mathbf{c}_i\hat{\mathbf{x}}_{k_i} \tag{8.41}$$

where $\mathbf{c}_i$ is an g vector. Thus, in this case the test statistic

$$\tau_i = \frac{\gamma_{k_i}^2}{\left(\mathbf{c}_i\boldsymbol{\Sigma}_{k/k-1}\mathbf{c}_i^{\mathrm{T}} + R_i\right)} \tag{8.42}$$

is tested against a chi square distribution with one degree of freedom. Note that only the first component of γ is equal to the first innovation generated by the sequential

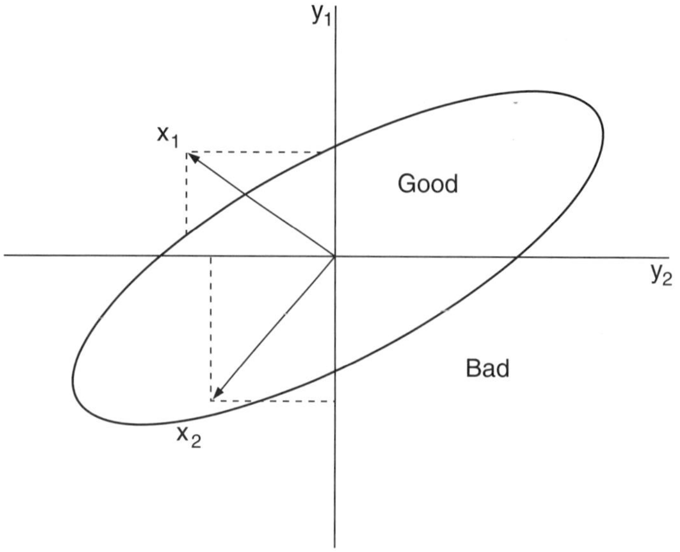

FIGURE 1 Innovation vector.

test. However, an advantage of performing the hypothesis test on the whole innovation vector is that the test is more sensitive to the detection of a failure, because the probability density function is n-dimensional as is shown in Fig. 1 for $n = 2$. In this figure vector **x** is outside of the region of acceptance, but with the first test it would be declared good, whereas in the case of processing the whole innovation vector it would be declared faulty. On the other hand, the vector $\mathbf{x}_2$, which is inside of the region with the first test, would be declared bad.

Finally, the selection of the critical value for the test is a trade-off between the sensitivity to fault detection and the probability of giving a false alarm.

Identification of Faulty Sensors

One of the disadvantages of replacing fault detection tests on components of the innovation vector by a test on the vector itself is that there is no longer a simple indication of the origin of the fault. In order to solve this problem the following procedure was implemented.

During the standard Kalman filter calculations, the matrix **M** has been evaluated from Eq. (8.38). When the test gives an alarm of malfunction, one or more elements of the innovation sequence vector is supposed to be at fault. In order to satisfy the abnormal situation, if one element is assumed to be at fault, the corresponding term in the matrix **M** would be greater than under normal circumstances. Thus,

$$\mathbf{M} = \underset{\displaystyle l-1 \qquad\quad 1}{\begin{bmatrix} \bar{\mathbf{M}} & \bar{\mathbf{m}} \\ \bar{\mathbf{m}}^{\mathrm{T}} & \mathbf{M}_{ll} + \delta m \end{bmatrix}} \begin{matrix} l-1 \\ 1 \end{matrix}, \tag{8.43}$$

where $\bar{\mathbf{M}}$ and $\bar{\mathbf{m}}$ contain the original elements of **M** (without change) and all the change is assumed in M_{ll}. δm is the increase in the variance due to the fault. This can be seen as considering the corresponding measurement in fault resulting in a change, δR, to the R_{ll} element. Since $\mathbf{R}$ is a diagonal matrix, it will affect $\mathbf{M}$ only in the corresponding diagonal elements. If we subdivide the innovation sequence vector, as well as the matrix $\mathbf{M}^{-1}$, as follows,

$$\tau = [\gamma_g \quad \gamma_f] \begin{bmatrix} \mathbf{M} & \\ & \delta_n \end{bmatrix}^{-1} \begin{bmatrix} \gamma_g \\ \gamma_f \end{bmatrix}, \tag{8.44}$$

where g and f mean good and faulty, respectively, then with $\delta m \to \infty$, the following recursive formula can be obtained:

$$\mathbf{M}_{\text{new}}^{-1} = \mathbf{M}_{\text{old}}^{-1}\left[\mathbf{I} - \mathbf{c}_i \left(\mathbf{c}_i^{\mathrm{T}}\mathbf{M}_{\text{old}}^{-1}\mathbf{c}_i\right)^{-1}\mathbf{c}_i^{\mathrm{T}}\mathbf{M}_{\text{old}}^{-1}\right], \tag{8.45}$$

where $\mathbf{c}_i$ is the ith row of the matrix $\mathbf{C}$ of measurements. In this way we can select the element of the vector γ that leads to a minimum value of τ in a sequential form. Since the number of elements in the innovation sequence are considered to be suspect one at a time, Eq. (8.45) provides a simple recursive formula for computing the variance of the innovation sequence when the effects of the various elements are analyzed. In this case, the recursive formula requires the inversion of a scalar quantity.

Bias Estimation

Once the occurrence of bad data is detected (through the previous procedure), we may either eliminate the sensor or we may assume simply that it has suffered degradation in the form of a bias. In the latter case, estimates of the bias may allow continued use of the sensor. That is, once the existence of a systematic error in one of the sensors is ascertained, its effect is modeled functionally.

Consequently, in the following discussion a sensor failure that affects one or more sensors will be assumed to add a constant bias of magnitude $\delta\mathbf{y}$ to the measurement vector, $\mathbf{y}$. In the presence of a sensor failure, let us consider the following models for the process and measurement:

$$\mathbf{x}_{k+1} = \mathbf{A}\mathbf{x}_k + \mathbf{w}_k \tag{8.46}$$

$$\mathbf{y}_k = \mathbf{C}\mathbf{x}_k + \mathbf{G}\delta\mathbf{y} + \varepsilon_k, \tag{8.47}$$

where $\mathbf{G}$ is a matrix of all zeros or ones, depending on whether the corresponding measurement is faulty or not. The constant $\delta\mathbf{y}$ represents the sign and magnitude of the failure. Once a bad sensor is identified, by any of the previous algorithms, the matrix $\mathbf{G}$ can be constructed.

Note: This is a more restricted formulation than the one posed in Eqs. (8.35) and (8.36), since only bias in the measurements is considered and an autonomous system is assumed. Also, here vector $\delta\mathbf{y}$ stands for vector $\mathbf{g}$ in Eq. (8.36). ♣

Defining a new state vector

$$\mathbf{z}_k = \begin{bmatrix} \mathbf{x}_k \\ \delta\mathbf{y}_k \end{bmatrix}, \tag{8.48}$$

we can rewrite Eqs. (8.46) and (8.47) as

$$\begin{aligned} \mathbf{z}_{k+1} &= \boldsymbol{\mu}\mathbf{z}_k + \mathbf{w}_k \\ \mathbf{y}_k &= \mathbf{L}\mathbf{z}_k + \boldsymbol{\varepsilon}_k, \end{aligned} \tag{8.49}$$

where

$$\boldsymbol{\mu} = \begin{bmatrix} \mathbf{A} & \mathbf{0} \\ \mathbf{0} & \mathbf{I} \end{bmatrix}, \quad \mathbf{L} = [\mathbf{C} \quad \mathbf{G}]. \tag{8.50}$$

The solution of this general state-bias estimation problem can be obtained through the application of the classical Kalman filtering technique. However, an efficient and alternative algorithm can be constructed for on-line implementation. This algorithm is based on the idea of a parallel processing technique that uses a decoupling procedure for the state-bias estimation (Friedland, 1969; Romagnoli and Gani, 1983).

By defining the auxiliary matrices $\mathbf{U}$ and $\mathbf{V}$, it is possible to obtain the corrected values of the error covariance and the estimate of the state as follows:

$$\begin{aligned} \boldsymbol{\Sigma}_k &= \boldsymbol{\Sigma}_k^0 + \delta\boldsymbol{\Sigma}_k \\ \mathbf{x}_k &= \mathbf{x}_k^0 + \delta\mathbf{x}_k, \end{aligned} \tag{8.51}$$

where $\boldsymbol{\Sigma}_k^0$ and $\boldsymbol{x}_k^0$ are the normal original elements of the filter in the absence of bias. Each element $\delta\mathbf{x}_k$ and $\delta\boldsymbol{\Sigma}_k$ corresponds to the correction (for the original filter) due to the bias estimation. They are calculated as

$$\begin{aligned} \delta\mathbf{x}_k &= \mathbf{V}_k^x \delta\mathbf{y}_k \\ \delta\boldsymbol{\Sigma}_k &= \mathbf{U}_k^x \boldsymbol{\Sigma}_k^{\delta y} \left(\mathbf{V}_k^x\right)^{\mathrm{T}}, \end{aligned} \tag{8.52}$$

while the estimate of the bias can be obtained from the recursion

$$\delta\mathbf{y}_k = \delta\mathbf{y}_{k-1} + \mathbf{K}_k^{\delta y}\left[\mathbf{m}_k - \left(\mathbf{C}\mathbf{U}_k^x + \mathbf{G}\right)\delta\mathbf{y}_{k-1}\right], \tag{8.53}$$

where $\mathbf{m}_k$ is the residual calculated from the standard Kalman filter, without the bias estimation. All of the recursion equations, as well as the complete procedure, can be found in the aforementioned publications.

This scheme may be viewed as a two-channel calculation procedure, where one channel corresponds to the filter in the normal way, and the other corresponds to the corrective part. This parallel computation will generate a corrective term (due to the bias estimation) that will affect the final results of the original normal filter. Since the state and parameter estimates are decoupled, the corrective term can be activated only when necessary, that is, when an anomaly occurs.

EXAMPLE 8.1

The results of an experimental application given in Porras and Romagnoli (1987) serve to display the features of this approach. The same experimental setup (see Fig. 2) used to illustrate the on-line implementation of a multichannel estimator (Bortolotto *et al.*, 1985) is considered here. It basically consists of a solid cylindrical rod with a heater housed in a hole longitudinally drilled at one end of the rod. An energy balance on the rod yields

$$\frac{\partial^2 T}{\partial \zeta^2} = \frac{2}{\lambda}\frac{h}{r}(T - T_o) - \frac{\rho c}{\lambda}\frac{\partial T}{\partial t} = 0,$$

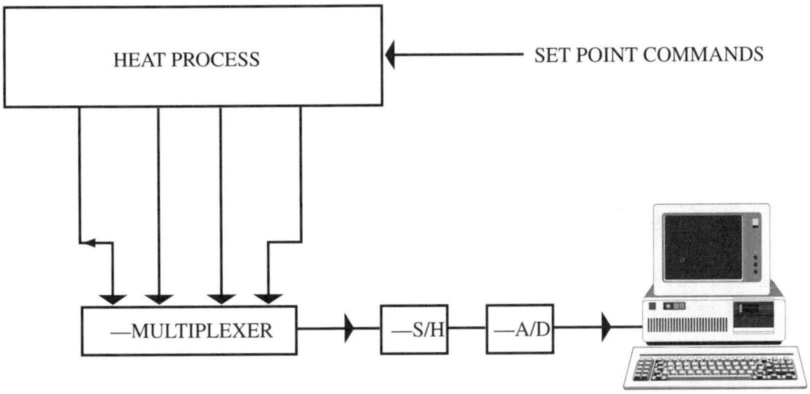

FIGURE 2 Process and data acquisition system for Example 8.1 (from Bortolotto *et al.*, 1985).

with boundary conditions

$$q = \frac{Q}{S} = -\lambda \left.\frac{dT}{d\zeta}\right|_{\zeta=0} = \text{constant}$$

$$\frac{Q_{\mathrm{L}}}{\mathrm{S}} = q_L = -\lambda \left.\frac{dT}{d\zeta}\right|_{\zeta=L} = h(T_L - T_o)$$

or

$$\left.\frac{dT}{d\zeta}\right|_{\zeta=L} = -\frac{h}{\lambda}(T_L - T_o),$$

where T is the air temperature and ζ the axial coordinate. For more details of the process and model description the reader is referred to the already-mentioned publications. A lumped version of the model was obtained by replacing the partial derivatives in these equations by their finite difference analogs.

The behavior of the detection algorithm is illustrated by adding a bias to some of the measurements. Curves A, B, C, and D of Fig. 3 illustrate the absolute values of the innovation sequences, showing the simulated error at different times and for different measurements. These errors can be easily recognized in curve E when the chi-square test is applied to the whole innovation vector ($n = 4$ and $\alpha = 0.01$). Finally, curves F, G, H, and I display the ratio between the critical value of the test statistic, τ, and the chi-value that arises from the source when the variance of the ith innovation (suspected to be at fault) has been substantially increased. This ratio, which is approximately equal to 1 under no-fault conditions, rises sharply when the discarded innovation is the one at fault.

This behavior allows easy detection of the source of the anomaly. In practice, under normal operation, the only curve processed in real time is E. But whenever the chi-square test detects a global faulty operation, the sequential process leading to curves F, G, H, and I starts, and thc innovation at fault is identified.

For the case of the bias estimation, several runs were performed simulating a known bias in one of the measurements, assuming different initial values of the variance of the bias. From Fig. 4 it can be seen that the convergence and the speed of response are heavily dependent on the initial values of Σ_{δ_y}. To help in the interpretation

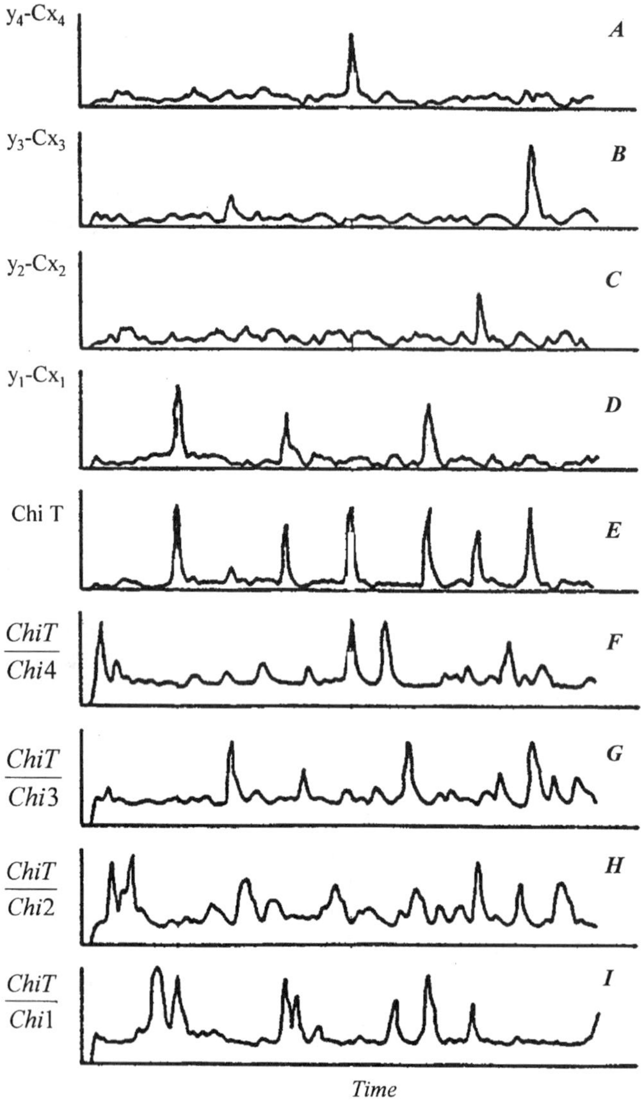

FIGURE 3 Tests for the innovation vector (from Porras and Romagnoli, 1987).

of the results, a band was drawn around a nominal value of the bias (about 12% of the bias). An estimate is considered good when the values predicted by the filter are inside the band.

8.3. DYNAMIC DATA RECONCILIATION: USING NONLINEAR PROGRAMMING TECHNIQUES

8.3.1. Problem Statement

As was previously shown, Kalman filtering techniques can be, and have been, successfully used on dynamic process data, to smooth measurement data recursively and

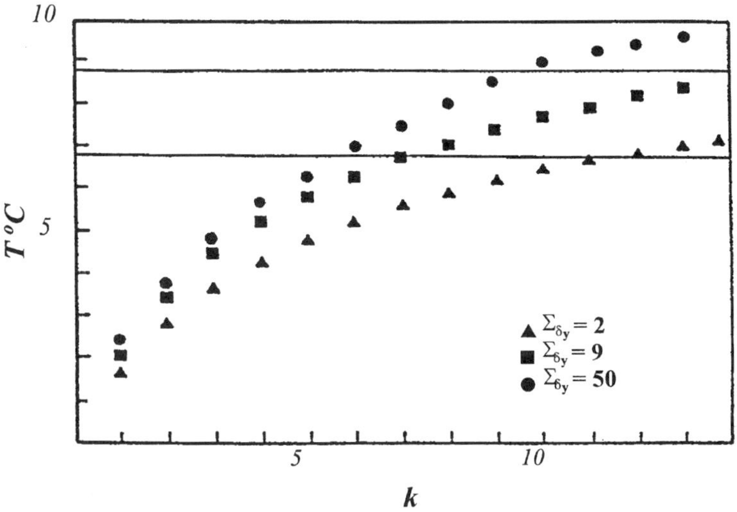

FIGURE 4 Speed of response (from Porras and Romagnoli, 1987).

to estimate parameters for linear systems. Modifications have been developed to handle nonlinear systems. These modifications typically involve replacing the nonlinear equations that represent the system with first-order approximations. For processes operating in highly nonlinear regions, linear approximations may not be satisfactory.

In this section, the use of nonlinear programming techniques for solving dynamic data reconciliation problems is discussed, as well as some existing methods for solving the resulting nonlinear differential/algebraic programming (NDAP) problem. The discussion follows closely the work performed by Edgar and co-workers in this area. The general nonlinear dynamic data reconciliation (NDDR) formulation can be written as (Liebman *et al.*, 1992; McBrayer and Edgar, 1995):

$$\begin{aligned}
&\underset{\hat{\mathbf{y}}}{\text{Min}}\ \Phi[\mathbf{y}, \hat{\mathbf{y}}, \sigma] \\
&\text{s.t.} \\
&\qquad \mathbf{f}\left[\frac{d\hat{\mathbf{y}}}{dt}, \hat{\mathbf{y}}(t)\right] = \mathbf{0} \\
&\qquad \eta[\hat{\mathbf{y}}(t)] = \mathbf{0} \\
&\qquad \omega[\hat{\mathbf{y}}(t)] \geq \mathbf{0},
\end{aligned} \tag{8.54}$$

where

- $\hat{\mathbf{y}}(t)$: estimate functions
- $\mathbf{y}$: discrete measurements
- σ: measurement noise standard deviations
- $\mathbf{f}$: differential equation constraints
- η: algebraic equality constraints
- ω: inequality constraints

Note that the measurements and estimates include both measured state variables and measured input variables. The inclusion of the input variables among those to be estimated establishes the error-in-variable nature of the data reconciliation problem.

The lengths of $\hat{\mathbf{y}}(t)$ and $\boldsymbol{\sigma}$ are equal to the total number of variables (state and input). The vector $\mathbf{y}$ comprises all $\mathbf{y}_k$, where $\mathbf{y}_k$ represent the measurements at discrete time t_k. The lengths of vectors $\mathbf{f}$, $\boldsymbol{\eta}$, and $\boldsymbol{\omega}$ are problem specific.

For most applications, the objective function is simply the weighted least squares (Liebman *et al.*, 1992)

$$\boldsymbol{\Phi}[\mathbf{y}, \hat{\mathbf{y}}(t), \boldsymbol{\sigma}] = \sum_{k=0}^{c} \frac{1}{2}[\hat{\mathbf{y}}(t_k) - \mathbf{y}_k]^{\mathrm{T}} \boldsymbol{\Psi}^{-1} [\hat{\mathbf{y}}(t_k) - \mathbf{y}_k], \tag{8.55}$$

where $\hat{\mathbf{y}}(t_k)$ represents the values of the estimate functions at discrete time t_k and $\boldsymbol{\Psi}$ is the variance–covariance matrix with $\boldsymbol{\Psi}_{ii} = \sigma_{ii}^2$. Variables t_0 and t_c represent the initial and current times, respectively.

As in the classical steady-state data reconciliation formulation, the optimal estimates are those that are as close as possible (in the least squares sense) to the measurements, such that the model equations are satisfied exactly.

8.3.2. Solution of the Differential/Algebraic Optimization Problem

The problem to be solved by the estimator may be stated as

$$\begin{aligned} &\min_{\hat{y}} \boldsymbol{\Phi}[\mathbf{y}, \hat{\mathbf{y}}, \boldsymbol{\sigma}] \\ &\text{s.t.} \\ &\qquad \frac{d\hat{\mathbf{y}}}{dt} - \mathbf{f}(\hat{\mathbf{y}}) = \mathbf{0} \\ &\qquad \boldsymbol{\eta}[\hat{\mathbf{y}}(t)] = \mathbf{0} \\ &\qquad \boldsymbol{\omega}[\hat{\mathbf{y}}(t)] \geq \mathbf{0}. \end{aligned} \tag{8.56}$$

There are two methods for computing solutions to problem (8.56):

1. Sequential solution and optimization
2. Simultaneous solution and optimization

1. The classic methods use an ODE solver in combination with an optimization algorithm and solve the problem sequentially. This solution strategy is referred to as a sequential solution and optimization approach, since for each iteration the optimization variables are set and then the differential equation constraints are integrated. Though straightforward, this approach is generally inefficient because it requires the accurate solution of the model equations at each iteration within the optimization, even when iterates are far from the final optimal solution.

2. In order to avoid solving the model equations to an unnecessary degree of accuracy at each step, the differential equations are approximated by a set of algebraic equations using a weighted residual method (Galerkin's method, orthogonal collocation, etc.). The model equations are then solved simultaneously with the other constraints within an unfeasible path optimization algorithm, such as successive quadratic programming. Using this approach, the differential equations are treated in the same manner as the other constraints and are not solved accurately until the final iteration of the optimization. Unfortunately, this method results in a relatively large NLP problem.

The original problem is restated as

$$\begin{aligned} &\underset{\hat{\mathbf{y}}}{\text{Min}}\ \boldsymbol{\Phi}[\mathbf{y}, \hat{\mathbf{y}}, \sigma] \\ &\text{s.t.} \\ &\qquad \mathbf{N}\mathbf{y} - \mathbf{f}(\hat{\mathbf{y}}) = \mathbf{0} \\ &\qquad \eta[\hat{\mathbf{y}}(t)] = \mathbf{0} \\ &\qquad \omega[\hat{\mathbf{y}}(t)] \geq \mathbf{0}, \end{aligned} \tag{8.57}$$

where **N** is a matrix of collocation weights (Villadsen and Stewart, 1967). Any derivatives with respect to spatial coordinates may be handled in a similar manner, and integral terms may be efficiently included by using appropriate quadrature formulas.

The simultaneous solution strategy offers several advantages over the sequential approach. A wide range of constraints may be easily incorporated and the solution of the optimization problem provides useful sensitivity information at little additional cost. On the other hand, the sequential approach is straightforward to implement and also has the advantage of well-developed error control. Error control for numerical integrators (used in the sequential approach) is relatively mature when compared, for example, to that of orthogonal collocation on finite elements (a possible technique for a simultaneous approach).

As pointed out by Liebman *et al.*, given a perfect model, an ideal data reconciliation scheme would use all information (process measurements) from the startup of the process until the current time. Unfortunately, such a scheme would necessarily result in an optimization problem of ever-increasing dimension. For practical implementation we can use a moving time window to reduce the optimization problem to manageable dimensions. A window approach was presented by Jang *et al.* (1986) and extended later by Liebman *et al.* (1992).

If the most recent available measurements are at time step c, then a history horizon $H\Delta t$ can be defined from $(t_c - H\Delta t)$ to t_c, where Δt is the time step size. In order to obtain enough redundant information about the process, it is important to choose a horizon length appropriate to the dynamic of the specific system (Liebman *et al.*, 1992). As shown in Fig. 5, only data measurements within the horizon will be reconciled during the nonlinear dynamic data reconciliation run.

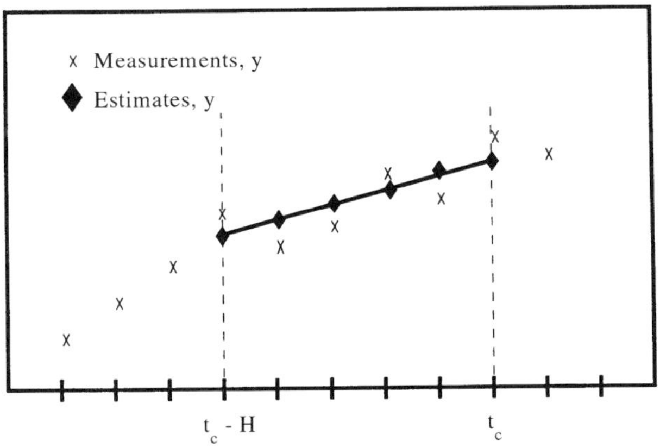

FIGURE 5 History horizon (from Liebman *et al.*, 1992).

A possible implementation algorithm comprises several steps (Liebman *et al.*, 1992), as shown below:

Step 1: Obtain process measurements
Step 2: Optimize $\mathbf{\Phi}$ for all $\hat{\mathbf{y}}(t)$ over $(t_c - H\Delta t) \leq t \leq t_c$
Step 3: Save the result of $\hat{\mathbf{y}}(t)$
Step 4: Repeat at next time t_{c+1}

An implementation of this algorithm, using the sequential procedure within the MATLAB environment, was proposed by Figueroa and Romagnoli (1994). To solve step 2, the *constr* function from the MATLAB Optimization Toolbox has been used. The numerical integration necessary in this step has been performed via the function *ode45* for the solution of ordinary differential equations.

EXAMPLE 8.2

We now proceed to demonstrate the application of the NDDR technique using a simulated CSTR with a first-order, exothermic reaction. The example was taken from Liebman *et al.* (1992). The dynamic model is given by

$$\frac{dA}{dt} = \frac{q}{V}(A_0 - A) - \alpha_{\mathrm{d}} K A$$

$$\frac{dT}{dt} = \frac{q}{V}(T_0 - T) + \alpha_{\mathrm{d}} \frac{-\Delta H_r}{\rho C_p} K A - \frac{U A_R}{\rho C_p V}(T - T_c)$$

where

$K = K_0 \exp(-E_{\mathrm{A}}/T)$, an Arrhenius rate expression
A_0: feed concentration, gmol cm^{-3}
T_0: feed temperature, K
A: tank concentration, gmol cm^{-3}
T: tank temperature, K

The parameter α_{d} is included to allow for catalyst deactivation as shown by Liebman. The data for the example (physical constants) are shown in Table 1. All temperatures and concentrations were scaled using a nominal reference concentration ($A_{\mathrm{r}} = 1 \times 10^{-6}$ gmol cm^{-3}) and a nominal reference temperature ($T_r = 100.0$ K).

TABLE 1 Data for the Example (Physical Constraints) (from Liebman *et al.*, 1992)

Parameter	Value	Units
q	10.0	cm^{3}s^{-1}
V	1000.0	cm^3
ΔH_{r}	−27,000.0	cal gmol^{-1}
ρ	0.001	g cm^{-3}
C_{p}	1.0	cal (g K)$^{-1}$
U	5.0×10^{-4}	cal (cm^2 s K)$^{-1}$
A_{R}	10.0	cm^2
T_{c}	340.0	K
K_0	7.86×10^{12}	s^{-1}
E_{A}	14,090.0	K
α_{d}	1.0	—

Measurements for both state variables, A and T, and both input variables, A_0 and T_0, were simulated at time steps of 2.5 s by adding Gaussian noise to the "true" values obtained through numerical integration of the dynamic equations. A measurement error with a standard deviation of 5% of the correspoding reference value was considered and the reconciliation of all measured variables (two states and two inputs) was carried out.

The proposed NDDR algorithm was applied with a history horizon of five time steps. The variance matrix was defined as

$$\Psi = \text{diag}\{0.325^2 \quad 0.175^2 \quad 0.0075^2 \quad 0.205^2\}.$$

Input variables were treated as constants over the entire horizon, allowing a decrease in the dimensionality of the optimization problem and also improving the performance of the reconciliation.

The steady-state simulation was initialized at a steady-state operating point of $A_0 = 6.5$, $T_0 = 3.5$, $A = 0.1531$, and $T = 4.6091$. At time 30 s, the feed concentration was stepped from 6.5 to 7.5.

The results are shown in Figs. 6 to 9. In these figures the stars correspond to the measured values, the dotted line to the simulated (free noise) values, and the solid line corresponds to the estimated values of the measurements. The estimate values for the states contain far less noise than the simulated measurements, as is shown in Figs. 6 and 7. Figures 8 and 9 show the estimate of the inputs. For the estimate of feed concentration, a lag is observed in the transient.

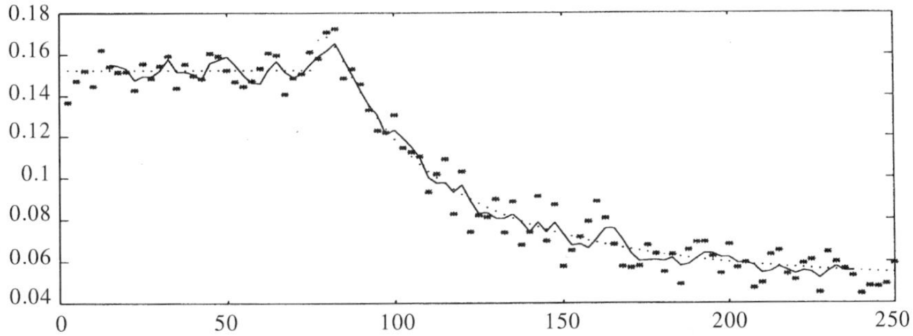

FIGURE 6 Concentration estimate response to step change in feed concentration.

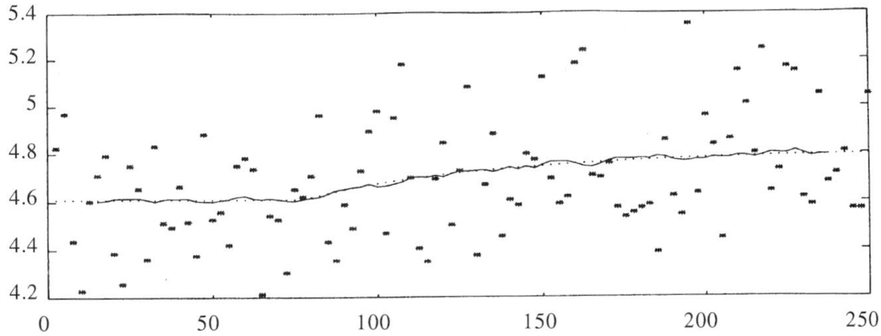

FIGURE 7 Temperature estimate response to step change in feed concentration.

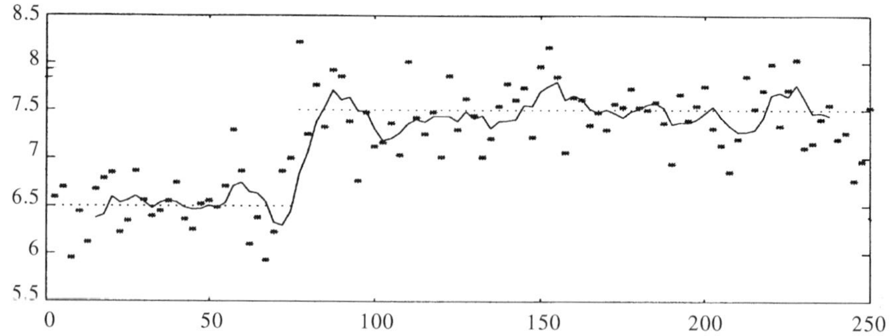

FIGURE 8 Step change in feed concentration.

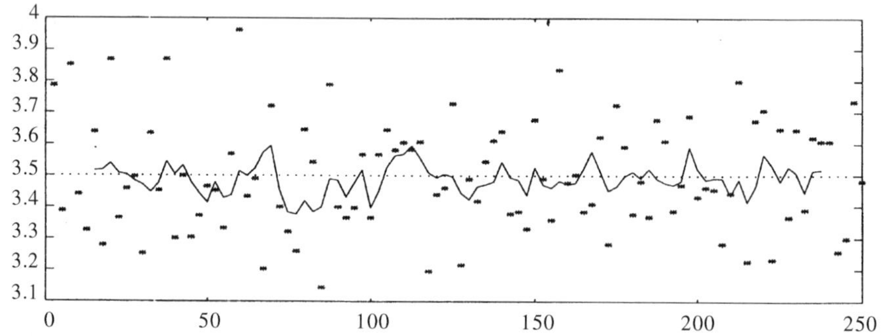

FIGURE 9 Feed temperature estimate.

8.3.3. Bias Detection and Estimation

The reconciliation method outlined before requires gross errors to be absent from the data before the rectification is carried out. If these error types are not absent, the reconciled values will exhibit "smearing" when compared with the true values.

Bias estimation in variables that are known a priori to be biased can be incorporated into the previous problem by incorporating the bias as a parameter (McBrayer and Edgar, 1995). The new Objective Function in Eq. (8.55) in the presence of bias becomes

$$\boldsymbol{\Phi}_b[\mathbf{y}, \hat{\mathbf{y}}(t), \boldsymbol{\sigma}, \hat{\mathbf{b}}] = \sum_{k=0}^{c} \frac{1}{2}[\hat{\mathbf{y}}(t_k) - (\mathbf{y}_k - \hat{\mathbf{b}}_j)]^{\mathrm{T}} \boldsymbol{\Psi}^{-1} [\hat{\mathbf{y}}(t_k) - (\mathbf{y}_k - \hat{\mathbf{b}}_j)], \tag{8.58}$$

where $\hat{\mathbf{b}}_j$ is the bias estimate for variable j.

The problem to be solved by the estimator may now be stated as

$$\begin{aligned} &\min_{\hat{\mathbf{y}}, \hat{\mathbf{b}}_j} \boldsymbol{\Phi}[\mathbf{y}, \hat{\mathbf{y}}, \boldsymbol{\sigma}, \hat{\mathbf{b}}_j] \\ &\text{s.t.} \\ &\qquad \frac{d\hat{\mathbf{y}}}{dt} - \mathbf{f}(\hat{\mathbf{y}}) = \mathbf{0} \\ &\qquad \boldsymbol{\eta}[\hat{\mathbf{y}}(t)] = \mathbf{0} \\ &\qquad \boldsymbol{\omega}[\hat{\mathbf{y}}(t)] \geq \mathbf{0}. \end{aligned} \tag{8.59}$$

By including $\hat{\mathbf{b}}_j$ in the inequality constraints, physical limits can be put on the range of admissible biases.

Remark. Although this method of estimation exhibits some oscillation, bias is a systematic parameter and the variance of the estimate is small relative to the measurement variance (McBrayer and Edgar, 1995). ♣

The same authors proposed an algorithmic approach for detecting the presence of a bias and for identifying the faulty sensor. Briefly, the main steps of the algorithm are as follows:

1. Using the model, a set of "base case" data is generated by adding Gaussian noise to the calculated measurements, using the same variance as the physical measurement.
2. These data are then reconciled and the test statistics calculated. These results correspond to the correct answer to which the measurements are compared.
3. The data reconciliation is then performed using the actual data, and these statistics are compared with those of the base case to determine the presence or absence of a bias.
4. If the test is positive, a bias term is added to a suspected measurement and estimated using the procedure outlined before.
5. The bias is then subtracted from the appropriate measurement, the reconciliation repeated, and the statistic compared with that of the base case. If the comparison is favorable, then the correct variable has been identified to contain a bias. If not, the procedure is repeated for a new suspected measurement, until the correct one is identified.

The complete procedure, together with a simulation application for a continuous stirred tank reactor, can be found in McBrayer and Edgar (1995).

Albuquerque and Biegler (1996) followed a different approach to incorporating bias into the dynamic data reconciliation, by taking into account the presence of a bias from the very beginning through the use of contaminated error distributions. This approach is fully discussed in Chapter 11.

8.4. CONCLUSIONS

In this chapter different aspects of data processing and reconciliation in a dynamic environment were briefly discussed. Application of the least square formulation in a recursive way was shown to lead to the classical Kalman filter formulation. A simpler situation, assuming quasi-steady-state behavior of the process, allows application of these ideas to practical problems, without the need of a complete dynamic model of the process.

Modification of the normal operation of the filter, to add the capability of detecting and compensating for abrupt changes, was discussed through the formulation of a "failure" model. It was shown that the solution of the general state-bias estimation problem can be obtained through the application of the classical Kalman filtering technique. A decoupling strategy was also introduced, based on the idea of parallel processing techniques.

Finally, an approach for nonlinear dynamic data reconciliation using nonlinear programming techniques was discussed. This formulation involves the optimization of an objective function through the adjustment of estimate functions constrained by differential and algebraic equalities and inequalities.

NOTATION

$\mathbf{a}$	vector defined by Eq. (8.16)
$\mathbf{A}$	Jacobian matrix of $\mathbf{f}(\mathbf{x}, t)$
$\mathbf{B}$	matrix defined by Eq. (8.32)
$\mathbf{b}$	jump of imputs
c	index for current time
$\mathbf{C}$	measurement model matrix
$\mathbf{D}$	matrix defined by Eq. (8.36)
$\mathbf{E}$	matrix defined by Eq. (8.35)
$\mathbf{f}$	differential equations constraints
$\mathbf{F}$	transition matrix
$\mathbf{g}$	jump of outputs
$\mathbf{G}$	matrix defined by Eq. (8.36)
H	width of the window
J	least square objective function value
k	time index
l	number of measurements
$\mathbf{L}$	matrix defined by Eq. (8.49)
m	number of constraints
$\mathbf{M}$	covariance matrix of γ
$\mathbf{n}$	vector of process noise
$\mathbf{N}$	matrix of collocation weights
$p(\mathbf{x}, t)$	probability density function for $\mathbf{x}(t)$
$\mathbf{P}$	matrix defined by Eq. (8.35)
$\mathbf{q}$	jumps of states
$\mathbf{Q}$	spectral density function for $\mathbf{w}$
$\mathbf{R}$	spectral density function for ε
$\mathbf{S}$	covariance matrix of $\mathbf{n}$
t	time
$\mathbf{u}$	known imput
$\mathbf{U}_x^i$	auxiliary matrix
$\mathbf{V}_x^i$	auxiliary matrix
$\mathbf{w}$	vector of modeling errors
$\mathbf{x}$	vector of state variables
$\mathbf{y}$	vector of measurements
$\mathbf{z}$	vector defined by Eq. (8.48)

Greek Symbols

ϕ	vector of measurement functions
ε	vector of errors between measurements and predictions
δ	Dirac delta function
Σ	estimate error covariance matrix
Δ	sampling interval
γ	vector of measurements residuals
τ	test statistic
$\delta\mathbf{y}$	magnitude of the bias
μ	matrix defined by Eq. (8.35)

Φ general objective function
σ standard deviation
η equality constraints
ω inequality constraints
Ψ covariance matrix for the measurement errors

Subscripts

0 initial state

Superscripts

$\cdot$ derivative
$\hat{}$ estimated value

REFERENCES

Albuquerque, J. S., and Biegler, L. T. (1996). Data reconciliation and gross error detection in dynamic systems. *AIChE J.* **42**, 2841–2856.

Anderson, B., and Moore, J. (1979). "Optimal Filtering." Prentice-Hall, Englewood Cliffs, NJ.

Bancha-Reid, A. T. (1960). "Elements of the Theory of Markov Processes and their Applications." McGraw-Hill, New York.

Bortolotto, G., Urbicain, M., and Romagnoli, J. A. (1985). On-line implementation of a multichannel estimator. *Comput. Chem. Eng.* **9**, 351–357.

Caglayan, A. E. (1980). Necessary and sufficient conditions for detectability of jumps in linear systems. *IEEE Trans. Autom. Control* **AC-25**, 833–834.

Figueroa, J. L., and Romagnoli, J. A. (1994). A strategy for dynamic data reconciliation within Matlab environment. *Australas. Chem. Eng. Conf. 22nd*, Perth, Australia, pp. 819–826.

Friedland, B. (1969). Treatment of bias in recursive filtering. *IEEE Trans. Autom. Control* **AC-14**, 359–367.

Gelb, A. (1974). "Applied Optimal Estimation." MIT Academic Press, Cambridge, MA.

Jang, S. S., Josepth, B., and Mukai, H. (1986). Comparison of two approaches to on-line parameter and state estimation problem of non-linear systems. *Ind. Eng. Chem. Process Des. Dev.* **25**, 809–814.

Jazwinski, A. H. (1970). "Stochastic Processes and Filtering Theory." Academic Press, New York.

Liebman, M. J., Edgar, T. F., and Lasdon, L. S. (1992). Efficient data reconciliation and estimation for dynamic process using non-linear programming techniques. *Comput. Chem. Eng.* **16**, 963–986.

McBrayer, K. F., and Edgar, T. F. (1995). Bias detection and estimation on dynamic data reconciliation. *J. Proc. Control* **15**, 285–289.

Porras, J. A., and Romagnoli, J. A. (1987). Data processing for instrument failure analysis in dynamical systems. *In* "Applied Modelling and Simulation of Technological Systems" (P. Borne and S. Tzafestas, eds.). Elsevier, North-Holland IMACS, Amsterdam.

Romagnoli, J. A., and Gani, R. (1983). Studies of distributed parameter systems: Decoupling the state parameter estimation problem. *Chem. Eng. Sci.* **38**, 1831–1843.

Stanley, G., and Mah, R. S. (1977). Estimation of flows and temperatures in process networks. *AIChE J.* **23**, 642–650.

Villadsen, J., and Stewart, W. E. (1967). Solution of boundary-value problems by orthogonal collocation. *Chem. Eng. Sci.* **22**, 1483–1501.

Wilsky, A. S., and Jones, L. (1976). A generalized likelihood ratio approach to the detection and estimation of jumps in linear systems. *IEEE Trans. Autom. Control* **AC-21**, 108–112.

9 JOINT PARAMETER ESTIMATION–DATA RECONCILIATION

In this chapter, the general problem of joint parameter estimation and data reconciliation will be discussed. The more general formulation, in terms of the error-in-variable method (EVM), where measurement errors in all variables are considered in the parameter estimation problem, will be stated. Finally, joint parameter and state estimation in dynamic processes will be considered.

9.1. INTRODUCTION

The estimation of model parameters is an important activity in the design, evaluation, optimization, and control of a process. As discussed in previous chapters, process data do not satisfy process constraints exactly and they need to be rectified. The reconciled data are then used for estimating parameters in process models involving, in general, nonlinear differential and algebraic equations (Tjoa and Biegler, 1992).

In a classical regression approach, the measurements of the independent variables are assumed to be free of error (i.e., for explicit models), while the observations of the dependent variables, the responses of the system, are subject to errors. However, in some engineering problems, observations of the independent variables also contain errors (i.e., for implicit models). In this case, the distinction between independent and dependent variables is no longer clear.

As pointed out by Tjoa and Biegler, in the first case, the optimization needs only to be performed in the parameters spaces, which is usually small. In the second case, we need to minimize the errors for all variables, and the challenge here is that the degrees of freedom become large, and increase with the number of experimental runs (data sets).

In the error-in-variable method, measurement errors in all variables are treated in the calculation of the parameters. Thus, EVM provides both parameter estimates and reconciled data estimates that are consistent with respect to the model.

Deming (1943) was among the first to consider the parameter estimation method for implicit models, formulating the general problem of parameter estimation in models by taking into account the errors in all measured variables. Britt and Luecke (1973) presented general algorithms for EVM where the objective function was optimized using Lagrange multipliers and the constraints were successively linearized with respect to the parameters and the measured variables. In this approach both the parameter estimates and the reconciled measurements were obtained simultaneously.

Other methods were proposed by Peneloux *et al.* (1976) and Reilly and Patino-Leal (1981). Their main characteristic is that the data reconciliation routine is nested within the parameter estimation routine. The advantage is that explicit calculation of the reconciled measurements is unnecessary and the size of the estimation problem is greatly reduced. More recent studies in EVM have been made by Schwetlick and Tiller (1985) and Valko and Vadja (1987). The main difference, with respect to previous approaches, is the separation of the parameter estimation step and the data reconciliation step, resulting in a two-stage calculation. Liebman and Edgar (1988) demonstrated improved reconciliation estimates using nonlinear programming, since the data reconciliation portion of the EVM seems to be especially sensitive to linearization errors. Further studies on the effect of linearization and the use of NLP to perform the data reconciliation problem were done by Kim *et al.* (1990). In this paper, several algorithms were tested using different case studies.

In this chapter, the general problem of joint parameter estimation and data reconciliation is discussed. First, the typical parameter estimation problem, in which the independent variables are error-free, is analyzed. Aspects related to the processing of the information are considered. Later, the more general formulation in terms of the error-in-variable method, where measurement errors in all variables are considered in the parameter estimation problem, is stated. Alternative solution techniques are briefly discussed; some special attention is given to the method of Valko and Vadja (1987) because it facilitates programming. This is because it can be implemented by putting together existing programs, especially those used in classical reconciliation routines. Finally, joint parameter and state estimation in dynamic processes will be considered and two different approaches, based on filtering techniques and nonlinear programming techniques, will be discussed.

9.2. THE PARAMETER ESTIMATION PROBLEM

Parameter estimation problems result when we attempt to match a model of known form to experimental data by an optimal determination of unknown model parameters. The exact nature of the parameter estimation problem will depend on the mathematical model. An important distinction has to be made at this point. A model will contain both state variables (concentrations, temperatures, pressures, etc.) and parameters (rate constants, dispersion coefficients, activation energies, etc.).

A further distinction still is possible by decomposing the state variables into two groups: independent and dependent variables. This decomposition will lead us to two different problems, as we will discuss later.

Let us now specify the model we will be considering. The following variables are defined:

θ: n-dimensional column vector of parameters whose numerical values are unknown $[\theta_1, \theta_2, \ldots, \theta_n]^T$.
$\mathbf{x}$: r-dimensional column vector of state variables $[x_1, x_2, \ldots, x_r]^T$.
$\mathbf{y}$: g-dimensional column vector of observed variables. These are the process variables that are actually measured in the experiments $[y_1, y_2, \ldots, y_g]^T$.

A single experiment consists of the measurement of each of the g observed variables for a given set of state variables (dependent, independent). Now if the independent state variables are error-free (explicit models), the optimization need only be performed in the parameter space, which is usually small.

So first we will concentrate on the problem in which the $\mathbf{x}$ independent variables are error-free and only errors in the dependent variables $\mathbf{y}$ need to be minimized. In fact $\mathbf{y}$ can be expressed as an explicit function of $\mathbf{x}$ and parameters θ, that is,

$$\mathbf{y} = \mathbf{f}(\mathbf{x}, \theta), \tag{9.1}$$

where $\mathbf{f}$ is a g-vector function of known form.

Note: When the independent variables, $\mathbf{x}$, are also subject to errors (implicit models), the distinction between independent and dependent variables is no longer clear and we need to minimize the errors for all variables. This will be the subject of Section 9.3. ♣

Coming back to problem (9.1), since we want to determine n parameters, we must perform at least n observations. If we assume that M different experiments are performed, then for each individual experiment we have a set of y's and x's. For the jth experiment we have

$$\mathbf{y}_j = \mathbf{f}(\mathbf{x}_j, \boldsymbol{\theta}), \quad j = 1, 2, \ldots, M. \tag{9.2}$$

It must be noted that the cases where the states and parameters are related by ordinary differential equations can be included in this general formulation. Suppose

$$\frac{dx_s}{dt} = f_s(t, \mathbf{x}(t), \boldsymbol{\theta}), \quad x_s(0) = x_{0s}, \quad s = 1, 2, \ldots, r. \tag{9.3}$$

The simplest case of this parameter estimation problem results if all state variables $x_s(t)$ and their derivatives $\dot{x}_s(t)$ are measured directly. Then the estimation problem involves only r algebraic equations. On the other hand, if the derivatives are not available by direct measurement, we need to use the integrated forms, which again yield a system of algebraic equations. In a study of a chemical reaction, for example, $\mathbf{y}$ might be the conversion and the independent variables might be the time of reaction, temperature, and pressure. In addition to quantitative variables we could also include qualitative variables as the type of catalyst.

EXAMPLE 9.1
Consider the isothermal rate equations

$$\frac{dC_A}{dt} = K_1 C_A$$

$$\frac{dC_B}{dt} = K_1 C_A - K_2 C_B$$

corresponding to the reactions

$$A \xrightarrow{K_1} B \xrightarrow{K_2} C$$

If the initial conditions are such that the total numbers of initial moles are known, that is,

$$C_{A_0} = 1, \quad C_{B_0} = 0, \quad C_{C_0} = 0, \quad t = 0,$$

we have for the integrated forms

$$C_A = e^{-K_1 t}$$

$$C_B = \frac{K_1}{K_2 - K_1}\left(e^{-K_1 t} - e^{-K_2 t}\right).$$

If we measure the concentrations of both A and B, then according to the previous definitions,

$$\mathbf{y} = [C_A; C_B]^T, \quad \boldsymbol{\theta} = [K_1, K_2]^T, \quad x = t,$$

and we have the form (9.1). If we measure C_A and C_B at M discrete values of time, then we have (9.2).

Because of error in the measurements and inaccuracies in the model, it is impossible to hope for an exact fit of (9.2). Instead, we will try to find values of these parameters that minimize some appropriate measure of the error. That is,

$$\mathbf{y}_j = \mathbf{f}(\mathbf{x}_j, \boldsymbol{\theta}) + \boldsymbol{\mu}_j, \quad j = 1, 2, \ldots, M, \tag{9.4}$$

where $\boldsymbol{\mu} = [\mu_1, \mu_2, \ldots, \mu_M]^{\mathrm{T}}$ is the vector of errors or residuals, between the observations and the predicted values. Now the parameter estimation problem, regardless of the form of the model, seeks the global minimum of

$$\begin{aligned} &\operatorname*{Min}_{\theta}\, (\mathbf{y} - \hat{\mathbf{y}})^{\mathrm{T}} \mathbf{W} (\mathbf{y} - \hat{\mathbf{y}}) \\ &\text{s.t.} \\ &\quad \hat{\mathbf{y}} - \mathbf{f}(\mathbf{x}, \boldsymbol{\theta}) = \mathbf{0}, \end{aligned} \tag{9.5}$$

where $\mathbf{W}$ is the weighting matrix. There are different ways to solve this problem. One approach is to solve it as a general NLP problem. In Tjoa and Biegler (1991), a hybrid SQP (HSQP) method was developed for nonlinear parameter estimation when the independent variables are error-free.

The first phase of the parameter estimation problem consists of choosing the measurements in such a way that the necessary conditions for estimability are satisfied. This means that we have to design the experiment such that if the measurements were totally without error, it would be possible to recover the desired parameters. These conditions were defined in Chapter 2. For instance, in the previous example, it would be theoretically impossible to recover K_1 and K_2 by measuring only C_A. In order to have the estimability conditions verified we should measure C_B, or C_A and C_B, or C_A and C_C.

The identification of the minimum set of observations required for a unique solution of a given problem is also included in this phase. That is, let n represent the number of parameters to be estimated and g_0 the minimum set of observations, with $g_0 \geq n$. Once these questions have been answered, the next step is to estimate the parameters. The analysis of the accuracy of the estimates is very important, in order to know along which variable we should move, when to take the samples, etc. This is what is called "design of the experiments" and will be discussed later.

In the following we describe the sequential processing of the information applied to parameter estimation. This allows us to analyze the accuracy of the estimate at each step in the procedure.

9.2.1. Sequential Processing of the Information for Experiment Design

In this section the sequential approach discussed in Chapter 6 will be extended to parameter estimation. An initial estimate of θ can be obtained from a minimal data set for which the number of observations is equal to the number of components of θ. Let $\mathbf{y}_0$ denote an observation at time $t = t_0$; if θ has n components, the minimal data set is symbolized by the vector

$$\mathbf{y}_0 = \begin{bmatrix} y_{01} \\ \cdot \\ \cdot \\ y_{0n} \end{bmatrix}. \tag{9.6}$$

The initial estimate, $\hat{\boldsymbol{\theta}}_0$, can be computed from the system of equations

$$\mathbf{f}_0(\hat{\boldsymbol{\theta}}_0) = \begin{bmatrix} f_{01}(\boldsymbol{\theta}_0) \\ \cdot \\ \cdot \\ f_{0n}(\boldsymbol{\theta}_0) \end{bmatrix} + \varepsilon_0 = \mathbf{y}_0, \tag{9.7}$$

where each element of $\mathbf{f}_0$ represents a prediction of an observed quantity at time $t = t_j$, $j = 1, 2, \ldots, M$. Now it is necessary to obtain the covariance matrix of the initial estimate error. Expanding the prediction function, $\mathbf{f}_0(\hat{\boldsymbol{\theta}}_0)$, about the true value of the parameter set, $\boldsymbol{\theta}$, and retaining only the constant and linear terms, we have

$$\mathbf{f}_0(\hat{\boldsymbol{\theta}}_0) = \mathbf{f}_0(\boldsymbol{\theta}) + \mathbf{D}_0(\boldsymbol{\theta})(\hat{\boldsymbol{\theta}}_0 - \boldsymbol{\theta}), \tag{9.8}$$

where

$$\mathbf{D}_0(\boldsymbol{\theta}) = \nabla \mathbf{f}_0(\boldsymbol{\theta}). \tag{9.9}$$

Since

$$\boldsymbol{\Delta}_0 = \hat{\boldsymbol{\theta}}_0 - \boldsymbol{\theta}, \tag{9.10}$$

we can write

$$\mathbf{D}_0(\boldsymbol{\theta})\boldsymbol{\Delta}_0 = \mathbf{f}_0(\hat{\boldsymbol{\theta}}_0) - \mathbf{f}_0(\boldsymbol{\theta}) = \varepsilon_0, \tag{9.11}$$

where ε_0 is the measurement error corresponding to the minimal data set. Thus,

$$\boldsymbol{\Psi} = \mathbf{E}\left[\varepsilon_0 \varepsilon_0^{\mathrm{T}}\right] = \mathbf{E}\left[\mathbf{D}_0 \boldsymbol{\Delta}_0 \boldsymbol{\Delta}_0^{\mathrm{T}} \mathbf{D}_0^{\mathrm{T}}\right]. \tag{9.12}$$

If we denote by $\boldsymbol{\Sigma}_0 = \mathbf{E}[\boldsymbol{\Delta}_0 \boldsymbol{\Delta}_0^{\mathrm{T}}]$ the covariance of the estimation error $\boldsymbol{\Delta}_0$, we have

$$\boldsymbol{\Sigma}_0 = \left[\mathbf{D}_0^{\mathrm{T}} \boldsymbol{\Psi}^{-1} \mathbf{D}_0\right]^{-1}. \tag{9.13}$$

We have now completed the preliminary step of obtaining the covariance for the initial estimate as a function of the known covariance matrix of the measurement errors. At time $t = t_1$, following the initial estimation, a new observation $\mathbf{y}_1$ is completed. Now the objective to be minimized by combining the new data with the previous estimate can be written as

$$\begin{aligned} J(\theta/\boldsymbol{\Delta}_0, \varepsilon_1) &= \boldsymbol{\Delta}_0^{\mathrm{T}} \boldsymbol{\Sigma}_0^{-1} \boldsymbol{\Delta}_0 + \varepsilon_1^{\mathrm{T}} \boldsymbol{\Psi}^{-1} \varepsilon_1 \\ &= (\hat{\boldsymbol{\theta}}_0 - \boldsymbol{\theta})^{\mathrm{T}} \boldsymbol{\Sigma}_0^{-1} (\hat{\boldsymbol{\theta}}_0 - \boldsymbol{\theta}) + [\mathbf{y}_1 - \mathbf{f}_1(\boldsymbol{\theta})]^{\mathrm{T}} \boldsymbol{\Psi}^{-1} [\mathbf{y}_1 - \mathbf{f}_1(\boldsymbol{\theta})], \end{aligned} \tag{9.14}$$

where $\mathbf{f}_1(\boldsymbol{\theta})$ is the predicted value of the observation $\mathbf{y}_1$ at time $t = t_1$ based on the value of the system parameters. Note the similarities between this formulation and the one described in Chapter 6 for data reconciliation using the general formulation from estimation theory, Eq. (6.61). In this case $\hat{\boldsymbol{\theta}}_0$, is the a priori estimate of the system parameters and $\boldsymbol{\Sigma}_0$ is the covariance of the a priori estimate error; however, there is no term related to the constraints (model) equations. The estimate error covariance, as we have seen in Chapter 6, is now given by

$$\boldsymbol{\Sigma}_1 = \left[\boldsymbol{\Sigma}_0^{-1} + \mathbf{D}_1^{\mathrm{T}} \boldsymbol{\Psi}^{-1} \mathbf{D}_1\right]^{-1}. \tag{9.15}$$

That is,

$$\boldsymbol{\Sigma}_1^{-1} = \left[\boldsymbol{\Sigma}_0^{-1} + \mathbf{D}_1^{\mathrm{T}} \boldsymbol{\Psi}^{-1} \mathbf{D}_1\right]. \tag{9.16}$$

This gives us the new error covariance when the measurement at time $t = t_1$ is processed as a function of the previous one. Note again that this formula is equivalent to the expressions in Chapter 6 when considering different blocks of information. The process can be iterated at each step with the introduction or deletion of new observations. By induction, the covariance for the error estimate at the ith step can be written as

$$\boldsymbol{\Sigma}_{i+1}^{-1} = \boldsymbol{\Sigma}_i^{-1} + \mathbf{D}_{i+1}^{\mathrm{T}} \boldsymbol{\Psi}^{-1} \mathbf{D}_{i+1}, \tag{9.17}$$

where positive signs stand for the addition of information.

Until now we have only considered moving along the variable time. This procedure, however, can be extended in general to other kinds of variables such as temperature, pressure, or some other qualitative variable such as the type of catalyst. In this way we can cover a whole range of operating conditions, processing the information at each stage in a sequential manner, using the information available from previous calculations.

Now, since the purpose of an experimental program is to gain information, in attempting to design the experiment we will try to plan the measurements in such a way that the final error estimate covariance is minimal. In our case, this can be achieved in a sequential manner by applying the matrix inversion lemma to Eq. (9.17), as we have shown in previous chapters.

EXAMPLE 9.2

Let us consider the simple, but very illustrative, example consisting of the reactions

$$A \xrightarrow{K_1} B \xrightarrow{K_2} C,$$

where K_1 and K_2 are first-order rate constants. Suppose further that C_B, the concentration of compound B, is measured at various times t, and $C_{B_0} = 0$, $C_{C_0} = 0$ and $C_{A_0} = 1$ are the initial values at time $t = 0$. Because of the estimability condition, we need to consider a minimum set of measurements composed of at least two different values of C_B at two times. According to our definition we have

$$y_j = C_{B_j}, \quad \theta = [K_1, K_2]^T, \quad x = t,$$

and

$$y_j = \mathbf{P}_j\theta + \varepsilon_j, \quad j = 1, \ldots, M,$$

where

$$P_{11_j} = \frac{\{K_1(K_1 - K_2)t_j + K_2\}e^{-K_1t_j} - K_2e^{-K_2t_j}}{(K_1 - K_2)^2}$$

$$P_{12_j} = \frac{\{-K_1(K_1 - K_2)t_j + K_1\}e^{-K_1t_j} - K_2e^{-K_2t_j}}{(K_1 - K_2)^2}.$$

A routine to calculate the optimal minimum set (optimal sampling times) has been implemented, which gives the minimum variance of the estimate errors. We have considered possible samples at M discrete times (equally spaced in time), for example, $M = 6$. That is, six measurements spaced every $\Delta t = 1$ period of time. The algorithm then looks at all of the possible combinations of measurement number 1 (at time $t = 1$) with each of the remaining, choosing the best combination. Then, combinations of measurement number 2 (at time $t = 2$) with each of the others are examined, and so on. It can be seen (Table 1) that to obtain the optimal minimum set, corresponding to two optimally located samples, measurements should be taken at times close to $t = 1$ and $t = 6$. In Table 1 the results are also given for $M = 12$ (12 equally spaced points in time $\Delta t = 0.5$). In this case the minimal set is composed of measurements number 2 and 12 (corresponding to times $t = 1$ and $t = 6$ for $\Delta t = 0.5$). In the same way we can isolate exactly the optimal minimum set of sampling times by further increasing M, that is, the number of allowable sampling points.

TABLE 1 Optimal Sampling Times for Different Values of M for Example 9.2

	Optimal sampling times			
Values of M	**Two**	**Three**	**Four**	**Five**
6	1;6	1;6;5	1;6;5;4	1;6;5;4;2
12	1;6	1;6;5.5	1;6;5.5;5	1;6;5.5;5;1.5
20	0.9;6	0.9;6;5.7	0.9;6;5.7;5.4	0.9;6;5.7;5.4;1.2

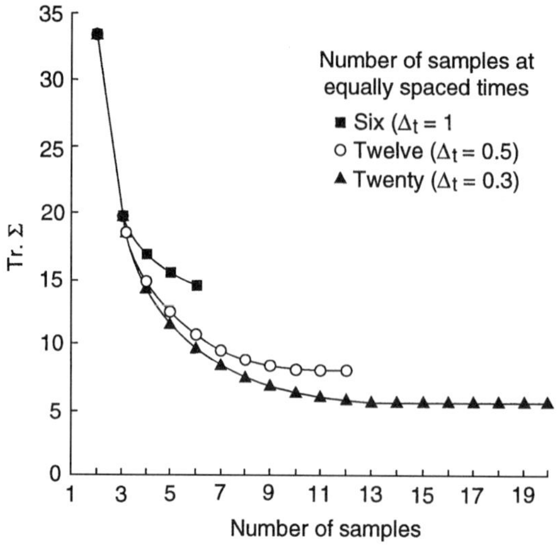

FIGURE 1 Trace of the error covariance as function of the number of optimal samples in time.

We have also considered the effect of increasing the number of optimally located samples on the final estimate. In this case, we start with the optimal minimum set and increase by one (optimally located) the number of samples. Figure 1 shows the results for different values of $M = 6, 12, 20$. As expected, the quality of the estimate is better when the number of optimally located samples is increased. However, after a few optimally located samples, the quality of the estimate does not improve significantly. Table 1 shows the corresponding optimal samples for different values of M and for different numbers of optimally located samples. In each case, after the minimum optimal set is determined, the next optimal samples are located around the times corresponding to the values for this optimal minimum set. The quality of the estimate is also directly affected by the variance of the measurements.

9.3. JOINT PARAMETER ESTIMATION–DATA RECONCILIATION PROBLEM

In the error-in-variable method (EVM), measurement errors in all variables are treated in the parameter estimation problem. EVM provides both parameter estimates and reconciled data estimates that are consistent with respect to the model. The regression models are often implicit and undetermined (Tjoa and Biegler, 1992), that is,

$$\mathbf{f}(\mathbf{y}, \mathbf{x}, \theta) = \mathbf{0}, \tag{9.18}$$

where $\mathbf{x}$ are the independent variables and $\mathbf{y}$ are the dependent ones. As stated by Tjoa and Biegler, "the estimation problem for these systems is often referred to as parameter estimation with implicit models or orthogonal distance regression (ODR), and we directly minimise a weighted least squares function that includes independent as well as dependent variables."

Now defining

$$\mathbf{z} = \begin{bmatrix} \mathbf{y} \\ \mathbf{x} \end{bmatrix},$$

an implicit model comprising b equations, p process variables, and n parameters can be described as

$$\mathbf{f}(\mathbf{z}, \boldsymbol{\theta}) = \mathbf{0}, \tag{9.19}$$

where

$$\begin{aligned} \mathbf{f} &= (f_1, \ldots, f_b)^{\mathrm{T}} \\ \mathbf{z} &= (z_1, \ldots, z_p)^{\mathrm{T}} \\ \boldsymbol{\theta} &= (\theta_1, \ldots, \theta_n)^{\mathrm{T}}. \end{aligned} \tag{9.20}$$

In experiments we observe the measurements values, $\tilde{\mathbf{z}}$, of the variables, $\mathbf{z}$, and allow for errors in all of them. Thus,

$$\tilde{\mathbf{z}}_j = \mathbf{z}_j + \varepsilon_j, \quad j = 1, \ldots, M. \tag{9.21}$$

Assuming that ε_j are normally distributed and uncorrelated, with zero mean and known positive definite covariance matrix $\boldsymbol{\Psi}_j$, the parameter estimation problem can be formulated as minimizing with respect to $\hat{\mathbf{z}}_j$ and $\hat{\boldsymbol{\theta}}$:

$$\begin{aligned} &\underset{z_j,\theta}{\text{Min}} \sum_{j=1}^{M} (\tilde{\mathbf{z}}_j - \mathbf{z}_j)^{\mathrm{T}} \boldsymbol{\Psi}_j^{-1} (\tilde{\mathbf{z}}_j - \mathbf{z}_j) \\ &\text{s.t.} \\ &\qquad \mathbf{f}(\mathbf{z}_j, \boldsymbol{\theta}) = \mathbf{0}, \quad j = 1, \ldots, M. \end{aligned} \tag{9.22}$$

In the following sections, different approaches to the solution of the preceding problem are briefly described. Special attention is devoted to the two-stage nonlinear EVM, and a method proposed by Valko and Vadja (1987) is described that allows the use of existing routines for data reconciliation, such as those used for successive linearization.

9.3.1. Solution Strategies for EVM

Several alternative algorithms have been proposed in the literature for dealing efficiently with EVM systems. The main approaches can be readily distinguished:

1. Simultaneous solution methods
2. Nested EVM
3. Two-stage EVM

1. Simultaneous Approach

The most straightforward approach for solving nonlinear EVM problems is to use nonlinear programming to estimate $\mathbf{z}_j$ and $\boldsymbol{\theta}$ simultaneously. In the traditional weighted least squares parameter estimation formulation there are only n optimization variables corresponding to the number of unknown parameters. In contrast, the simultaneous parameter estimation and data reconciliation formulation has $(pM + n)$

optimization variables. The dimensionality of the problem increases directly with the number of data sets and can become large.

A feasible path optimization approach can be very expensive because an iterative calculation is required to solve the undetermined model. A more efficient way is to use an unfeasible path approach to solve the NLP problem; however, many of these large-scale NLP methods are only efficient in solving problems with few degrees of freedom. A decoupled SQP method was proposed by Tjoa and Biegler (1991) that is based on a globally convergent SQP method.

2. Nested EVM

Kim *et al.* (1990) proposed a nested, nonlinear EVM, following ideas similar to those of Reilly and Patino-Leal (1981). In this approach, the parameter estimation is decoupled from the data reconciliation problem; however, the reconciliation problem is optimized at each iteration of the parameter estimation problem.

The algorithm is as follows:

Step 1: At $i = 1, \hat{\mathbf{z}} = \tilde{\mathbf{z}}$ and $\hat{\boldsymbol{\theta}} = \boldsymbol{\theta}_0$.
Step 2: Find the minimum of the function for $\hat{\mathbf{z}}$ and $\hat{\boldsymbol{\theta}}$:

$$\begin{aligned} & J_1 = \underset{\theta}{\text{Min}} \sum_{j=1}^{M} (\tilde{\mathbf{z}}_j - \mathbf{z}_j)^{\mathrm{T}} \boldsymbol{\Psi}_j^{-1} (\tilde{\mathbf{z}}_j - \mathbf{z}_j) \\ & \text{s.t.} \\ & J_2 = \underset{\hat{z}_j}{\text{Min}} \sum_{j=1}^{M} (\tilde{\mathbf{z}}_j - \mathbf{z}_j)^{\mathrm{T}} \boldsymbol{\Psi}_j^{-1} (\tilde{\mathbf{z}}_j - \mathbf{z}_j) \\ & \text{s.t.} \\ & \qquad \mathbf{f}(\mathbf{z}_j, \boldsymbol{\theta}) = \mathbf{0}, \quad j = 1, 2, \ldots, M. \end{aligned} \tag{9.23}$$

As pointed out by Kim *et al.* (1990), the difference between this algorithm and that of Patino-Leal is that the successive linearization solution is replaced with the nonlinear programming problem in Eq. (9.23). The nested NLP is solved as a set of decoupled NLPs, and the size of the largest optimization problem to be solved is reduced to the order of n.

3. Two-Stage EVM

To define the two-stage algorithm we will follow the work of Valko and Vadja (1987), which basically decouples the two problems. Starting from the definition of the general EVM problem in Eq. (9.22), the vectors $\mathbf{z}_1, \ldots, \mathbf{z}_M$ minimize the constrained problem at fixed θ, if and only if each $\mathbf{z}_j$, $j = 1, 2, \ldots, M$, is the solution to

$$\begin{aligned} & \text{Min } J(\mathbf{z}) = (\tilde{\mathbf{z}} - \mathbf{z})^{\mathrm{T}} \boldsymbol{\Psi}^{-1} (\tilde{\mathbf{z}} - \mathbf{z}) \\ & \text{s.t.} \\ & \qquad \mathbf{f}(\mathbf{z}, \theta) = \mathbf{0}, \end{aligned} \tag{9.24}$$

where the index j is dropped, since problem (9.24) is solved for each $j = 1, 2, \ldots, M$ separately.

Now, linearizing the constraints in (9.24) about some estimate $\hat{\mathbf{z}}$, and defining the equation error in terms of this linear approximation,

$$\mathbf{e} = \mathbf{f}(\hat{\mathbf{z}}, \boldsymbol{\theta}) + \mathbf{N}(\hat{\mathbf{z}}, \boldsymbol{\theta})[\tilde{\mathbf{z}} - \hat{\mathbf{z}}], \tag{9.25}$$

where $\mathbf{N}$ is the Jacobian matrix of $\mathbf{f}$ with respect to the variables $\mathbf{z}$. Thus, keeping $\hat{\mathbf{z}}$ fixed, we need to solve the typical linear (linearized) data reconciliation problem

$$\begin{aligned} &\text{Min } \mathbf{d}^{\mathrm{T}}\boldsymbol{\Psi}^{-1}\mathbf{d} \\ &\text{s.t.} \\ &\qquad -\mathbf{N}(\hat{\mathbf{z}}, \boldsymbol{\theta})\mathbf{d} + \mathbf{f}(\hat{\mathbf{z}}, \boldsymbol{\theta}) + \mathbf{N}(\hat{\mathbf{z}}, \boldsymbol{\theta})(\tilde{\mathbf{z}} - \hat{\mathbf{z}}) = \mathbf{0}, \end{aligned} \tag{9.26}$$

whose solution is given by

$$\hat{\mathbf{d}} = \boldsymbol{\Psi}\mathbf{N}^{\mathrm{T}}(\mathbf{N}\boldsymbol{\Psi}\mathbf{N}^{\mathrm{T}})^{-1}[\mathbf{f} + \mathbf{N}(\tilde{\mathbf{z}} - \hat{\mathbf{z}})], \tag{9.27}$$

where $\mathbf{d} = \tilde{\mathbf{z}} - \mathbf{z}$.

The problem can be solved using the successive linearization technique until convergence is achieved. The fixed point in the iteration is denoted by $\hat{\mathbf{z}}$. It is the solution of (9.24) and satisfies

$$\tilde{\mathbf{z}} - \hat{\mathbf{z}} = \boldsymbol{\Psi}\mathbf{N}^{\mathrm{T}}(\mathbf{N}\boldsymbol{\Psi}\mathbf{N}^{\mathrm{T}})^{-1}[\mathbf{f} + \mathbf{N}(\tilde{\mathbf{z}} - \hat{\mathbf{z}})] \tag{9.28}$$

Setting this equation into Eq. (9.24), we can compute the minimum of J in an alternative way:

$$J(\hat{\mathbf{z}}) = [\mathbf{f} + \mathbf{N}(\tilde{\mathbf{z}} - \hat{\mathbf{z}})]^{\mathrm{T}}(\mathbf{N}\boldsymbol{\Psi}\mathbf{N}^{\mathrm{T}})^{-1}[\mathbf{f} + \mathbf{N}(\tilde{\mathbf{z}} - \hat{\mathbf{z}})]. \tag{9.29}$$

This equation contains explicit information about the effect of the parameters on the objective function J. This has been exploited in the following algorithm.

Algorithm

Step 1: At $i = 1$ select $\hat{\mathbf{z}}_j(i) = \tilde{\mathbf{z}}_j$, $j = 1, 2, \ldots, M$.
Step 2: Find the minimum $\boldsymbol{\theta}^{i+1}$ of the function

$$J(\theta) = \sum_{j=1}^{M}[\mathbf{f}_j + \mathbf{N}_j(\tilde{\mathbf{z}}_j - \hat{\mathbf{z}}_j(i))]^{\mathrm{T}}\left(\mathbf{N}_j\boldsymbol{\Psi}\mathbf{N}_j^{\mathrm{T}}\right)^{-1}[\mathbf{f}_j + \mathbf{N}_j(\tilde{\mathbf{z}}_j - \hat{\mathbf{z}}_j(i)], \tag{9.30}$$

where $\mathbf{f}_j = \mathbf{f}(\mathbf{z}_j, \boldsymbol{\theta})$ and $\mathbf{N}_j = \partial\mathbf{f}(\mathbf{z}_j, \boldsymbol{\theta})/\partial\mathbf{z}$ computed at $(\hat{\mathbf{z}}_j(i), \boldsymbol{\theta})$. If $i > 1$ and $\|\boldsymbol{\theta}^i - \boldsymbol{\theta}^{i+1}\| \leq T_b$, then finish; otherwise, proceed to step 3.

Step 3: At fixed θ^{i+1} perform the data reconciliation for each $j = 1, 2, \ldots, M$ using successive linearization:

$$\hat{\mathbf{z}}_j^{\text{new}} = \tilde{\mathbf{z}}_j - \boldsymbol{\Psi}_j\mathbf{N}_j^{\mathrm{T}}\left(\mathbf{N}_j\boldsymbol{\Psi}_j\mathbf{N}_j^{T}\right)^{-1}[\mathbf{f}_j + \mathbf{N}_j(\tilde{\mathbf{z}}_j - \hat{\mathbf{z}}_j)]. \tag{9.31}$$

Denote by $\hat{\mathbf{z}}(i+1)$ the result of repeating Eq. (9.31) until convergence. Replace i by $i + 1$ and return to Step 2.

This type of algorithm facilitates programming, since the entire procedure can be implemented by putting together existing programs, such as the case of the successive linearization step, which can be implemented using the already available data reconciliation strategy. An alternative formulation was proposed by Kim *et al.* (1990), the difference being that the successive linearization solution is replaced with a nonlinear programming problem in Step 3.

EXAMPLE 9.3

To demonstrate the application of the two-stage EVM algorithm on a simulated example, let us consider the flowsheet of Fig. 2 taken from Bahri *et al.* (1996). It

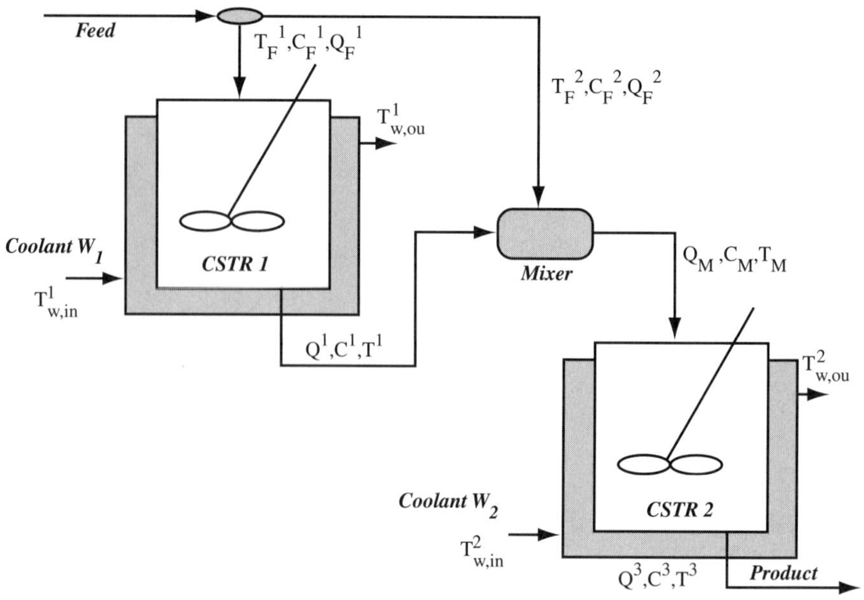

FIGURE 2 The two CSTRs flowsheet (from Bahri *et al.*, 1996).

consists of two continuous stirred tank reactors (CSTRs) in series with an intermediate mixer for the introduction of the second feed. A single, irreversible, exothermic, first-order reaction ($A \rightarrow B$) takes place in both reactors.

For our analysis we will just consider steady-state operation. The stage variables are the temperatures and concentrations in the reactors and the mixer. The four independent variables are input composition (C_F), two input flowrates (Q_F^1, Q_F^2), and input temperature (T_F). A CSTR steady-state model was used to simulate the measurements of the dependent variables. Details of the model equations representing the material and energy balances can be found in de Henning and Perkins (1991). The total number of measurements considered in this study is 21. The reference data set was generated adding a 1% Gaussian noise to each measurement. The parameters requiring estimation are the heat transfer coefficient in each reactor. As they are considered to be equal, a single parameter estimation problem results. The actual value of this parameter, used in the simulations that generated the measurements, is 0.35.

Different runs were performed by changing the noise level on the measurements as well as initial value of the parameter. Only a single set of experimental data was considered in our calculations. Table 2 gives a summary of the results, that is, the final value of the parameter obtained from the application of the algorithm, as a function of the initial value and the measurement noise. We can clearly see the effect of the noise level on the parameter estimator's accuracy, as well as its effect on the number of iterations.

Cases 4 and 5 deserve some special consideration. They were performed under the same conditions in terms of noise and initial parameter value, but in case 5 the covariances (weights) of the temperature measurements were increased with respect to those in the remaining measurements. For case 4 it was noticed that, although a normal random distribution of the errors was considered in generating the measurements, some systematic errors occur, especially in measurement numbers 6, 8,

TABLE 2 Summary of Results for the Two CSTRs Example

Case	Initial value of parameter	Final value of parameter	Noise level	Number of iterations	Iterations for parameter estimation
1	0.5	0.3209	1%	5	14,7,7,7,7
2	0.5	0.3363	0.5%	4	11,7,7,7
3	0.5	0.3473	0.1%	3	12,7,7
4	0.2	0.3249	1%	4	7,7,7,7
5	0.2	0.3565	1%*	22	12,8, . . . ,8,3,3

TABLE 3 Comparison of Actual, Measured, and Reconciled Data for Case 4 in Table 2

Variable (from Fig. 2)	Actual value	Measured value	Reconciled value
C^1	0.0307	0.0311	0.0282
T^1	385.0744	387.4882	387.4414
C^2	0.0509	0.0509	0.0421
T^2	360.7426	362.0110	362.4886
C_M	7.3654	7.3141	6.8828
T_M	372.1917	378.5046	377.0245
C_f^1	20.0000	20.0118	20.2201
T_F^1	350.000	356.2898	356.7557
C_F^2	20.0000	20.0528	20.1560
T_F^2	350.0000	353.0509	353.1371
$T_{w,ou}^1$	319.3400	314.7218	314.8330
$T_{w,ou}^2$	283.7160	281.7267	281.7218
W_1	0.3318	0.3359	0.3722
W_2	0.7996	0.7945	0.8048
$T_{w,in}^1$	250.0000	251.4434	251.4409
$T_{w,in}^2$	250.0000	249.0999	249.1162
Q_F^1	0.3552	0.3547	0.3556
Q_F^2	0.2062	0.2034	0.1828
Q^1	0.3552	0.3507	0.3556
Q_M	0.5614	0.5669	0.5384
Q^2	0.5614	0.5611	0.5384

and 11. Consequently, in case 5 a larger covariance for these measurements was also considered. Under these conditions, the accuracy of the parameter estimation and data reconciliation further improved. This shows the importance of a good knowledge of the covariance matrix as well as the importance of being able to deal with systematic errors. Tables 3 and 4 and Figs. 3 and 4 illustrate the comparisons, in terms of the amount of correction on each data point after the joint parameter estimation and data reconciliation procedure. This joint approach for parameter estimation and data reconciliation is further investigated in Chapter 12 in the context of estimating the heat transfer coefficient for an industrial furnace from available plant data.

TABLE 4 Comparison of Actual, Measured, and Reconciled Data for Case 5 in Table 2

Variable (from Fig. 2)	Actual value	Measured value	Reconciled value
C^1	0.0307	0.0311	0.0300
T^1	385.0744	387.4882	385.8322
C^2	0.0509	0.0509	0.0496
T^2	360.7426	362.0110	361.1329
C_M	7.3654	7.3141	7.3149
T_M	372.1917	378.5046	373.9660
C_f^1	20.0000	20.0118	20.0114
T_F^1	350.0000	356.2898	353.2848
C_F^2	20.0000	20.0528	20.0528
T_F^2	350.0000	353.0509	350.9374
$T_{w,ou}^1$	319.3400	314.7218	320.6448
$T_{w,ou}^2$	283.7160	281.7267	283.8052
W_1	0.3318	0.3359	0.3358
W_2	0.7996	0.7945	0.7944
$T_{w,in}^1$	250.0000	251.4434	251.4434
T_{w,in^2}	250.0000	249.0999	249.1000
Q_F^1	0.3552	0.3547	0.3563
Q_F^2	0.2062	0.2034	0.2045
Q^1	0.3552	0.3507	0.3563
Q_M	0.5614	0.5669	0.5608
Q^2	0.5614	0.5611	0.5608

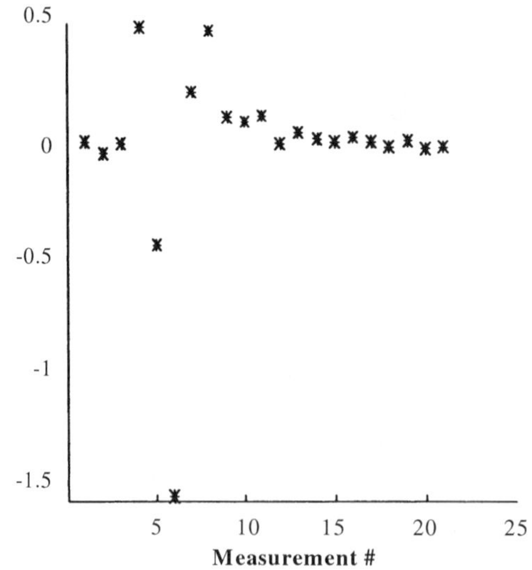

FIGURE 3 Amount of correction on the measurements after joint parameter and data reconciliation for case 4 in Table 2.

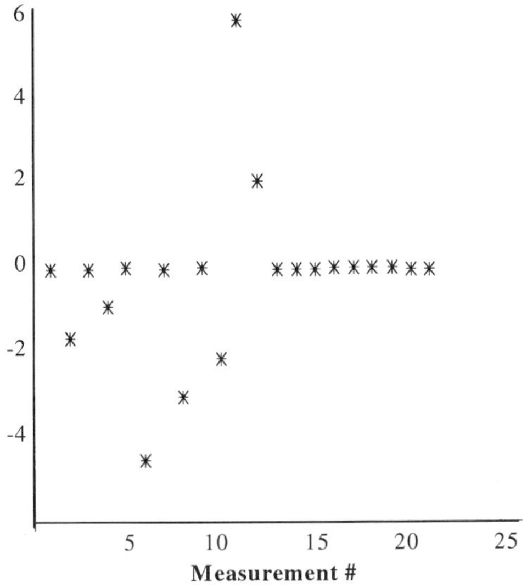

FIGURE 4 Amount of correction on the measurements after joint parameter and data reconciliation for case 5 in Table 2.

9.4. DYNAMIC JOINT STATE–PARAMETER ESTIMATION: A FILTERING APPROACH

9.4.1. Problem Statement and Definition

The problem of state–parameter estimation in dynamic systems is considered in terms of decoupling the estimation procedure. By using the extended Kalman filter (EKF) approach, the state–parameter estimation problem is defined and a decoupling procedure developed that has several advantages over the classical approach.

Following Bortolotto *et al.* (1985), let us consider the dynamic system governed by the following stochastic model:

$$\begin{aligned} \mathbf{x}_{k+1} &= \mathbf{F}\mathbf{x}_k + \mathbf{w}_k \\ \mathbf{y}_k &= \mathbf{C}\mathbf{x}_k + \boldsymbol{\varepsilon}_k, \end{aligned} \tag{9.32}$$

where $\mathbf{w}_k$ and ε_k represent the forcing random noise entering the system and the random noise that corrupts the measurements, respectively. The statistical characterization of these Gaussian white noises is represented as

$$\begin{aligned} E[\mathbf{w}\mathbf{w}^{\mathrm{T}}] &= \mathbf{R}_1, \quad E[\boldsymbol{\varepsilon}\boldsymbol{\varepsilon}^{\mathrm{T}}] = \mathbf{R}_2 \\ E[\mathbf{x}(0)] &= \mathbf{x}_0, \quad E\left[\mathbf{x}_0\mathbf{x}_0^{\mathrm{T}}\right] = \boldsymbol{\Pi}_0. \end{aligned} \tag{9.33}$$

Furthermore, matrices $\mathbf{F}$, $\mathbf{C}$, and $\boldsymbol{\Pi}_0$ may depend, in general, on a finite-dimensional vector $\boldsymbol{\theta}$ that must be determined. The EKF approach for determining the unknown vector $\mathbf{a}$ involves extending the state vector $\mathbf{x}$ with the parameter $\boldsymbol{\theta}$; thus,

$$\mathbf{a} = \begin{bmatrix} \mathbf{x} \\ \boldsymbol{\theta} \end{bmatrix} \tag{9.34}$$

with the model

$$\mathbf{a}_{k+1} = \mathbf{f}(\mathbf{a}_k) + \begin{bmatrix} \mathbf{w}_k \\ \mathbf{0} \end{bmatrix}$$
$$\mathbf{y}_k = \mathbf{c}(\mathbf{a}_k) + \varepsilon_k, \tag{9.35}$$

where

$$\mathbf{f}(\mathbf{a}_k) = \begin{bmatrix} \mathbf{f}(\boldsymbol{\theta}_k)\mathbf{x}_k \\ \boldsymbol{\theta}_k \end{bmatrix}$$
$$\mathbf{c}(\mathbf{a}_k) = \mathbf{C}(\boldsymbol{\theta}_k)\mathbf{x}_k. \tag{9.36}$$

9.4.2. Decoupling the State–Parameter Estimation Problem

According to the previous section, in order to deal with the state–parameter estimation problem we have to solve a nonlinear set of filtering equations. The extended Kalman filter leads to the following equations (Ursin, 1980):

$$\hat{\mathbf{a}}_{k+1} = \mathbf{G}_k\hat{\mathbf{a}}_k + \mathbf{B}_k[\mathbf{y}_{k+1} - \mathbf{L}_{k+1}\mathbf{G}_k\hat{\mathbf{a}}_k] \tag{9.37}$$

$$\mathbf{B}_k = \boldsymbol{\Sigma}_k\mathbf{L}_k^T\mathbf{S}_k^{-1} \tag{9.38}$$

$$\boldsymbol{\Sigma}_{k+1} = \mathbf{G}_k[\mathbf{I} - \mathbf{B}_k\mathbf{L}]\boldsymbol{\Sigma}_k\mathbf{G}_k^T + \mathbf{Q}, \tag{9.39}$$

where

$$\mathbf{S}_k = \left(\mathbf{L}_k\boldsymbol{\Sigma}_k\mathbf{L}_k^T + \mathbf{R}\right) \tag{9.40}$$

and

$$\hat{\mathbf{a}}_{0/-1} = \hat{\mathbf{a}}(0) = \begin{bmatrix} \mathbf{0} \\ \hline \hat{\boldsymbol{\theta}}_0 \end{bmatrix} \tag{9.41}$$

$$\boldsymbol{\Sigma}_0 = \begin{bmatrix} \boldsymbol{\Pi}_0 & \mathbf{0} \\ \mathbf{0} & \boldsymbol{\Psi}_0 \end{bmatrix} \tag{9.42}$$

$$\mathbf{G}_k = \left.\frac{\partial \mathbf{f}}{\partial \mathbf{a}^{\mathrm{T}}}\right|_{\mathbf{a}=\hat{\mathbf{a}}_k} = \begin{bmatrix} \mathbf{F}_k & \mathbf{H}_k^1 & \cdot & \cdot & \mathbf{H}_k^n \\ 0 & \mathbf{I} & & & 0 \\ \cdot & & & \cdot & \cdot \\ \cdot & & & \cdot & \cdot \\ 0 & & \cdot & \cdot & \mathbf{I} \end{bmatrix} \tag{9.43}$$

$$\mathbf{L}_k = \left.\frac{\partial \mathbf{c}}{\partial \mathbf{a}^{\mathrm{T}}}\right|_{\mathbf{a}=\hat{\mathbf{a}}_k} = \begin{bmatrix} A_k & A_k^1 & \ldots & A_k^n \end{bmatrix} \tag{9.44}$$

$$\mathbf{H}_k^i = \left.\frac{\partial(\mathbf{G}(\boldsymbol{\theta})\mathbf{x}_k)}{\partial \boldsymbol{\theta}^i}\right|_{\theta=\hat{\theta}_k}^{x=\hat{x}_k} \tag{9.45}$$

$$\mathbf{A}_k^i = \left.\frac{\partial(\mathbf{C}_k(\boldsymbol{\theta}_k))}{\partial \boldsymbol{\theta}^i}\right|_{\mathbf{a}=\hat{\mathbf{a}}_k} \tag{9.46}$$

$$\mathbf{Q} = \begin{bmatrix} \mathbf{R}_1 & \mathbf{0} \\ \mathbf{0} & \mathbf{0} \end{bmatrix} \tag{9.47}$$

$$\mathbf{R} = \begin{bmatrix} \mathbf{R}_2 & \mathbf{0} \\ \mathbf{0} & \mathbf{0} \end{bmatrix}. \tag{9.48}$$

In view of the aforementioned augmented problem, the error covariance matrix for the augmented system will be of the form

$$\boldsymbol{\Sigma} = \begin{bmatrix} \boldsymbol{\Sigma}_x & \cdot & \cdot & \boldsymbol{\Sigma}_{x\theta_m} \\ \cdot & \cdot & \cdot & \cdot \\ \cdot & & & \cdot \\ \boldsymbol{\Sigma}_{\theta_m x} & \cdot & \cdot & \boldsymbol{\Sigma}_{\theta_m} \end{bmatrix}, \tag{9.49}$$

which shows the interaction between the state–parameter estimation problems, where

$$\boldsymbol{\Sigma}_{\mathbf{x}} = \mathbf{E}\{(\mathbf{x} - \hat{\mathbf{x}})(\mathbf{x} - \hat{\mathbf{x}})^{\mathrm{T}}\} \tag{9.50}$$

and

$$\begin{aligned} \boldsymbol{\Sigma}_{\mathbf{x}\theta_i} &= \mathrm{E}\{(\mathbf{x} - \hat{\mathbf{x}})\boldsymbol{\theta}_i^{\mathrm{T}}\} \\ \boldsymbol{\Sigma}_{\theta_i} &= \mathrm{E}\{\boldsymbol{\theta}_i \boldsymbol{\theta}_i^{\mathrm{T}}\}. \end{aligned} \tag{9.51}$$

Romagnoli and Gani (1983) and Bortolotto *et al.* (1985) developed and implemented on-line a methodology for decoupling the state–parameter estimation problem. It was shown that the estimates of the states and error covariance matrix could be expressed as

$$\begin{aligned} \mathbf{x}_k &= \mathbf{x}_k^0 + \sum_{i=1}^{n} \delta \mathbf{x}^i \\ \boldsymbol{\Sigma}_k &= \boldsymbol{\Sigma}_k^0 + \sum_{i=1}^{n} \delta \boldsymbol{\Sigma}_k^i, \end{aligned} \tag{9.52}$$

where $\boldsymbol{\Sigma}_k^0$ and $\mathbf{x}_k^0$ are the normal (original) elements for the filter, in the absence of the parameter estimation. Each element $\delta\mathbf{x}^i$ and $\delta\boldsymbol{\Sigma}^i$ corresponds to the correction (on the original estimate) due to the parameter estimation. They are evaluated as

$$\begin{aligned} \delta \mathbf{x}^i &= \mathbf{V}_{\mathrm{x}}^{\mathrm{i}}(k)\hat{\boldsymbol{\theta}}_k^i \\ \delta \boldsymbol{\Sigma}^i &= \mathbf{U}_{\mathrm{x}}^{\mathrm{i}}(k)\mathbf{E}_{\mathrm{k}}^{\mathrm{i}}\left(\mathbf{U}_{\mathrm{x}}^{\mathrm{i}}(k)\right)^{\mathrm{T}}. \end{aligned} \tag{9.53}$$

The individual estimates for the parameter can be obtained from the recursion

$$\hat{\boldsymbol{\theta}}_{k+1}^i = \hat{\boldsymbol{\theta}}_k^i + \mathbf{B}_k^i \left[n_k - (\mathbf{C}_k \mathbf{U}_x^i(k)\hat{\boldsymbol{\theta}}_k^i\right], \tag{9.54}$$

where $\mathbf{U}^i$, $\mathbf{V}^i$, and $\mathbf{E}^i$ are auxiliary matrices. The complete procedure, with all the recursions, can be found in Romagnoli and Gani (1983).

EXAMPLE 9.4

The experimental setup, described in Example 8.1, for calculating the bias in a dynamic environment will be used here to discuss the parameter estimation methodology. In this case both the surface heat transfer coefficient (h) and the thermal conductivity (λ) of the body in the condition of natural convection in air are considered (Bortolotto *et al.*, 1985).

The extended state vector is now defined as

$$\mathbf{a} = \begin{bmatrix} \mathbf{x} \\ h \\ \lambda \end{bmatrix},$$

and a multichannel scheme implemented for the joint state–parameter estimation.

The algorithm was tested under conditions:

1. The channel for state estimation opened first and after some period of time (when a "good" reference trajectory was obtained) the two channels for the parameters were opened simultaneously.
2. The channel for state estimation opened first and the two other channels were opened sequentially at different periods of time.
3. The algorithm was tested at different operating conditions.

As a sample, the performance of the algorithm for different initial values of Σ_h and Σ_λ is given in Fig. 5. Figure 6 shows the improvement in convergence characteristics when channels are opened one at a time rather than simultaneously. Some of the runs, along with the specified statistical characterization, are given in Table 5. Additional information, as well as results, can be obtained from Bortolotto *et al.* (1985).

The multichannel procedure introduces an alternative approach to the problem of dynamic state–parameter estimation. The decoupling of the state estimator from

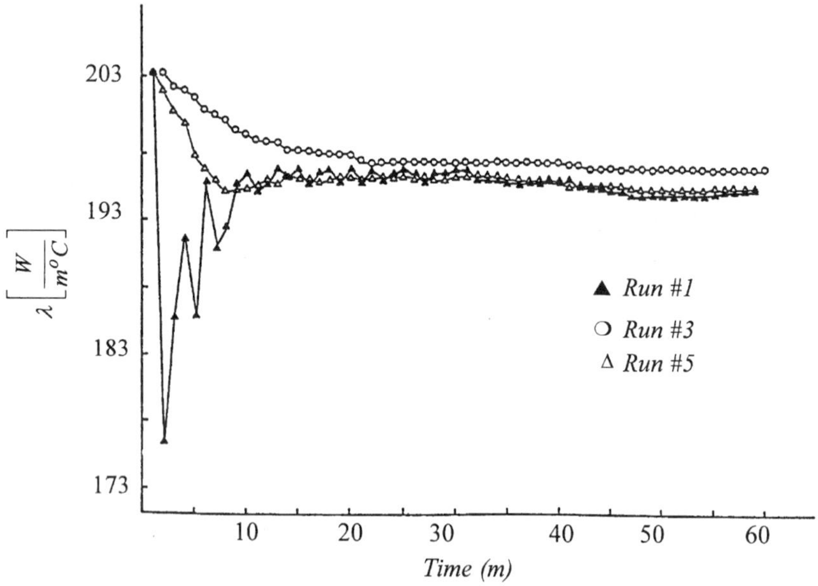

FIGURE 5 Behavior of the filter with different initial values of Σ_i. (From Bortolotto *et al.*, 1985.)

TABLE 5 Results for Different Initial Values of Σ_h and Σ_λ (from Bortolotto *et al.*, 1985)

	Initial parameter values		Weights for covariance matrices		Final parameter values	
Run #	λ	h	Σ_h	Σ_λ	h	λ
1	203	8.04	10	1000	7.38	196.1
2	203	8.04	1	100	7.31	195.8
3	203	8.04	0.1	10	7.34	197.2
4	203	8.04	0.01	30	7.38	196.9
5	203	8.04	0.01	30	7.41	196.5
6	173	6.81	10	1000	8.23	190.7
7	173	6.81	1	100	8.23	191.3
8	173	6.81	0.1	10	8.20	199.5
9	173	6.81	0.01	1	8.20	180.3
10	173	6.81	0.3	50	8.23	191.3
11	187	7.38	10	1000	7.82	195.00
12	187	7.38	1	100	7.85	195.00
13	187	7.38	0.1	10	7.85	194.7
14	187	7.38	0.01	1	7.56	190.5
15	187	7.38	0.2	40	7.85	194.5

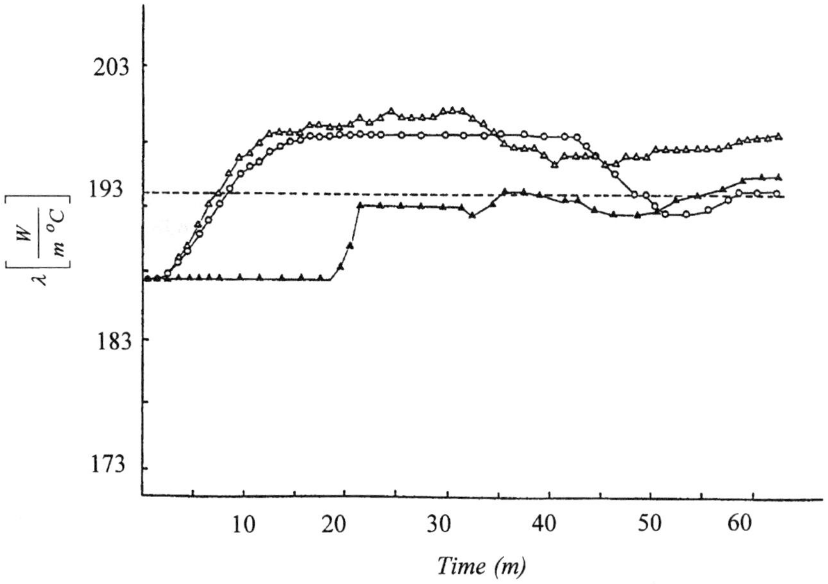

FIGURE 6 Convergence characteristics of the algorithm selecting the channel: △, both parameters estimated simultaneously; ▲, channel for parameter h opened first; O, Channel for parameter λ opened first (from Bortolotto *et al.*, 1985).

the parameter estimator results in the manipulation of smaller vectors and matrices, leading to efficient practical application. The data handling is reduced, as the information is taken only when required, in accordance with the expected variation of the particular parameter with time. In this way, the size of the problem becomes independent of the number of parameters to be estimated, since such estimation is performed sequentially and whenever desired.

9.5. DYNAMIC JOINT STATE–PARAMETER ESTIMATION: A NONLINEAR PROGRAMMING APPROACH

In this section the extension of the use of nonlinear programming techniques to solve the dynamic joint data reconciliation and parameter estimation problem is briefly discussed. As shown in Chapter 8, the general nonlinear dynamic data reconciliation (NDDR) formulation can be written as:

$$\begin{aligned}
&\min_{\hat{\mathbf{y}}} \boldsymbol{\Phi}[\mathbf{y}, \hat{\mathbf{y}}, \sigma] \\
&\text{s.t.} \\
&\qquad \mathbf{f}\left[\frac{d\hat{\mathbf{y}}}{dt}, \hat{\mathbf{y}}(t)\right] = 0 \\
&\qquad \eta[\hat{\mathbf{y}}(t)] = 0 \\
&\qquad \omega[\hat{\mathbf{y}}(t)] \geq 0.
\end{aligned} \tag{9.55}$$

Note: The measurements and estimates include both measured state variables and measured input variables. The inclusion of the input variables among those to be estimated establishes the error-in-variable nature of the data reconciliation problem. ♣

If the measurement noises are normally distributed and independent across the data sets, the objective function is simply the weighted least squares

$$\boldsymbol{\Phi}[\mathbf{y}, \hat{\mathbf{y}}(t), \sigma] = \sum_{k=0}^{c} \frac{1}{2}[\hat{\mathbf{y}}(t_k) - \mathbf{y}_k]^{\mathrm{T}} \boldsymbol{\Psi}^{-1} [\hat{\mathbf{y}}(t_k) - \mathbf{y}_k]. \tag{9.56}$$

By discretizing the differential algebraic equations model using some standard discretization techniques (Liebman *et al.*, 1992; Alburquerque and Biegler, 1996) to convert the differential constraints to algebraic constraints, the NDDR problem can be solved as the following NLP problem:

$$\begin{aligned}
&\min_{\hat{\mathbf{y}}} \boldsymbol{\Phi}[\mathbf{y}, \hat{\mathbf{y}}, \sigma] \\
&\text{s.t.} \\
&\qquad \eta[\hat{\mathbf{y}}(t)] = 0 \\
&\qquad \omega[\hat{\mathbf{y}}(t)] \geq 0,
\end{aligned} \tag{9.57}$$

where η now includes all the equality constraints.

Parameters and unmeasured variables can be incorporated into this formulation (Alburquerque and Biegler, 1996). Let us consider that we have measured variables, $\mathbf{x}$, and unmeasured variables, $\mathbf{u}$, and also suppose that they are constrained by some physical model and are dependent on unknown parameters, θ. Then we can represent the equalities in Eq. (9.57) by the general form

$$\eta(\mathbf{x}, \mathbf{u}, \boldsymbol{\theta}) = \mathbf{0}. \tag{9.58}$$

The general problem is then to estimate $\boldsymbol{\theta}$ and $\mathbf{u}$ knowing the values of the measurements, $\mathbf{y}$, and the probability distribution function of ε (measurement error). If $P(\varepsilon)$ is the error distribution, then $\mathbf{y}$ will be distributed according to $P(\mathbf{y} - \mathbf{x} \mid \boldsymbol{\theta}, \mathbf{u})$. Thus, according to Bayes' theorem, (Alburquerque and Biegler, 1996), the posterior distribution of the parameter and the unmeasured variables, given the data, will be

$$P(\boldsymbol{\theta}, \mathbf{u} \mid \mathbf{y}) = L(\boldsymbol{\theta}, \mathbf{u})\Pi(\boldsymbol{\theta}, \mathbf{u}), \tag{9.59}$$

where $\Pi(\boldsymbol{\theta}, \mathbf{u})$ is the a priori distribution function of the parameters, $\boldsymbol{\theta}$, and the unmeasured variables, $\mathbf{u}$. Following Alburquerque and Biegler and using the maximum a posteriori method, we minimize the negative of the log posterior subject to the model, that is,

$$\begin{aligned} &\operatorname*{Min}_{\theta,\mathrm{u}} -\log P(\boldsymbol{\theta}, \mathbf{u} \mid \mathbf{y}) \\ &\text{s.t.} \\ &\qquad \eta[\mathbf{x}, \mathbf{u}, \boldsymbol{\theta}] = \mathbf{0} \\ &\qquad \omega[\mathbf{x}, \mathbf{u}, \boldsymbol{\theta}] \geq \mathbf{0}. \end{aligned} \tag{9.60}$$

If the measurement noise is normally distributed and independent across the data set, and if we use a flat prior distribution, problem (9.60) become a nonlinear least squares problem and can be solved using the techniques discussed in Section 9.3.

9.6. CONCLUSIONS

In this chapter, the general problem of joint parameter estimation and data reconciliation was discussed. First, the typical parameter estimation problem was analyzed, in which the independent variables are error-free, and aspects related to the sequential processing of the information were considered. Later, the more general formulation in terms of the error-in-variable method (EVM), where measurement errors in all variables are considered in the parameter estimation problem, was stated. Alternative solution techniques were briefly discussed. Finally, joint parameter–state estimation in dynamic processes was considered and two different approaches, based on filtering techniques and nonlinear programming techniques, were discussed.

NOTATION

$\mathbf{a}$	vector defined by Eq. (9.34)
$\mathbf{A}$	matrix defined by Eq. (9.46)
b	number of model equations
$\mathbf{B}$	matrix defined by Eq. (9.38)
$\mathbf{C}$	matrix of measurement model
C_i	concentration of component i
$\mathbf{c}$	vector defined by Eq. (9.36)
$\mathbf{d}$	vector defined by Eq. (9.27)
$\mathbf{D}_0$	matrix defined by Eq. (9.9)
$\mathbf{e}$	vector defined by Eq. (9.25)
$\mathbf{E}_k^i$	auxiliary matrix
$\mathbf{f}$	vector function that comprises the process model
$\mathbf{F}$	transition matrix
$\mathbf{G}$	matrix defined by Eq. (9.43)
g	number of measurements
g_0	minimum set of observations
$\mathbf{H}$	matrix defined by Eq. (9.45)

h	heat transfer coefficient
j	index of experiments
J	objective function value
K	constant of reaction rate
$\mathbf{L}$	matrix defined by Eq. (9.44)
M	number of experiments
$\mathbf{N}$	Jacobian matrix of $\mathbf{f}(\hat{\mathbf{z}}, \boldsymbol{\theta})$
n	number of parameters
p	number of process variables for implicit models
Q	flowrate
$\mathbf{Q}$	matrix defined by Eq. (9.47)
r	number of state variables
$\mathbf{R}$	matrix defined by Eq. (9.48)
$\mathbf{R}_1$	covariance matrix of $\mathbf{w}$
$\mathbf{R}_2$	covariance matrix of ε
s	index for state variables
$\mathbf{S}$	matrix defined by Eq. (9.40)
T	temperature
T_b	threshold value
t	time
$\mathbf{U}_x^i$	auxiliary matrix
$\mathbf{u}$	unmeasured variables
$\mathbf{V}_x^i$	auxiliary matrix
$\mathbf{w}$	forcing random noise entering the system
$\mathbf{W}$	weighting matrix
$\mathbf{x}$	vector of state variables
$\mathbf{y}$	vector of measurements
$\mathbf{z}$	vector of process variables for implicit models

Greek Symbols

ε	vector of measurement errors
$\boldsymbol{\Psi}$	covariance matrix of measurement errors
$\boldsymbol{\theta}$	vector of parameters
$\boldsymbol{\Delta}$	parameter estimation error
$\boldsymbol{\Pi}_0$	covariance matrix of $\mathbf{x}_0$
λ	thermal conductivity
$\boldsymbol{\Sigma}$	covariance of the parameter estimation error
μ	vector of errors between measurements and predictions
$\boldsymbol{\Phi}$	general objective function
δ	correction due to parameter estimation
σ	standard deviation
η	equality constraints
ω	inequality constraints

Subscripts

0	initial state

Superscripts

$\tilde{}$	measured value
$\cdot$	derivative
$\hat{}$	estimated value

REFERENCES

Alburquerque, J. S., and Biegler, L. T. (1996). Data reconciliation and gross error detection for dynamic systems. *AIChE J.* **42**, 2841–2856.

Bahri, P., Bandoni, A., and Romagnoli, J. A. (1996). Effect of disturbances in optimizing control: Steady-state open-loop backoff problem. *AIChE J.* **42**, 983–995.

Bortolotto, G., Urbicain, M. J., and Romagnoli, J. A. (1985). On-line implementation of a multichannel estimator. *Comput. Chem. Eng.* **9**, 351–357.

Britt, H. I., and Luecke, R. H. (1973). The estimation of parameters in non-linear implicit models. *Technometrics* **15**, 223–247.

de Henning, S. R., and Perkins, J. D. (1991). "Structural Decisions in On-line Optimisation," Tech. Rep. B93-37. Imperial College, London.

Deming, W. E. (1943). "Statistical Adjustment of Data." Wiley, New York.

Kim, I. W., Liebman, M. J., and Edgar, T. F. (1990). Robust error-in-variable estimation using non-linear programming techniques. *AIChE J.* **36**, 985–993.

Liebman, M. J., and Edgar, T. F. (1988). Data reconciliation for non-linear processes. *Prepr. AIChE Annu. Meet.*, Washington, DC.

Liebman, M. J., Edgar, T. F., and Lasdon, L. S. (1992). Efficient data reconciliation and estimation for dynamic process using non-linear programming techniques. *Comput. Chem. Eng.* **16,** 963–986.

Peneloux, A. R., Deyrieux, E., and Neau, E. (1976). The maximum likelihood test and the estimation of experimental inaccuracies: Application of data reduction for vapour-liquid equilibrium. *J. Phys.* **7**, 706.

Reilly, P. M., and Patino-Leal, H. (1981). A Bayesian study of the error-in-variable models. *Technometrics* **23**, 221–231.

Romagnoli, J. A., and Gani, R. (1983). Studies of Distributed parameter systems: Decoupling the state parameter estimation problem. *Chem. Eng. Sci.* **38**, 1831–1843.

Schwetlick, H., and Tiller, V. (1985). Numerical methods for estimating parameters in non-linear models. *Technometrics* **27**, 17–24.

Tjoa, I. B., and Biegler, L. T. (1991). Reduced successive quadratic programming strategy for error-in-variables estimation. *Comput. Chem. Eng.* **16**, 523–533.

Ursin, B. (1980). Asymptotic convergence properties of the extended Kalman filter using filtered state estimates. *IEEE Trans. Autom. Control* **AC-25**, 1207–1211.

Valko, P., and Vadja, S. (1987). An extended Marquardt-type procedure for fitting error-in-variable models. *Comput. Chem. Eng.* **11**, 37–43.

10 ESTIMATION OF MEASUREMENT ERROR VARIANCES FROM PROCESS DATA

Most techniques for process data reconciliation start with the assumptions that the measurement errors are random variables obeying a known statistical distribution and that the covariance matrix of measurement errors ($\boldsymbol{\Psi}$) is given. In contrast, in this chapter we discuss direct and indirect approaches for estimating the variances of measurement errors from sample data. Furthermore, a robust strategy is presented for dealing with the presence of outliers in the data set.

10.1. INTRODUCTION

Most of the available techniques for data reconciliation are based on the assumption that the measurement errors are random variables obeying a known statistical distribution, namely, a normal distribution. Furthermore, they start with the assumption of a known covariance matrix of the measurement errors. Very little has been said about how such information is available, and in most practical situations, this matrix is unknown or known only approximately. Added to this, the covariance matrix, $\boldsymbol{\Psi}$, will vary when the process is at different operating conditions. That is, for one measurement device, the measurements' error level may be different from range to range. For instance, a normally operating pressure gauge will have different errors when working in different scale ranges, say 20% and 50%. In such cases, changes in the measurements' error level indicate that the magnitude of the covariance matrix elements will be different. Even if the covariance is already known, it still needs to be validated. Consequently, the definition or calculation of this statistical property of the data is of prime importance in implementing a reliable and accurate reconciliation procedure.

Only a few publications in the literature have dealt with this problem. Almasy and Mah (1984) presented a method for estimating the covariance matrix of measured errors by using the constraint residuals calculated from available process data. Darouach *et al.* (1989) and Keller *et al.* (1992) have extended this approach to deal with correlated measurements. Chen *et al.* (1997) extended the procedure further, developing a robust strategy for covariance estimation, which is insensitive to the presence of outliers in the data set.

The next sections describe direct and indirect techniques, developed in these publications, for estimating the variances of process data, as well as a robustification procedure for dealing with contaminated information.

10.2. DIRECT METHOD

As pointed out by Keller *et al.* (1992), if the process is truly at steady state, then estimation by the so-called direct method using the sample variance and covariance is adequate and simple to use. Let y_i be the ith element in a vector of measured variables, then the sample variance of the r repeated measurements of y_i is given by

$$\operatorname{var}(y_i) = \frac{1}{r-1}\sum_{k=1}^{r}(y_{ik} - \bar{y}_i)^2 \tag{10.1}$$

and the covariance of y_i and y_j is given by

$$\operatorname{cov}(y_i, y_j) = \frac{1}{r-1}\sum_{k=1}^{r}(y_{ik} - \bar{y}_i)(y_{jk} - \bar{y}_j), \tag{10.2}$$

where

$$\bar{y}_i = \frac{1}{r}\sum_{k=1}^{r} y_{ik}. \tag{10.3}$$

r is the size of the window, which is a function of the time during which the process is truly at steady state. This is an unbiased maximum likelihood estimator, based on the assumption of an independent error distribution with constant variance.

This procedure (based on sample variance and covariance) is referred to as the *direct method* of estimation of the covariance matrix of the measurement errors. As it stands, it makes no use of the inherent information content of the constraint equations, which has proved to be very useful in process data reconciliation. One shortcoming of this approach is that these r samples should be under steady-state operation, in order to meet the independent sampling condition; otherwise, the direct method could give incorrect estimates.

As pointed out by Almasy and Mah (1984), in real practical situations, the process conditions are continuously undergoing changes, and steady state is almost never attained. Practically speaking, steady state has meaning only within the framework of the time interval and the range of variation allowed by the analyst. Variance and covariance estimated by the foregoing technique will be very poor if the variation is comparable in magnitude to, or exceeds, the measurement errors. This drawback can be eliminated through the use of an indirect method, which is discussed next.

10.3. INDIRECT METHOD

As discussed before, in a strict sense, there is always some degree of dependence between the sample data. An alternative approach is to make use of the covariance matrix of the constraint residuals to eliminate the dependence between sample data (or the influence of unsteady-state behavior of the process during sampling periods). This is the basis of the so-called *indirect approach.*

Consider a g-dimensional process sample data vector at a time k,

$$\mathbf{y}_k = \mathbf{x}_k + \varepsilon_k, \tag{10.4}$$

where $\mathbf{x}_k$ is the vector of process variables at time k and ε_k is the vector of measurement errors, also at time k. Usually, $\mathbf{x}_k$ is called to satisfy the m-dimensional ($m \leq g$) set of algebraic equations (e.g., multicomponent mass and energy balances)

$$\mathbf{A}\mathbf{x}_k = \mathbf{0}, \tag{10.5}$$

where $\mathbf{A} = (a_{ij})_{m\times g}, i = 1, \ldots, m; j = 1, \ldots, g.$

From Eqs. (10.4) and (10.5), the residuals $\mathbf{r}_k$ in the balance equations can be calculated by

$$\mathbf{r}_k = \mathbf{A}\mathbf{y}_k = \mathbf{A}\mathbf{x}_k + \mathbf{A}\varepsilon_k = \mathbf{A}\varepsilon_k. \tag{10.6}$$

Assuming that ε_k is Gaussian with zero mean and positive definite covariance matrix $\mathbf{\Psi}$ then

$$E(\varepsilon_k) = \mathbf{0} \tag{10.7a}$$

$$E\left(\varepsilon_k\varepsilon_k^{\mathrm{T}}\right) = \mathbf{\Psi} \tag{10.7b}$$

$$E(\mathbf{r}_k) = E(\mathbf{A}\varepsilon_k) = \mathbf{A}E(\varepsilon_k) = \mathbf{0}, \tag{10.7c}$$

where $\mathbf{\Psi} = (\Psi_{ij})_{g\times g}, i = 1, \ldots, g; j = 1, \ldots, g.$

From Eqs. (10.6) and (10.7) we can obtain the covariance matrix of the residuals, $\mathbf{\Phi}$, as follows:

$$\mathbf{\Phi} = \mathrm{cov}(\mathbf{r}_k) = E\left(\mathbf{r}_k\mathbf{r}_k^{\mathrm{T}}\right) = E\left(\mathbf{A}\varepsilon_k\varepsilon_k^{\mathrm{T}}\mathbf{A}^{\mathrm{T}}\right) = \mathbf{A}E\left(\varepsilon_k\varepsilon_k^{\mathrm{T}}\right)\mathbf{A}^{\mathrm{T}} = \mathbf{A}\mathbf{\Psi}\mathbf{A}^{\mathrm{T}}, \tag{10.8}$$

where $\mathbf{\Phi} = (\Phi_{ij})_{m\times m}, i = 1, \ldots, m; j = 1, \ldots, m.$

Using the Kronecker product of matrices and the vec(o) operator (Almasy and Mah, 1984; see also Appendix A), Eq. (10.8) can be restated as

$$\mathrm{vec}(\Phi) = (\mathbf{A} \otimes \mathbf{A})\, \mathrm{vec}(\mathbf{\Psi}). \tag{10.9}$$

The indirect method uses Eq. (10.9) to estimate $\mathbf{\Psi}$. This procedure requires the value of the covariance matrix, $\mathbf{\Phi}$, which can be calculated from the residuals using the balance equations and the measurements.

10.3.1. Computational Procedure

Two alternative procedures have been suggested in the literature to solve the problem and they will be discussed next. *Alternative 1* was proposed by Almasy and Mah (1984). They attempt to minimize the sum of the squares of the off-diagonal elements of a measurement error covariance matrix subject to the relation deduced from the statistical properties of the residuals constraints. *Alternative 2* was proposed by Keller *et al.* (1992) to deal with the diagonal and non-diagonal case. They presented an analytical procedure (following Dorouach *et al.*, 1989) based on the relation deduced from the statistical properties of the material constraints.

Alternative 1 (Almasy and Mah, 1984)

From Eq. (10.9), with reference to $\boldsymbol{\Psi}$, let $\mathbf{d}$ be the vector of $g(\mathbf{y})$ diagonal elements, and $\mathbf{t}$ the vector of $g(\mathbf{t})$ off-diagonal elements. In general,

$$g(\mathbf{t}) \le g(\mathbf{y})\frac{[g(\mathbf{y}) - 1]}{2}. \tag{10.10}$$

The remaining off-diagonal elements are zero by specification. Vectors $\mathbf{d}$ and $\mathbf{t}$ may be obtained from vec $(\boldsymbol{\Psi})$ by premultiplying by the appropriate mapping matrices $\mathbf{D}$ and $\mathbf{T}$, that is,

$$\mathbf{d} = \mathbf{D}\ \mathrm{vec}(\boldsymbol{\Psi}) \tag{10.11}$$

$$\mathbf{t} = \mathbf{T}\ \mathrm{vec}(\boldsymbol{\Psi}), \tag{10.12}$$

It is then possible to recover Ψ from

$$\mathrm{vec}(\boldsymbol{\Psi}) = \mathbf{D}^{+}\mathbf{d} + \mathbf{T}^{+}\mathbf{t} \tag{10.13}$$

with appropriate matrices $\mathbf{D}^{+}$ and $\mathbf{T}^{+}$.

As before, a vector $\mathbf{p}$ of $g(\mathbf{p})$ elements can be defined, associated with $\boldsymbol{\Phi}$, such that

$$g(\mathbf{p}) = g(\mathbf{r})\frac{[g(\mathbf{r}) + 1]}{2} \tag{10.14}$$

and

$$\mathbf{p} = \mathbf{P}\ \mathrm{vec}(\boldsymbol{\Phi}). \tag{10.15}$$

Substituting Eqs. (10.9) and (10.13) into Eq. (10.15) yields

$$\begin{aligned}\mathbf{p} &= \mathbf{P}(\mathbf{A} \otimes \mathbf{A}) \cdot (\mathbf{D}^{+}\mathbf{d} + \mathbf{T}^{+}\mathbf{t}) \\ &= \mathbf{M}\mathbf{d} + \mathbf{N}\mathbf{t}.\end{aligned} \tag{10.16}$$

The problem of estimating the covariance matrix of the measurement errors may now be formulated as

$$\min_{\mathrm{d,t}} \mathbf{t}^{\mathrm{T}}\mathbf{t}, \tag{10.17}$$

subject to the constraints given by Eq. (10.16).

The solution of this problem is given by

$$\mathbf{d} = \mathbf{S}\mathbf{M}^{\mathrm{T}}\mathbf{R}\mathbf{p} \tag{10.18}$$

and

$$\mathbf{t} = \mathbf{N}^{\mathrm{T}}(\mathbf{R} - \mathbf{R}\mathbf{M}\mathbf{S}\mathbf{M}^{\mathrm{T}}\mathbf{R})\mathbf{p}, \tag{10.19}$$

where

$$\mathbf{R} = (\mathbf{N}\mathbf{N}^{\mathrm{T}})^{-1} \tag{10.20}$$

$$\mathbf{S} = (\mathbf{M}^{\mathrm{T}}\mathbf{R}\mathbf{M})^{-1}. \tag{10.21}$$

The necessary conditions for the existence of the inverse matrices in Eqs. (10.20) and (10.21) are

$$g(\mathbf{t}) \geq g(\mathbf{p}) \geq g(\mathbf{y}). \tag{10.22}$$

According to Almasy and Mah:

1. The conditions imposed by Eq. (10.22) by themselves do not guarantee the existence of **R** and **S**.
2. We need to compute matrices $\mathbf{SM}^{\mathrm{T}}\mathbf{R}$ and $\mathbf{N}^{\mathrm{T}}(\mathbf{R} - \mathbf{RMSM}^{\mathrm{T}}\mathbf{R})$ only once and apply the results of Eqs. (10.18) and (10.19) each time a **p** is obtained.

A procedure for computing matrices **M** and **N** is given in Almasy and Mah (1984).

Alternative 2 (Keller *et al.*, 1992)

In this approach we have two possible situations:

1. *Diagonal case*. If the measurement errors are not correlated, $\boldsymbol{\Psi}$ is a diagonal matrix. Let us consider

$$\mathbf{d} = [\Psi_{11}, \Psi_{22}, \ldots, \Psi_{gg}]^{\mathrm{T}}. \tag{10.23}$$

This vector contains all of the parameters that need to be estimated. Equation (10.9) can be written as

$$\mathrm{vec}(\boldsymbol{\Phi}) = \mathbf{Fd}, \tag{10.24}$$

where

$$\mathrm{vec}(\boldsymbol{\Phi}) = [\Phi_{11}, \Phi_{12}, \ldots, \Phi_{1m}, \Phi_{21}, \Phi_{22}, \ldots, \Phi_{2m}, \ldots, \Phi_{m1}, \Phi_{m2}, \ldots, \Phi_{mm}]^{\mathrm{T}} \tag{10.25}$$

$$\mathbf{F} = \begin{bmatrix} a_{11}\mathbf{A}_1 & a_{12}\mathbf{A}_2 & \ldots & a_{1g}\mathbf{A}_g \\ a_{21}\mathbf{A}_1 & a_{22}\mathbf{A}_2 & \ldots & a_{2g}\mathbf{A}_g \\ & & \ldots & \\ a_{m1}\mathbf{A}_1 & a_{m2}\mathbf{A}_2 & \ldots & a_{mg}\mathbf{A}_g \end{bmatrix} \tag{10.26}$$

and $\mathbf{A}_j$ is the jth column of matrix **A**; $\mathrm{vec}(\boldsymbol{\Phi})$ is an m^2 vector, **d** is a g vector, and **F** is a (m^2, g) matrix.

Now, if $(m^2 \geq g)$, the solution of Eq. (10.24), under the assumption of an independent and normal error distribution with constant variance can be obtained as the maximum likelihood estimator of **d** and is given by

$$\hat{\mathbf{d}} = (\mathbf{F}^{\mathrm{T}}\mathbf{F})^{-1}\mathbf{F}^{T}\mathrm{vec}(\boldsymbol{\Phi}). \tag{10.27}$$

EXAMPLE 10.1

To illustrate the application of the strategy for the diagonal case, we consider the process system taken from Ripps (1965). As indicated in Chapter 5, it consists of a chemical reactor with four streams, two entering and two leaving the process. All the stream flowrates are assumed to be measured, and their true values are $\mathbf{x} = [0.1739 \;\; 5.0435 \;\; 1.2175 \;\; 4.00]^{\mathrm{T}}$. The corresponding system matrix, **A**, and the covariance of the measurement errors, $\boldsymbol{\Psi}$, are also known and given by

$$\mathbf{A}_1 = \begin{bmatrix} 0.1 & 0.6 & -0.2 & -0.7 \\ 0.8 & 0.1 & -0.2 & -0.1 \\ 0.1 & 0.3 & -0.6 & -0.2 \end{bmatrix}, \quad \boldsymbol{\Psi} = \begin{bmatrix} 0.000289 & & & \\ & 0.0025 & & \\ & & 0.000576 & \\ & & & 0.04 \end{bmatrix}.$$

A multivariate normal distribution data set with the variance and mean given by this $\boldsymbol{\Psi}$ and $\mathbf{x}$ was generated by the Monte Carlo method to simulate the process sampling data. The data size was 1000 and it was used to investigate the performance of the indirect method.

In the estimation of $\boldsymbol{\Psi}$ from thc convcntional indirect approach, using the Keller *et al.* (1992) method, ($\boldsymbol{\Psi}_C$) is given by

$$\boldsymbol{\Psi}_C = \begin{bmatrix} 0.0002678 & 0 & 0 & 0 \\ 0 & 0.00246 & 0 & 0 \\ 0 & 0 & 0.000512 & 0 \\ 0 & 0 & 0 & 0.0416 \end{bmatrix}.$$

2. *General case*. If some measurement errors are correlated, the covariance matrix is not diagonal. It is assumed that we know the sensors which are subjected to correlated measurement errors, for example because they share some common elements (e.g., power supplies). There are then s off-diagonal elements of $\boldsymbol{\Psi}$, $(\Psi_{pq}, \ldots, \Psi_{bl})$, that need to be estimated in addition to the diagonal ones.

Let $\mathbf{f} = (\Psi_{11}, \Psi_{22}, \ldots, \Psi_{gg}, \Psi_{pq}, \ldots, \Psi_{bl})^T$. Then Eq. (10.9) can be written as

$$\text{vec}(\boldsymbol{\Phi}) = \mathbf{G}\mathbf{f}, \tag{10.28}$$

where

$$\mathbf{G} = \begin{bmatrix} a_{11}\mathbf{A}_1 & a_{12}\mathbf{A}_2 & \ldots & a_{1g}\mathbf{A}_g & a_{1p}\mathbf{A}_p + a_{1q}\mathbf{A}_q & \ldots & a_{1b}\mathbf{A}_b + a_{1l}\mathbf{A}_l \\ a_{21}\mathbf{A}_1 & a_{22}\mathbf{A}_2 & & a_{2g}\mathbf{A}_g & a_{2p}\mathbf{A}_p + a_{2q}\mathbf{A}_q & \ldots & a_{2b}\mathbf{A}_b + a_{2l}\mathbf{A}_l \\ & & & & & \ldots & \\ a_{m1}\mathbf{A}_1 & a_{m2}\mathbf{A}_2 & \ldots & a_{mg}\mathbf{A}_g & a_{mp}\mathbf{A}_p + a_{mq}\mathbf{A}_q & \ldots & a_{mb}\mathbf{A}_b + a_{ml}\mathbf{A}_l \end{bmatrix}, \tag{10.29}$$

where $\mathbf{A}_j$ is the jth column of matrix $\mathbf{A}$, $\mathbf{A} = (\mathbf{A}_1, \mathbf{A}_2, \ldots, \mathbf{A}_g)$.

From Eq. (10.28), the maximum likelihood estimator of $\mathbf{f}$ is given by

$$\mathbf{f} = (\mathbf{G}^T\mathbf{G})^{-1}\mathbf{G}^T\text{vec}(\boldsymbol{\Phi}). \tag{10.30}$$

The estimability conditions of variance are now rank($\mathbf{G}$) $= g + s$.

EXAMPLE 10.2

This example was taken from Keller *et al.* (1992). It consists of the network of 7 nodes and 12 streams given in Fig. 1. In this case, we make a change about the position of the correlated sensors to make sure that the target covariance conforms

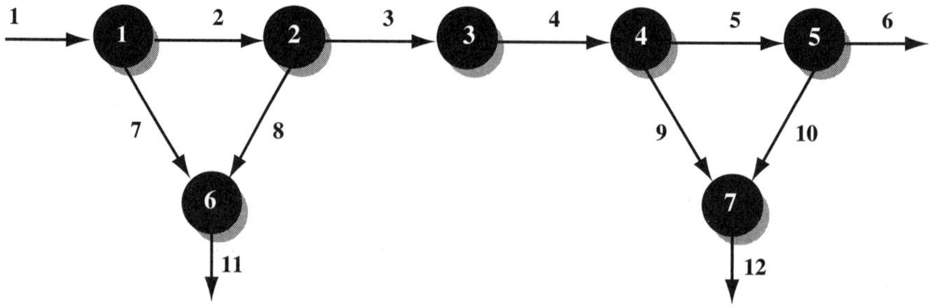

FIGURE 1 Process network for Example 10.2 (from Keller *et al.*, 1992).

to the common assumption on the property of positive definiteness. Thus, the measurement errors of sensors 2 and 5 and sensors 6 and 11 are to be correlated. The simulation condition is similar to the uncorrelated case. The system matrix, $\mathbf{A}$, and target covariance, $\boldsymbol{\Psi}$, for this case are

$$\mathbf{A} = \begin{bmatrix} 1 & -1 & 0 & 0 & 0 & 0 & -1 & 0 & 0 & 0 & 0 & 0 \\ 0 & 1 & -1 & 0 & 0 & 0 & 0 & -1 & 0 & 0 & 0 & 0 \\ 0 & 0 & 1 & -1 & 0 & 0 & 0 & 0 & 0 & 0 & 0 & 0 \\ 0 & 0 & 0 & 1 & -1 & 0 & 0 & 0 & -1 & 0 & 0 & 0 \\ 0 & 0 & 0 & 0 & 1 & -1 & 0 & 0 & 0 & -1 & 0 & 0 \\ 0 & 0 & 0 & 0 & 0 & 0 & 1 & 1 & 0 & 0 & -1 & 0 \\ 0 & 0 & 0 & 0 & 0 & 0 & 0 & 0 & 1 & 1 & 0 & -1 \end{bmatrix}$$

$$\boldsymbol{\Psi} = \begin{bmatrix} 30.00 & 0 & 0 & 0 & 0 & 0 & 0 & 0 & 0 & 0 & 0 & 0 \\ 0 & 30.00 & 0 & 0 & 6.00 & 0 & 0 & 0 & 0 & 0 & 0 & 0 \\ 0 & 0 & 20.00 & 0 & 0 & 0 & 0 & 0 & 0 & 0 & 0 & 0 \\ 0 & 0 & 0 & 20.00 & 0 & 0 & 0 & 0 & 0 & 0 & 0 & 0 \\ 0 & 6.00 & 0 & 0 & 7.50 & 0 & 0 & 0 & 0 & 0 & 0 & 0 \\ 0 & 0 & 0 & 0 & 0 & 15.00 & 0 & 0 & 0 & 0 & 9.00 & 0 \\ 0 & 0 & 0 & 0 & 0 & 0 & 10.00 & 0 & 0 & 0 & 0 & 0 \\ 0 & 0 & 0 & 0 & 0 & 0 & 0 & 10.00 & 0 & 0 & 0 & 0 \\ 0 & 0 & 0 & 0 & 0 & 0 & 0 & 0 & 10.00 & 0 & 0 & 0 \\ 0 & 0 & 0 & 0 & 0 & 0 & 0 & 0 & 0 & 8.10 & 0 & 0 \\ 0 & 0 & 0 & 0 & 0 & 9.00 & 0 & 0 & 0 & 0 & 20.00 & 0 \\ 0 & 0 & 0 & 0 & 0 & 0 & 0 & 0 & 0 & 0 & 0 & 20.00 \end{bmatrix}$$

A comparison with the target covariance shows that the *conventional indirect approach* gives a very good estimation of the covariance in this case:

$$\boldsymbol{\Psi}_c = \begin{bmatrix} 29.52 & 0 & 0 & 0 & 0 & 0 & 0 & 0 & 0 & 0 & 0 & 0 \\ 0 & 30.01 & 0 & 0 & 4.92 & 0 & 0 & 0 & 0 & 0 & 0 & 0 \\ 0 & 0 & 19.72 & 0 & 0 & 0 & 0 & 0 & 0 & 0 & 0 & 0 \\ 0 & 0 & 0 & 19.24 & 0 & 0 & 0 & 0 & 0 & 0 & 0 & 0 \\ 0 & 4.92 & 0 & 0 & 7.09 & 0 & 0 & 0 & 0 & 0 & 0 & 0 \\ 0 & 0 & 0 & 0 & 0 & 12.52 & 0 & 0 & 0 & 0 & 9.51 & 0 \\ 0 & 0 & 0 & 0 & 0 & 0 & 11.25 & 0 & 0 & 0 & 0 & 0 \\ 0 & 0 & 0 & 0 & 0 & 0 & 0 & 11.25 & 0 & 0 & 0 & 0 \\ 0 & 0 & 0 & 0 & 0 & 0 & 0 & 0 & 11.44 & 0 & 0 & 0 \\ 0 & 0 & 0 & 0 & 0 & 0 & 0 & 0 & 0 & 8.24 & 0 & 0 \\ 0 & 0 & 0 & 0 & 0 & 9.51 & 0 & 0 & 0 & 0 & 20.17 & 0 \\ 0 & 0 & 0 & 0 & 0 & 0 & 0 & 0 & 0 & 0 & 0 & 21.51 \end{bmatrix}.$$

10.4. ROBUST COVARIANCE ESTIMATOR

The performances of the indirect conventional methods described previously are very sensitive to outliers, so they are not robust. The main reason for this is that they use a direct method to calculate the covariance matrix of the residuals ($\boldsymbol{\Phi}$). If outliers are present in the sampling data, the assumption about the error distribution will be violated. In such situations, the final estimation of the covariance of the measurements will not be a maximum likelihood estimator. In other words, this may lead to an incorrect estimation. For this reason an alternative robust procedure is needed.

DEFINITION

An estimator is called robust if it is insensitive to mild departures from the underlying assumptions and is also only slightly more inefficient relative to conventional approaches when these assumptions are satisfied.

10.4.1. The M-Estimator

One type of common robust estimator is the so-called M-estimator or generalized maximum likelihood estimator, originally proposed by Huber (1964). The basic idea of an M-estimator is to assign weights to each vector, based on its own Mahalanobis distance, so that the amount of influence of a given point decreases as it becomes less and less characteristic.

Let $\mathbf{z}_1, \ldots, \mathbf{z}_r$ be samples of j dimension. Then, the M-estimator of the location (vector $\mathbf{m}$) and the scatter (matrix $\mathbf{V}$) are obtained by solving the following set of simultaneous equations (Maronna, 1976; Huber, 1981, p. 212):

$$\frac{1}{r}\sum_{i=1}^{i=r} u_1(w_i)(\mathbf{z}_i - \mathbf{m}) = \mathbf{0} \tag{10.31}$$

$$\frac{1}{r-1}\sum_{i=1}^{i=r} u_2\left(w_i^2\right)(\mathbf{z}_i - \mathbf{m})(\mathbf{z}_i - \mathbf{m})^{\mathrm{T}} = \mathbf{V}, \tag{10.32}$$

where u_1 and u_2 are arbitrary weights and $w_i^2 = (\mathbf{z}_i - \mathbf{m})^{\mathrm{T}}\mathbf{V}^{-1}(\mathbf{z}_i - \mathbf{m})$, $i = 1, \ldots, r$, are the squared distances of the observation vectors from the current estimate of location $\mathbf{m}$. Note that when $u_1 = u_2 = 1$, Eqs. (10.31) and (10.32) become

$$\frac{1}{r}\sum_{i=1}^{i=r} (\mathbf{z}_i - \mathbf{m}) = \mathbf{0}$$

$$\frac{1}{r-1}\sum_{i=1}^{i=r} (\mathbf{z}_i - \mathbf{m})(\mathbf{z}_i - \mathbf{m})^{\mathrm{T}} = \mathbf{V}.$$

In this case, $\mathbf{V}$ is the classical sample covariance matrix of $\mathbf{z}$.

In Chen *et al.* (1997) the following simultaneous estimation strategy is adopted to construct the M-estimator:

- Assuming the covariance $\mathbf{V}$ is positive definite, the Cholesky decomposition of $\mathbf{V}$ is

$$\mathbf{V} = \mathbf{C}\mathbf{C}^{\mathrm{T}}. \tag{10.33}$$

Then Eqs. (10.31) and (10.32) can be written as

$$E\{u_1(\|\mathbf{C}^{-1}(\mathbf{z} - \mathbf{m})\|)\mathbf{C}^{-1}(\mathbf{z} - \mathbf{m})\} = \mathbf{0} \tag{10.34}$$

$$E\{u_2(\|\mathbf{C}^{-1}(\mathbf{z} - \mathbf{m})\|^2)[\mathbf{C}^{-1}(\mathbf{z} - \mathbf{m})][\mathbf{C}^{-1}(\mathbf{z} - \mathbf{m})]^{\mathbf{T}}\} = \mathbf{I}, \tag{10.35}$$

where $\|\cdot\|$ denotes the norm of vectors or matrixes and $\mathbf{I}$ stands for the identity matrix.

• Let $n = \|\mathbf{C}^{-1}(\mathbf{z} - \mathbf{m})\|$. Then the functions u_1 and u_2 will be determined by the following minimax problem:

$$\begin{aligned} \varphi(n, c) &= \max\{-c, \min(n, c)\} \\ \text{with} \quad \varphi_i &= nu_i(n), \quad i = 1, 2, \end{aligned} \tag{10.36}$$

where $\varphi(n, c)$ is known as the Huber's psi function; c is a constant specified by the user to take into account the acceptable loss in efficiency of the Gaussian model in exchange for resistance to outliers.

Taking

$$\varphi_1 = \varphi(n, c) \tag{10.37}$$

$$\varphi_2 = \varphi(n, c^2)/\beta, \tag{10.38}$$

where β is chosen to make $\mathbf{V}$ an asymptotically unbiased estimator of the covariance matrix in a multivariate normal situation, the following algorithm can be devised to estimate the covariance matrix of the residuals, $\mathbf{\Phi}$.

10.4.2. The Algorithm (Chen *et al.*, 1997)

Let $\mathbf{y}_1, \mathbf{y}_2, \ldots, \mathbf{y}_r$ be the g-dimensional vector of data with $\mathbf{y}_i = [y_{1i}, y_{2i}, \ldots, y_{gi}]^{\mathrm{T}}$, $\mathbf{A} = (a_{ij})_{m\times g}$ is the matrix for the constraint equations. The algorithm for robust covariance estimation can be implemented as follows.

Step 1: Calculation of the residuals, $\mathbf{r}_i$, by means of Eq. (10.6).
Step 2: Calculation of the covariance matrix of residuals:

• *Variable initialization.* Initialize the location (mean) m_j, scale b_j, and covariance matrix $\mathbf{\Phi}$ with

$$m^*_{j,0} = \underset{i}{\text{median}}(r_{ji}); \quad j = 1, 2, \ldots, m; \quad i = 1, 2, \ldots, r \tag{10.39}$$

$$b_j = \underset{i}{\text{median}}(|r_{ji} - m_j|)/0.6745 \tag{10.40}$$

$$\mathbf{\Phi}^*_0 = \text{diag}\left(b_1^2, b_2^2, \ldots, b_m^2\right). \tag{10.41}$$

• *Iterative calculation.* The solution of the M-estimator Eqs. (10.31) and (10.32) is obtained by means of an iterative procedure that, for iteration q, is made up of the following steps:
(a) Estimation of the weight function u_1, u_2:

$$u_1(w_i) = \begin{cases} 1 & w_i < c \\ c/w_i & \text{otherwise} \end{cases} \tag{10.42}$$

$$u_2\left(w_i^2\right) = (u_1(w_i))^2/\beta, \tag{10.43}$$

with

$$\beta = G(c^2/2,1.5) + 2c^2(1 - F(c)), \tag{10.44}$$

where F represents the multivariate normal cumulative distribution function and $G(a, f)$ is the gamma distribution with f degrees of freedom. In this work, c^2 is taken to be the 90% point of the χ^2 distribution.

The Mahalanobis distances

$$w_i^2 = \left(\mathbf{r}_i - \mathbf{m}_{q-1}^*\right)^{\mathrm{T}} \left(\mathbf{\Phi}_{q-1}^*\right)^{-1} \left(\mathbf{r}_i - \mathbf{m}_{q-1}^*\right), \quad i = 1, 2, \ldots, r, \tag{10.45}$$

of the sample vectors from the current estimate of location $\mathbf{m}^*$ are measured in the metric of $\mathbf{\Phi}^*$, the current estimate of the covariance matrix.

(b) Estimation of the location $\mathbf{m}^*$:

$$\mathbf{m}_q^* = \mathbf{m}_{q-1}^* + \sum_{i=1}^{r} u_1(w_i)\left(\mathbf{r}_i - \mathbf{m}_{q-1}^*\right) \Bigg/ \sum_{i=1}^{r} u_i(w_i). \tag{10.46}$$

(c) Estimation of Φ^*:

$$\mathbf{\Phi}_q^* = \frac{1}{r} \sum_{i=1}^{r} u_2\left(w_i^2\right)\left(\mathbf{r}_i - \mathbf{m}_q^*\right)\left(\mathbf{r}_i - \mathbf{m}_q^*\right)^{\mathrm{T}}. \tag{10.47}$$

The iteration process terminates when the maximum difference in the elements of $\mathbf{\Phi}$ between two successive iterations is smaller than a prespecified threshold, in this work taken to be 10^{-6}.

Step 3: Robust maximum likelihood estimation of $\mathbf{\Psi}$:
First calculate the maximum likelihood estimator of vec ($\mathbf{\Psi}$):

$$\mathrm{vec}(\mathbf{\Psi}) = \mathbf{f} = (\mathbf{G}^{\mathrm{T}}\mathbf{G})^{-1}\mathbf{G}^{\mathrm{T}}\mathrm{vec}(\mathbf{\Phi}). \tag{10.48}$$

Then the robust covariance $\mathbf{\Psi}$ can be obtained by reshaping vec($\mathbf{\Psi}$) as follows:

$$\Psi = \mathrm{vec}^{-1}(\mathrm{vec}(\mathbf{\Psi})) = \mathrm{vec}^{-1}(\mathbf{f}). \tag{10.49}$$

End

The following remarks are in order:

Remark 1: The selection of weight functions u_1 and u_2 is basically arbitrary. However, these functions have to satisfy certain conditions in order to ensure the existence and uniqueness of a solution. The weight functions u_1 and u_2 utilized in this approach have been verified as meeting all these conditions (Maronna, 1976). Since u_1 and u_2 solve the minimax problem (10.36), they are also known as Huber type weights.

Remark 2: It is worthwhile to mention that the breaking point of the M-estimator, the fraction of outliers that can be tolerated without the estimator potentially breaking down, is about $1/(g+1)$, where g is the dimension of the data (Maronna, 1976). In data reconciliation, outliers refer to the abnormal observations and are a small fraction of the whole data set. In our view, the breaking point property is reasonably good for practical implementations.

Actually this approach follows a general recipe that works for many robust procedures: *Clean the data by pulling outliers towards their fitted values, refitting iteratively until convergence is obtained.* By assigning a weight to each of the observations, the individual observation's contribution to covariance can be adjusted in order to avoid outliers governing the estimation of the covariance. The empirical convergence behavior of the iterative calculation procedure is good (Maronna, 1976), but there is not yet theoretical proof and more research is needed. Besides the M-estimator, there are several simultaneous estimation strategies, but none of them have been proved. It is worth mentioning that when referring to Eqs. (10.40), (10.41) and (10.47), the iterative algorithm will maintain the positive definiteness of the covariance matrix.

To investigate the performance of the proposed robust estimator, Monte Carlo studies were performed on the two previous examples used for the case without outliers.

EXAMPLE 10.3 (Chen *et al.*, 1997)

As before, all the stream flowrates are assumed to be measured in the process flowsheet presented by Ripps (1965). The corresponding system matrix, **A**, and the covariance of the measurement errors, $\boldsymbol{\Psi}$, are given in Section 10.3.

A multivariate normal distribution data set was generated by the Monte Carlo method using the values of variances and true flowrates in order to simulate the process sampling data. The data, of sample size 1000, were used to investigate the performance of the robust approach in the two cases, with and without outliers.

Without outliers. The estimation of $\boldsymbol{\Psi}$ both from the conventional indirect approach $\boldsymbol{\Psi}_{\mathrm{C}}$ and from the robust approach $\boldsymbol{\Psi}_{\mathrm{R}}$ gave similar results when compared with the target covariance matrix $\boldsymbol{\Psi}$:

$$\boldsymbol{\Psi}_{\mathrm{C}} = \begin{bmatrix} 0.0002678 & 0 & 0 & 0 \\ 0 & 0.00246 & 0 & 0 \\ 0 & 0 & 0.000512 & 0 \\ 0 & 0 & 0 & 0.0416 \end{bmatrix}$$

$$\boldsymbol{\Psi}_{\mathrm{R}} = \begin{bmatrix} 0.0002712 & 0 & 0 & 0 \\ 0 & 0.002498 & 0 & 0 \\ 0 & 0 & 0.000507 & 0 \\ 0 & 0 & 0 & 0.0411 \end{bmatrix}.$$

With outliers. For comparison, we still use the same data set as case 1 (without outliers), except one outlier is introduced to x_3 (from 1.2128 to 7.2128). The estimation results are

$$\boldsymbol{\Psi}_{\mathrm{C}} = \begin{bmatrix} 0.000306 & 0 & 0 & 0 \\ 0 & 0.00715 & 0 & 0 \\ 0 & 0 & 0.03637 & 0 \\ 0 & 0 & 0 & 0.03874 \end{bmatrix}$$

$$\boldsymbol{\Psi}_{\mathrm{R}} = \begin{bmatrix} 0.0002717 & 0 & 0 & 0 \\ 0 & 0.002493 & 0 & 0 \\ 0 & 0 & 0.000515 & 0 \\ 0 & 0 & 0 & 0.0411 \end{bmatrix}.$$

$\boldsymbol{\Psi}_{\mathrm{R}}$ from the robust estimator still gives the correct answer, as expected. However, the conventional approach fails to provide a good estimate of the covariance even for the case when only one outlier is present in the sampling data.

EXAMPLE 10.4 (Chen *et al.*, 1997)

The second case study is that of Example 10.2 given in Figure 10.1. As before, the measurement errors of sensors 2 and 5, and sensors 6 and 11 are to be correlated.

The system matrix, **A**, and target covariance matrix, Ψ, for this case are given in Section 10.3. The simulation condition is similar to the uncorrelated case; the two cases, with and without outliers, are tested.

Without outliers. As shown by comparison with the target covariance, both the *conventional indirect approach* and the *robust approach* give a very good estimation of the covariance in this case:

$$\Psi_C = \begin{bmatrix} 29.52 & 0 & 0 & 0 & 0 & 0 & 0 & 0 & 0 & 0 & 0 & 0 \\ 0 & 30.01 & 0 & 0 & 4.92 & 0 & 0 & 0 & 0 & 0 & 0 & 0 \\ 0 & 0 & 19.72 & 0 & 0 & 0 & 0 & 0 & 0 & 0 & 0 & 0 \\ 0 & 0 & 0 & 19.24 & 0 & 0 & 0 & 0 & 0 & 0 & 0 & 0 \\ 0 & 4.92 & 0 & 0 & 7.09 & 0 & 0 & 0 & 0 & 0 & 0 & 0 \\ 0 & 0 & 0 & 0 & 0 & 12.52 & 0 & 0 & 0 & 0 & 9.51 & 0 \\ 0 & 0 & 0 & 0 & 0 & 0 & 11.25 & 0 & 0 & 0 & 0 & 0 \\ 0 & 0 & 0 & 0 & 0 & 0 & 0 & 11.25 & 0 & 0 & 0 & 0 \\ 0 & 0 & 0 & 0 & 0 & 0 & 0 & 0 & 11.44 & 0 & 0 & 0 \\ 0 & 0 & 0 & 0 & 0 & 0 & 0 & 0 & 0 & 8.24 & 0 & 0 \\ 0 & 0 & 0 & 0 & 0 & 9.51 & 0 & 0 & 0 & 0 & 20.17 & 0 \\ 0 & 0 & 0 & 0 & 0 & 0 & 0 & 0 & 0 & 0 & 0 & 21.51 \end{bmatrix}$$

$$\Psi_R = \begin{bmatrix} 29.13 & 0 & 0 & 0 & 0 & 0 & 0 & 0 & 0 & 0 & 0 & 0 \\ 0 & 29.81 & 0 & 0 & 4.91 & 0 & 0 & 0 & 0 & 0 & 0 & 0 \\ 0 & 0 & 19.56 & 0 & 0 & 0 & 0 & 0 & 0 & 0 & 0 & 0 \\ 0 & 0 & 0 & 19.13 & 0 & 0 & 0 & 0 & 0 & 0 & 0 & 0 \\ 0 & 4.91 & 0 & 0 & 6.98 & 0 & 0 & 0 & 0 & 0 & 0 & 0 \\ 0 & 0 & 0 & 0 & 0 & 12.38 & 0 & 0 & 0 & 0 & 9.54 & 0 \\ 0 & 0 & 0 & 0 & 0 & 0 & 11.24 & 0 & 0 & 0 & 0 & 0 \\ 0 & 0 & 0 & 0 & 0 & 0 & 0 & 10.09 & 0 & 0 & 0 & 0 \\ 0 & 0 & 0 & 0 & 0 & 0 & 0 & 0 & 11.45 & 0 & 0 & 0 \\ 0 & 0 & 0 & 0 & 0 & 0 & 0 & 0 & 0 & 8.22 & 0 & 0 \\ 0 & 0 & 0 & 0 & 0 & 9.54 & 0 & 0 & 0 & 0 & 19.83 & 0 \\ 0 & 0 & 0 & 0 & 0 & 0 & 0 & 0 & 0 & 0 & 0 & 21.22 \end{bmatrix}.$$

With outliers. We still use the same data set as for the previous case, except that two outliers are introduced to x_2, x_5 (from 2.7683 and 4.4475 to 102.7682 and −95.5525). The results are as follows:

$$\Psi_C = \begin{bmatrix} 31.48 & 0 & 0 & 0 & 0 & 0 & 0 & 0 & 0 & 0 & 0 & 0 \\ 0 & 40.86 & 0 & 0 & -6.1 & 0 & 0 & 0 & 0 & 0 & 0 & 0 \\ 0 & 0 & 20.02 & 0 & 0 & 0 & 0 & 0 & 0 & 0 & 0 & 0 \\ 0 & 0 & 0 & 19.90 & 0 & 0 & 0 & 0 & 0 & 0 & 0 & 0 \\ 0 & -6.1 & 0 & 0 & 17.37 & 0 & 0 & 0 & 0 & 0 & 0 & 0 \\ 0 & 0 & 0 & 0 & 0 & 12.34 & 0 & 0 & 0 & 0 & 9.11 & 0 \\ 0 & 0 & 0 & 0 & 0 & 0 & 11.64 & 0 & 0 & 0 & 0 & 0 \\ 0 & 0 & 0 & 0 & 0 & 0 & 0 & 10.27 & 0 & 0 & 0 & 0 \\ 0 & 0 & 0 & 0 & 0 & 0 & 0 & 0 & 11.44 & 0 & 0 & 0 \\ 0 & 0 & 0 & 0 & 0 & 0 & 0 & 0 & 0 & 8.14 & 0 & 0 \\ 0 & 0 & 0 & 0 & 0 & 9.51 & 0 & 0 & 0 & 0 & 20.62 & 0 \\ 0 & 0 & 0 & 0 & 0 & 0 & 0 & 0 & 0 & 0 & 0 & 21.68 \end{bmatrix}$$

$$
\Psi_R = \begin{bmatrix}
28.94 & 0 & 0 & 0 & 0 & 0 & 0 & 0 & 0 & 0 & 0 & 0 \\
0 & 29.90 & 0 & 0 & 4.80 & 0 & 0 & 0 & 0 & 0 & 0 & 0 \\
0 & 0 & 19.56 & 0 & 0 & 0 & 0 & 0 & 0 & 0 & 0 & 0 \\
0 & 0 & 0 & 19.06 & 0 & 0 & 0 & 0 & 0 & 0 & 0 & 0 \\
0 & 4.80 & 0 & 0 & 7.12 & 0 & 0 & 0 & 0 & 0 & 0 & 0 \\
0 & 0 & 0 & 0 & 0 & 12.36 & 0 & 0 & 0 & 0 & 9.54 & 0 \\
0 & 0 & 0 & 0 & 0 & 0 & 11.18 & 0 & 0 & 0 & 0 & 0 \\
0 & 0 & 0 & 0 & 0 & 0 & 0 & 11.05 & 0 & 0 & 0 & 0 \\
0 & 0 & 0 & 0 & 0 & 0 & 0 & 0 & 11.46 & 0 & 0 & 0 \\
0 & 0 & 0 & 0 & 0 & 0 & 0 & 0 & 0 & 8.21 & 0 & 0 \\
0 & 0 & 0 & 0 & 0 & 9.54 & 0 & 0 & 0 & 0 & 19.90 & 0 \\
0 & 0 & 0 & 0 & 0 & 0 & 0 & 0 & 0 & 0 & 0 & 21.21
\end{bmatrix}.
$$

The robust estimator still provides a correct estimation of the covariance matrix; on the other hand, the estimate Ψ_C, provided by the conventional approach, is incorrect and the signs of the correlated coefficients have been changed by the outliers.

10.5. CONCLUSIONS

The covariance matrix of measurement errors is a very useful statistical property. Indirect methods can deal with unsteady sampling data, but unfortunately they are very sensitive to outliers and the presence of one or two outliers can cause misleading results. This drawback can be eliminated by using robust approaches via M-estimators. The performance of the robust covariance estimator is better than that of the indirect methods when outliers are present in the data set.

NOTATION

$\mathbf{A}$	matrix of linear constraints
b	initial standard deviations of residuals
$\mathbf{C}$	matrix from Cholesky decomposition of $\mathbf{V}$
c	constant
$\mathbf{d}$	vector of $g(\mathbf{x})$ diagonal elements of Ψ
$\mathbf{D}$	matrix that maps diagonal elements of Ψ into $\mathbf{d}$
$\mathbf{D}^+$	matrix that maps vector $\mathbf{d}$ into the diagonal elements of matrix Ψ
f	degrees of freedom
$\mathbf{f}$	vector of estimated elements of Ψ
$\mathbf{F}$	matrix defined by Eq. (10.26)
F	multivariate cumulative function distribution
g	number of measured variables
$\mathbf{G}$	matrix defined by Eq. (10.29)
G	gamma function
j	dimension of $\mathbf{z}$
k	sample index
m	number of process constraints
$\mathbf{m}$	location vector
$\mathbf{M}$	matrix defined by Eq. (10.16)
$\mathbf{N}$	matrix defined by Eq. (10.16)
$\mathbf{p}$	vector formed with elements of Φ

$\mathbf{P}$	matrix that maps elements of $\mathbf{\Phi}$ into vector $\mathbf{p}$
q	iteration index
r	number of repeated measurements
$\mathbf{r}$	residuum of process constraints
$\mathbf{R}$	matrix defined by Eq. (10.20)
s	number of off-diagonal elements of $\mathbf{\Psi}$ for the general case
$\mathbf{S}$	matrix defined by Eq. (10.21)
$\mathbf{t}$	off-diagonal elements of matrix $\mathbf{\Psi}$
$\mathbf{T}$	matrix that maps off-diagonal elements of matrix Ψ into $\mathbf{t}$
$\mathbf{T}^+$	matrix that maps vector $\mathbf{t}$ into off-diagonal elements of $\mathbf{\Psi}$
u	arbitrary weights
$\mathbf{V}$	scatter matrix
w	distance of the observation vectors from the current estimate of $\mathbf{m}$
$\mathbf{x}$	vector of true value of measured variables
$\mathbf{X}_d$	maximum likelihood estimator of $\mathbf{V}$
$\mathbf{y}$	vector of measurements
$\mathbf{z}$	general vector

Greek Symbols

ε	measurement random errors
$\mathbf{\Psi}$	covariance matrix of measurement errors
$\mathbf{\Phi}$	covariance matrix of residuum

Superscripts

$\hat{}$	least square estimation

Subscripts

C	conventional estimation
R	robust estimation

Other Symbols

cov( )	sample covariance of
var( )	sample variance of
vec( )	see Appendix A
n( )	the cardinality of
$\otimes$	the symbol of Kronecker product of matrices

REFERENCES

Almasy, G. A., and Mah, R. S. H. (1984). Estimation of measurement error variances from process data. *Ind. Eng. Chem. Process Des. Dev.* **23**, 779–784.

Chen, J., Bandoni, A., and Romagnoli, J. A. (1997). Robust estimation of measurements error variance/covariance from process sampling data. *Comput. Chem. Eng.* **21**, 593–600.

Darouach, M., Ragot, R., Zasadzinski, M., and Karzakala, G. (1989). Maximum likelihood estimator of measurement error variances in data reconciliation. *IFAC, AIPAC Symp.* **2**, 135–139.

Huber, P. J. (1964). A robust estimation of a location parameter. *Ann. Math. Stat.* **35**, 73–101.

Huber, P. J. (1981). "Robust Statistics." Wiley, New York.

Keller, J. Y., Zasadzinski, M., and Darouach, M. (1992). Analytical estimator of measurement error variances in data reconciliation. *Comput. Chem. Eng.* **16**, 185–188.
Maronna, R. A. (1976). Robust M-estimators of multivariate location and scatter. *Ann. Stat.* **4**, 51–67.
Neudecker, H. (1969). Some theorems on matrix differentiation with special reference to Kronecker matrix products. *Am. Stat. Assoc. J.* **64**, 953–963.
Ripps, D. L. (1965). Adjustment of experimental data. *Chem. Eng. Prog., Symp. Ser.* **61**, 8–13.

APPENDIX A[1]: THE KRONECKER PRODUCT OF MATRICES AND THE vec(o) OPERATOR

The Kronecker Product

Let $\mathbf{A} = \{a_{ij}\}$ be an $(n \times m)$ matrix and $\mathbf{B}$ an $(s \times t)$ matrix. Then the *Kronecker product* of $\mathbf{A}$ and $\mathbf{B}$, denoted as $\mathbf{A} \otimes \mathbf{B}$, is defined as the $[(n \cdot s) \times (m \cdot t)]$ matrix

$$\mathbf{A} \otimes \mathbf{B} = \{a_{ij} \cdot \mathbf{B}\}, \tag{A10.1}$$

where $a_{ij} \cdot \mathbf{B}$ denotes the product of matrix $\mathbf{B}$ by the scalar a_{ij}.

The vec(o) Operator

Let the jth column of matrix $\mathbf{A}$ be denoted by $\mathbf{a}_j$, so that

$$\mathbf{A} = (\mathbf{a}_1, \mathbf{a}_2, \ldots, \mathbf{a}_m). \tag{A10.2}$$

Neudecker (1969) defined an $(m \cdot n)$ column vector

$$\text{vec}(\mathbf{A}) = \begin{bmatrix} a_1 \\ a_2 \\ \vdots \\ a_m \end{bmatrix}, \tag{A10.3}$$

that is, vec($\mathbf{A}$) is obtained by writing the columns of $\mathbf{A}$ below each other.

[1]From Almasy and Mah (1984).

11 NEW TRENDS

This chapter discusses some recent approaches for dealing with different aspects of the data reconciliation problem. A more general formulation in terms of a probabilistic framework is first introduced, and its use in dealing with gross error is discussed in particular. Robust estimation approaches are then considered, in which the estimators are designed so that they are insensitive to outliers. Finally, a strategy that combines principal component analysis and steady-state data reconciliation will be discussed.

11.1. INTRODUCTION

In the previous chapters, data reconciliation schemes have been discussed extensively. These schemes, used to remove both noise and gross errors from measured plant data, provide a more accurate description of the true operational status of the plant.

As was shown, the conventional method for data reconciliation is that of weighted least squares, in which the adjustments to the data are weighted by the inverse of the measurement noise covariance matrix so that the model constraints are satisfied. The main assumption of the conventional approach is that the errors follow a normal Gaussian distribution. When this assumption is satisfied, conventional approaches provide unbiased estimates of the plant states. The presence of gross errors violates the assumptions in the conventional approach and makes the results invalid.

We have discussed, in Chapter 7, a number of auxiliary gross error detection/identification/estimation schemes, for identifying and removing the gross errors from the measurements, such that the normality assumption holds. Another approach is to take into account the presence of gross errors from the beginning, using, for example, contaminated error distributions. In this case, the regression accommodates the presence of the outliers and gross error detection can be performed simultaneously.

In addition, conventional approaches assume that the only available information about the process is the known model constraints. However, a wealth of information is available in the operating history of the plant. In this case, together with spatial redundancy, there is also temporal redundancy, that is, temporal redundancy exists when measurements at different past times are available. This temporal redundancy contains information about the measurement behavior such as the probability distribution. The methods discussed in the first two sections of this chapter try to exploit these ideas by formulating the reconciliation problem in a different way.

Correlations are inherent in chemical processes even where it can be assumed that there is no correlation among the data. Principal component analysis (PCA) transforms a set of correlated variables into a new set of uncorrelated ones, known as principal components, and is an effective tool in multivariate data analysis. In the last section we describe a method that combines PCA and the steady-state data reconciliation model to provide sharper, and less confounding, statistical tests for gross errors.

11.2. THE BAYESIAN APPROACH

As discussed before, in the conventional data reconciliation approach, auxiliary gross error detection techniques are required to remove any gross error before applying reconciliation techniques. Furthermore, the reconciled states are only the maximum likelihood states of the plant, if feasible plant states are equally likely. That is, $P\{\mathbf{x}\} = 1$ if the constraints are satisfied and $P\{\mathbf{x}\} = 0$ otherwise. This is the so-called binary assumption (Johnston and Kramer, 1995) or flat distribution.

To alleviate these assumptions the maximum likelihood rectification (MLR) technique was proposed by Johnston and Kramer. This approach incorporates the prior distribution of the plant states, P{**x**}, into the data reconciliation process to obtain the maximum likely rectified states given the measurements. Mathematically the problem can be stated (Johnston and Kramer, 1995) as

$$\underset{\mathrm{x}}{\mathrm{Max}}\, P\{\mathbf{x} \mid \mathbf{y}\}. \tag{11.1}$$

According to Bayes' theorem, the probability of the states, given the data, will be

$$\underset{\mathrm{x}}{\mathrm{Max}}\, P\{\mathbf{x} \mid \mathbf{y}\} = \underset{\mathrm{x}}{\mathrm{Max}}\, P\{\mathbf{y} \mid \mathbf{x}\} P\{\mathbf{x}\} / P\{\mathbf{y}\}, \tag{11.2}$$

where $P\{\mathbf{y}/\mathbf{x}\}$ indicates the probability distribution of the measurements given the states, $P\{\mathbf{x}\}$ is the prior probability distribution of the states, and $P\{\mathbf{y}\}$ is the probability distribution of the measurements. Since $P\{\mathbf{y}\}$ acts as a normalization constant for a given set of data and does not depends on x, it can be ignored. The states that maximize the conditional probability density of the plant states, given the data, are, finally,

$$\underset{\mathrm{x}}{\mathrm{Max}}\, P\{\mathbf{x} \mid \mathbf{y}\} = \underset{\mathrm{x}}{\mathrm{Max}}\, P\{\mathbf{y} \mid \mathbf{x}\} P\{\mathbf{x}\}. \tag{11.3}$$

The following three cases can be defined according to different assumptions about $P\{\mathbf{y}/\mathbf{x}\}$ and $P\{\mathbf{x}\}$ and are discussed later:

Case 1: Binary assumption on $P\{\mathbf{x}\}$ (flat distribution), and sensor errors are assumed to follow a normal distribution

Case 2: Binary assumption on $P\{\mathbf{x}\}$, and sensor errors do not follow a normal distribution

Case 3: Relaxed assumption on $P\{\mathbf{x}\}$, and sensor errors do not follow a normal distribution

Case 1. The implicit binary assumption on $P\{\mathbf{x}\}$ implies that $P\{\mathbf{x}\}$ is a binary variable, that is,

- $P\{\mathbf{x}\} = 1$ if and only if $\mathbf{x}$ satisfies the model equations (equality and inequality constraints)
- $P\{\mathbf{x}\} = 0$ if $\mathbf{x}$ fails to satisfy the model equations

Under this assumption the $P\{\mathbf{x}\}$ term is converted to a set of constraints and our original problem is converted to the following constrained optimization:

$$\begin{aligned} &\operatorname*{Max}_{\mathrm{x}} P\{\mathbf{y} \mid \mathbf{x}\} \\ &\text{s.t.} \\ &\qquad \varphi(\mathbf{x}) = \mathbf{0} \\ &\qquad \omega(\mathbf{x}) \leq \mathbf{0}. \end{aligned} \tag{11.4}$$

If in addition to the binary assumption on $P\{\mathbf{x}\}$, sensor errors are normally distributed and independent across the data sets, the problem becomes our typical nonlinear least squares data reconciliation problem:

$$\begin{aligned} &\operatorname*{Max}_{\mathrm{x}} (\mathbf{y} - \mathbf{x})^{\mathrm{T}} \boldsymbol{\Psi}^{-1} (\mathbf{y} - \mathbf{x}) \\ &\text{s.t.} \\ &\qquad \varphi(\mathbf{x}) = \mathbf{0} \\ &\qquad \omega(\mathbf{x}) \leq \mathbf{0}. \end{aligned} \tag{11.5}$$

Case 2. The assumption on $P\{\mathbf{x}\}$ still holds in this case, but a nonnormal model of $P\{\mathbf{y}/\mathbf{x}\}$ is considered. To take into account the presence of outliers, the contaminated normal distribution (Tjoa and Biegler, 1991) is considered in most cases.

Let us state the measurement model for the ith observation as the sum of the true state being measured and an additive error, that is,

$$y_i = x_i + \delta_i, \tag{11.6}$$

where δ_i is the measurement error. For the case where δ_i is independent of x_i,

$$P\{y_i \mid x_i\} = P\{\delta_i\}. \tag{11.7}$$

Here are two possible outcomes:

- $\mathrm{G} = \{$Gross error occurred$\}$ with probability η
- $\mathrm{R} = \{$Random error occurred$\}$ with probability $1 - \eta$

When G occurs, the error will follow a distribution $P\{\delta \mid \mathrm{G}\}$, and when R occurs, the error will follow $P\{\delta \mid \mathrm{R}\}$. Under these conditions the distribution of δ_i will be

$$P\{\delta_i\} = (1 - \eta_i) P\{\delta_i \mid \mathrm{R}\} + \eta_i P\{\delta_i \mid \mathrm{G}\}. \tag{11.8}$$

Assuming that the errors are uncorrelated, the joint distribution of the biases is the product of the individual biases, that is,

$$P\{\delta\} = \prod_i P\{\delta_i\}. \tag{11.9}$$

Normal distribution is commonly used for random and gross errors, but gross error distributions have a higher variance than the random error distributions. This leads to the n-contaminated normal distribution with normal contamination, that is,

$$P\{\delta_i\} = (1-\eta_i)\frac{1}{\sqrt{2\pi}\sigma_i}\exp\left(-\frac{\delta_i^2}{2\sigma_i^2}\right) + \eta_i\frac{1}{\sqrt{2\pi}rst_i\sigma_i}\exp\left(-\frac{\delta_i^2}{2rst_i^2\sigma_i^2}\right), \tag{11.10}$$

where σ_i is the standard deviation of the normal noise band in sensor i and rst_i ($rst_i > 1$) is the ratio of the standard deviation of the gross error distribution of sensor i to the standard deviation of the normal noise band for sensor i.

Under these assumptions, our original problem is converted to the following constrained optimization:

$$\begin{aligned} &\operatorname*{Max}_{\mathrm{x}} P\{\delta\} \\ &\text{s.t.} \\ &\qquad \varphi(\mathbf{x}) = \mathbf{0} \\ &\qquad \omega(\mathbf{x}) \le \mathbf{0}, \end{aligned} \tag{11.11}$$

where $P\{\delta\}$ is given by Eqs. (11.8) and (11.9).

Tjoa and Biegler (1991) used this formulation within a simultaneous strategy for data reconciliation and gross error detection on nonlinear systems. Albuquerque and Biegler (1996) used the same approach within the context of solving an error-in-all-variable-parameter estimation problem constrained by differential and algebraic equations.

Case 3. Taking into account the probability distribution of the sensors errors given by Eqs. (11.8) and (11.9) and the model of $P\{\mathbf{x}\}$, this more general formulation solves the following problem to obtain the most likely rectified states, given the measurements:

$$\begin{aligned} &\operatorname*{Max}_{\mathrm{x}} P\{\delta\}P\{\mathbf{x}\} \\ &\text{s.t.} \\ &\qquad \varphi(\mathbf{x}) = \mathbf{0} \\ &\qquad \omega(\mathbf{x}) \le \mathbf{0}. \end{aligned} \tag{11.12}$$

One problem encountered in solving Eq. (11.12) is the modeling of the prior distribution $P\{\mathbf{x}\}$. It is assumed that this distribution is not known in advance and must be calculated from historical data. Several methods for estimating the density function of a set of variables are presented in the literature. Among these methods are histograms, orthogonal estimators, kernel estimators, and elliptical basis function (EBF) estimators (see Silverman, 1986; Scott, 1992; Johnston and Kramer, 1994; Chen *et al.*, 1996). A wavelet-based density estimation technique has been developed by Safavi *et al.* (1997) as an alternative and superior method to other common density estimation techniques. Johnston and Kramer (1998) have proposed the recursive state density estimation (RSDE) method to find $P\{\mathbf{x}\}$ from noisy data. This method links the data reconciliation and probability density estimation through the principles of expected maximization.

Following Johnston and Kramer (1998) and using an EBF probability density estimator, the form of the estimator for $P\{\mathbf{x}\}$ is

$$P\{\mathbf{x}\} = \sum_{h=1}^{H} q_h \phi_h(\mathbf{x}, \mathbf{m}_h, \mathbf{Q}_h), \tag{11.13}$$

where

$$\phi(\mathbf{x}, \mathbf{m}_h, \mathbf{Q}_h) = \frac{1}{\sqrt{(2\pi)^g \det(\mathbf{Q}_h)}} \exp\left(-\frac{(\mathbf{x}-\mathbf{m}_h)^T \mathbf{Q}_h^{-1}(\mathbf{x}-\mathbf{m}_h)}{2}\right). \tag{11.14}$$

$\mathbf{m}_h$ are the unit centers and $\mathbf{Q}_h$ are the unit covariance matrices; q_h are the mixing proportions; and h is the window size. In this way, from Eq. (11.1), the resulting reconciliation problem requiring solution is

$$\begin{aligned} &\underset{\mathrm{x}}{\mathrm{Max}}\ P\{\delta\}\left[\sum_{h=1}^{H} q_h \phi_h(\mathbf{x}; \mathbf{m}_h; \mathbf{Q}_h)\right] \\ &\text{s.t.}\quad \varphi(\mathbf{x}) = \mathbf{0} \\ &\qquad\ \omega(\mathbf{x}) \leq \mathbf{0}. \end{aligned} \tag{11.15}$$

In general, it is computationally expensive to find the solution to problem (11.12). An alternative approach based on a surrogate objective function was developed by Johnston and Kramer (1998). This approach, for the unconstrained and linearly constrained cases, has an analytical solution thus simplifying the calculations. The complete procedure can be found in the aforementioned publication.

EXAMPLE 11.1

The same problem discussed in Example 5.1 is taken to illustrate the ideas described in this section. The system consists of a chemical reactor with four streams, two entering and two leaving the process. All the stream flowrates are assumed to be measured, and their true values are $\mathbf{x} = [0.1739\ \ 5.0435\ \ 1.2175\ \ 4.00]^{\mathrm{T}}$. The corresponding system matrix, $\mathbf{A}$, and the covariance matrix, $\boldsymbol{\Psi}$, of the measurement errors are also known and given by

$$\mathbf{A} = \begin{bmatrix} 0.1 & 0.6 & -0.2 & -0.7 \\ 0.8 & 0.1 & -0.2 & -0.1 \\ 0.1 & 0.3 & -0.6 & -0.2 \end{bmatrix}$$

$$\boldsymbol{\Psi} = \begin{bmatrix} 0.0029 & 0 & 0 & 0 \\ 0 & 0.0025 & 0 & 0 \\ 0 & 0 & 0.0006 & 0 \\ 0 & 0 & 0 & 0.04 \end{bmatrix}.$$

Monte Carlo data for $\mathbf{y}$ were generated according to $\boldsymbol{\Psi}$ with mean $\mathbf{x}$, to simulate process sampling data. A window size of 25 was used here and to demonstrate the performance of the Bayesian approach.

Following Tjoa and Biegler (1991) we have modeled $P\{\delta\}$ as a bivariate likelihood distribution, a contaminated normal distribution as shown in Eq. (11.10) with

$rsd = 10$ and $P\{\mathbf{x}\} = 1$ (flat distribution). Monte Carlo simulations have been performed on several distributions. All these generated distributions were contaminated by some isolated outliers randomly generated. Only the results for Cauchy distribution with 10% outliers are shown here.

Figures 1 to 4 illustrate the results of the reconciliation for the four variables involved. As can be seen, this approach does not completely eliminate the influence of the outliers. For some of the variables, the prediction after reconciliation is actually deteriorated because of the presence of outliers in some of the other measurements. This is in agreement with the findings of Albuquerque and Biegler (1996), in the sense that the results of this approach can be very misleading if the gross error distribution is not well characterized.

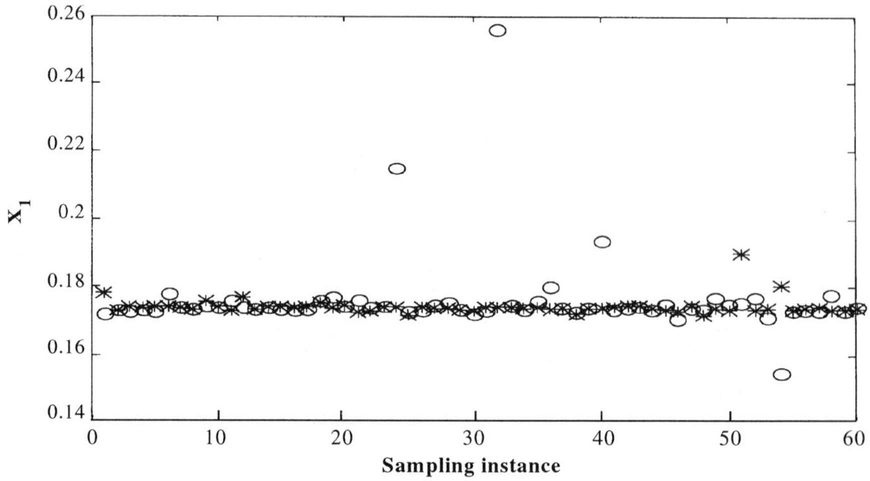

FIGURE 1 Reconciliation results for variable x_1 with Cauchy distribution and 10% of outliers ($*$, measured value; $\circ$, reconciled value).

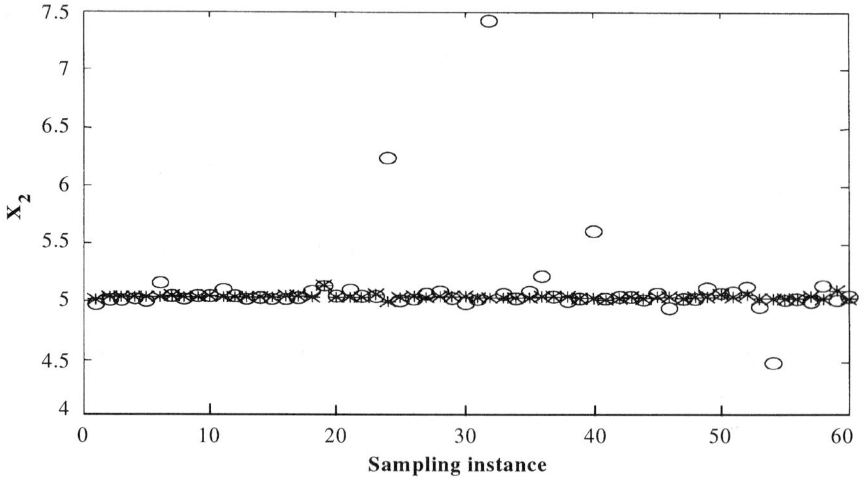

FIGURE 2 Reconciliation results for variable x_2 with Cauchy distribution and 10% of outliers ($*$, measured value; $\circ$, reconciled value).

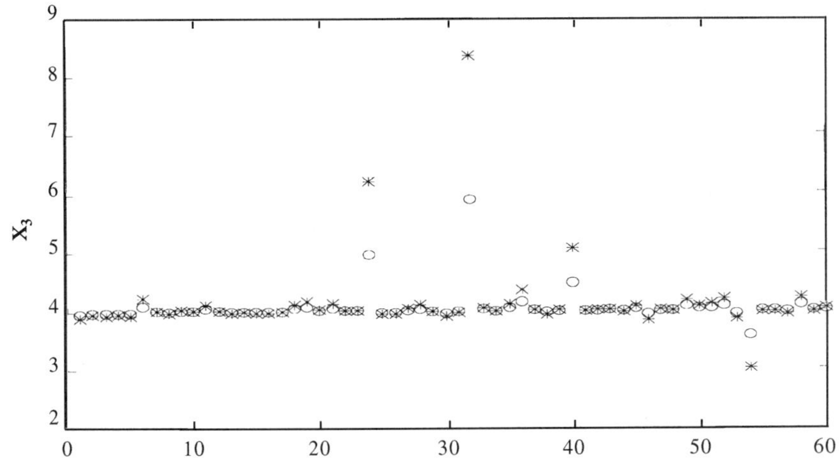

FIGURE 3 Reconciliation results for variable x_3 with Cauchy distribution and 10% of outliers ($*$, measured value; $\circ$, reconciled value).

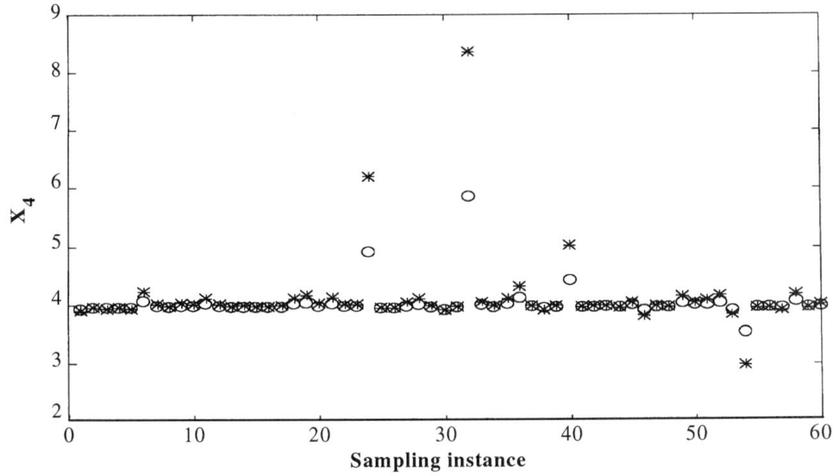

FIGURE 4 Reconciliation results for variable x_4 with Cauchy distribution and 10% of outliers ($*$, measured value; $\circ$, reconciled value).

11.3. ROBUST ESTIMATION APPROACHES

In robust statistics, rather than assuming an ideal distribution, an estimator is constructed that will give unbiased results in the presence of this ideal distribution, but that will try to minimize the sensitivity to deviations from ideality. Several approaches are described here:

- M-estimators approach
- QQ-plots approach
- The trust function approach

11.3.1. M-Estimators Approach

Robust system identification and estimation has been an important area of research since the 1990s in order to get more advanced and robust identification and estimation schemes, but it is still in its initial stages compared with the classical identification and estimation methods (Wu and Cinar, 1996). With the classical approach we assume that the measurement errors follow a certain statistical distribution, and all statistical inferences are based on that distribution. However, departures from all ideal distributions, such as outliers, can invalidate these inferences. In robust statistics, rather than assuming an ideal distribution, we construct an estimator that will give unbiased results in the presence of this ideal distribution, but will be insensitive to deviation from ideality to a certain degree (Alburquerque and Biegler, 1996).

Consider a nonlinear system with g inputs and l outputs described by the following model:

$$\mathbf{y} = \mathbf{f}(\mathbf{x}, \theta) + \mathbf{e}(t), \tag{11.16}$$

where $\mathbf{y}$ is an l-dimensional vector, $\mathbf{x}$ is a g-dimensional vector, $\mathbf{e}$ is the noise vector, and θ is the n-dimensional parameter to be estimated from the noisy data. The ordinary least squares (OLS) estimates of parameter θ are obtained by minimizing

$$\sum_{i=1}^{M} (y_i - f(x_i, \theta))^2, \tag{11.17}$$

or equivalently by solving the system of n equations that results from differentiating expression (11.17) with respect to θ and setting the derivative equal to zero. That is,

$$\sum_{i=1}^{M} \frac{\partial}{\partial \theta_j}(y_i - f(x_i, \theta))^2 = \mathbf{0}, \quad j = 1, \ldots, n. \tag{11.18}$$

If the errors are normally distributed, the OLS estimates are the maximum likelihood estimates of θ and the estimates are unbiased and efficient (minimum variance estimates) in the statistical sense. However, if there are outliers in the data, the underlying distribution is not normal and the OLS will be biased. To solve this problem, a more robust estimation methods is needed.

Several robust identification and estimation methods have been considered in the literature. One approach is the use of L_p estimators, in which we estimate the parameters of a given process by minimizing (Butler *et al.*, 1990)

$$\sum_{i=1}^{M} |y_i - f(x_i, \theta)|^p \tag{11.19}$$

instead of the sum of the error square. Least absolute deviation (LAD) and ordinary least squares estimation (OLS) correspond to $p = 1$ and $p = 2$, respectively. Values of p smaller than 2 are associated with tails that are thicker than the normal distribution. Forsythe (1972) suggested that setting $p = 1.5$ can give good robust features for a thicker tailed distribution. L_p estimators are, in fact, special cases of robust M-estimators (Butler *et al.*, 1990). Huber's M-estimates (Huber, 1981) provide the general framework for robust estimation. These estimates are based on the concept of the influence function (IF).

If we denote the residual values by

$$u_i(\theta) = y_i - f(x_i, \theta), \quad i = 1, \ldots, M, \tag{11.20}$$

the M-estimates of vector θ are

$$\hat{\theta} = \arg\min \sum_{i=1}^{M} \rho(u_i(\theta)), \tag{11.21}$$

where ρ is usually a convex function in order for the solution to be unique. Comparing Eq. (11.21) with Eq. (11.18), the estimates can be defined as the solution of the n equations

$$\sum_{i=1}^{M} \psi(u_i) \frac{\partial f(x_i, \theta)}{\partial \theta_j} = \mathbf{0}, \quad j = 1, \ldots, n, \tag{11.22}$$

where $\psi(u) = \rho'(u)$. If ρ is convex and continuous, the definitions expressed in (11.21) and (11.22) are equivalent. For all but the simplest f functions, the solution of (11.22) must be obtained numerically, generally by using some iterative process.

Occasionally it is convenient to refer to the ρ function in (11.21), but generally the form (11.22) is used in robust M-estimation. The use of the ψ form is due to Hampel's concept of the influence function (Hampel *et al.*, 1986). According to the IF concept, the value of ψ represents the effect of the residuals on the parameter estimation. If ψ is unbounded, it means that an outlier has an infinite effect on the estimation. Thus, the most important requirement for robustness is that ψ must be bounded and should have a small value when the residual is large. In fact, the value of the ψ function corresponds to the gross error sensitivity (Hampel *et al.*, 1986), which measures the worst (approximate) influence that a small amount of contamination of fixed size can have on the value of the estimator.

For least squares estimation, ρ is the errors squared function, and the influence function $\psi = u$. For a very large residual, $u \to \infty$, and ψ also grows to infinity; this means that a single outlier has a large influence on the estimation. That is, for least squares estimation, every observation is treated equally and has the same weight.

Now we must consider the problem in robust estimation of how to choose the ψ function properly so that it meets the crucial condition that it is bounded when the argument u is very large, while remaining a continuous function. In other words, unlike least squares, we do not treat the observations equally, but rather we give a limited (usually small) weight to large error observations. In some cases zero weight may be assigned to a large error observation in order to ignore its contribution to the estimation.

Moberg *et al.* (1980) use the residuals from the preliminary fit to classify the error distribution according to the tail-weight and skewness characteristics. They then use the classification to select a ψ function. The general form of the function for a skewed light-tailed distribution according to Moberg is

$$\psi(u) = \frac{a^2 u}{(b + u)^2 + c^2}, \tag{11.23}$$

where the parameters a, b, and c are determined empirically.

Within the context of data reconciliation and gross error detection, Alburquerque and Biegler (1996) used a ρ function given by

$$\rho(u) = c^2\left[\frac{|u|}{c} - \log\left(1 + \frac{|u|}{c}\right)\right], \tag{11.24}$$

where c is a tuning parameter. The corresponding influence function is given by

$$\psi(u) \propto \frac{u}{1 + \frac{|u|}{c}}, \tag{11.25}$$

which is bounded, since $\lim_{u\to\infty} \psi(u) \propto c$. This estimator behaves like the least squares estimator for small residuals, but like an absolute estimator for large residuals (Alburquerque and Biegler, 1996).

An alternative approach to these methods is to obtain the influence function directly from the error distribution. In this case, for the maximum likelihood estimation of the parameters, the ψ function can be chosen as follows:

$$\psi(u) = -\frac{f'(u)}{f(u)}, \tag{11.26}$$

where f is the density function of the true underlying distribution of $\{\mathbf{e}(t)\}$. This is an MLE of the parameters θ, and it possesses the smallest asymptotic variance.

Wu and Cinar (1996) use a polynomial approximation (ARMENSI) of the error density function f based on a generalized exponential family, such as

$$f(u) = \xi \exp\left[-\int \frac{v(u)}{w(u)}\, du\right], \tag{11.27}$$

where v and w are reasonable functions and ξ is a scalar constant. Generally, v and w are polynomials, but they can be allowed to include more general functions such as logarithms and trigonometric expressions. Detailed explanation can be found in Wu and Cinar (1996). The disadvantage of this approach is the predetermination of the orders of the polynomials. An alternative adaptive wavelet based robust estimator (WARME) approach has been proposed by Wang and Romagnoli (1998). This approach uses an embedded wavelet method for density estimation. These approaches have obvious advantages over the other approaches, because they do not assume that the errors belong to some generalized distribution. Consequently, we do no need to preselect a form for the ψ function as it can be obtained directly from the residuals. However, this is still an open and continuing area of research.

EXAMPLE 11.2

This simple example consists of the model of a lollipop dissolution in a mixing experiment, which is a typical experiment in a chemical engineering laboratory (Wang and Romagnoli, 1998). The power number $P/(\rho N^2 D^5)$ is related to the Reynolds number $D^2 N\rho/\mu$ by the following nonlinear model:

$$y(t) = \theta_1 + \frac{\theta_2}{u(t)} + e(t),$$

where $y(t) = P/(\rho N^3 D^5)$, $x(t) = D^2 N\rho/\mu$, $u(t)$ is uniformly distributed between [1, 15] and $e(t)$ is the error distribution mentioned earlier.

TABLE 1 Comparison of the Performance of Different Estimators for the Uniform and t^2 Distributions

	Relative error		Efficiency	
Method	**Uniform**	t^2	**Uniform**	t^2
OLS	0.4455	12.0906	0.0958	0.0091
Huber's	0.1009	0.2588	0.4229	0.4245
ARMENSI	0.0993	0.1205	0.4296	0.9120
Biegler's	0.1066	0.2341	0.4003	0.4692
WARME	0.0427	0.1099	1.0000	1.0000

A Monte Carlo study comparing five parameter estimation procedures using different error distributions was performed. The five estimators are OLS, Huber's minimax estimator, Biegler's fair function, ARMENSI, and the wavelet-based adaptive robust M-estimator (WARME). Various error distributions were generated, ranging from light-tailed to heavy-tailed and from symmetric to skewed. All these generated distributions were contaminated by some isolated, randomly generated outliers.

The empirical relative errors of the model parameter vector θ are calculated by

$$\sum_{i=1}^{2}\left(\frac{\theta_i - \hat{\theta}_i}{\theta_i}\right)^2,$$

where

$$\theta = [1.1232 \quad 49.0075], \text{ and } \hat{\theta} \text{ is the estimated value of the parameter.}$$

The relative efficiencies have also been obtained by comparing these relative errors to the smallest value of the other estimates for each case studied. The smaller the relative error, the better the model parameter estimation; the larger the relative efficiency, the better the estimator. Results are listed in Table 1 for the uniform and t^2 distributions.

This Monte Carlo study shows that, for this family of error distributions, the WARME method has the best performance. Also, as expected, the OLS performs the best for the normal distribution, but performs very poorly for the case when outliers are present in the data set. The Huber minimax estimator can decrease the influence of the outliers, but it only works well for the symmetric distribution. Biegler's fair function is also designed for symmetric distributions.

11.3.2. QQ-Plots Approach

Chen *et al.* (1998) developed an integrated approach, which can delete the influence of outliers in the data reconciliation problem, based on the idea of QQ-plots. The basic concept of this approach is to calibrate sampling data by means of its own main structure so that the influence of data decreases as it becomes less and less characteristic. In this way, the final data reconciliation procedure will be resistant to outliers. A limiting transformation, which operates on the data set, is defined to eliminate or reduce the influence of outliers on the performance of the conventional data reconciliation procedure.

Quantile probability plots (QQ-plots) are useful data structure analysis tools originally proposed by Wilk and Gnanadesikan (1968). By means of probability plots they provide a clear summarization and palatable description of data. A variety of application instances have been shown by Gnanadesikan (1977). Durovic and Kovacevic (1995) have successfully implemented QQ-plots, combining them with some ideas from robust statistics (e.g., Huber, 1981) to make a robust Kalman filter.

In comparing two distribution functions, a plot of the points whose coordinates are the quantiles $\{q_{z1}(p_c), q_{z2}(p_c)\}$ for different values of the cumulative probability p_c is a QQ-plot. If z_1 and z_2 are identically distributed variables, then the plot of z_1-quantiles versus z_2-quantiles will be a straight line with slope 1 and will point toward the origin.

A desired property (linear invariance property) of QQ-plots is that when the two distributions involved in the comparison are possibly different only in location and/or scale, the configuration of the QQ-plots will still be linear, with a nonzero intercept if there is a difference in location, and/or a slope different from unity if there is a difference in scale.

Following Durovic and Kovacevic (1995), let us now consider a measurement sample recorder $\{y_{(i)}\}, i = 1, \ldots, M$ (here, i can be considered as each sampling instance), from a distribution $F(y)$, corresponding to a probability density function $f(y)$. When the samples $\{y_{(i)}\}$ are rearranged in ascending order $\{y_i\}$, the probability that an observation y will have rank i in the ordered sequence $\{y_i\}$ follows from the Bernoulli experiment (Papoulis, 1991):

$$P(i/y) = \binom{M-1}{i-1} F^{i-1}(y)(1 - F(y))^{M-i}, \tag{11.28}$$

and the conditional expectation $E(i/y)$ is

$$E(i/y) = 1 + (M-1)F(y). \tag{11.29}$$

Rewriting Eq. (11.29), one obtains the QQ-plots expression as follows:

$$\begin{aligned} y_i &= F^{-1}\{[E(i/y) - 1]/(n-1)\} \\ &\approx F^{-1}\{(i-1)/(M-1)\}, \quad i = 1, 2, \ldots, M \\ &\approx F^{-1}\{(i-0.5)/M\}, \end{aligned} \tag{11.30}$$

where $F^{-1}(^\circ)$ is the inverse of the distribution function $F(^\circ)$.

The essence of a QQ-plot is to plot the ordered sample values against some representative values from a presumed null standard distribution $F(^\circ)$. These representative values are the quantiles of the distribution function $F(^\circ)$ corresponding to a cumulative probability p_{ci} [e.g., $(i - 0.5)/M$] and are determined by the expected values of the standard order statistics from the reference distribution. Thus, if the configuration of the QQ-plot in Eq. (11.30) is fairly linear, it indicates that the observations $(\{y_{(i)}\}, i = 1, \ldots, M)$ have the same distribution function as $F(^\circ)$, even in the tails.

Figures 5 and 6 show a typical noise record $\{y_i\}$ and the corresponding QQ-plots when the noise follows a standard Gaussian distribution $F_n(^\circ)$ with zero mean and unit variance.

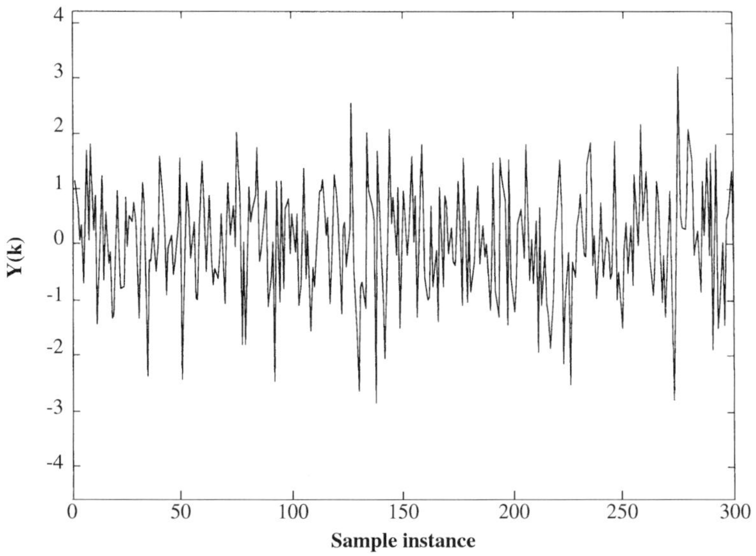

FIGURE 5 Normal data samples (from Chen *et al.*, 1998).

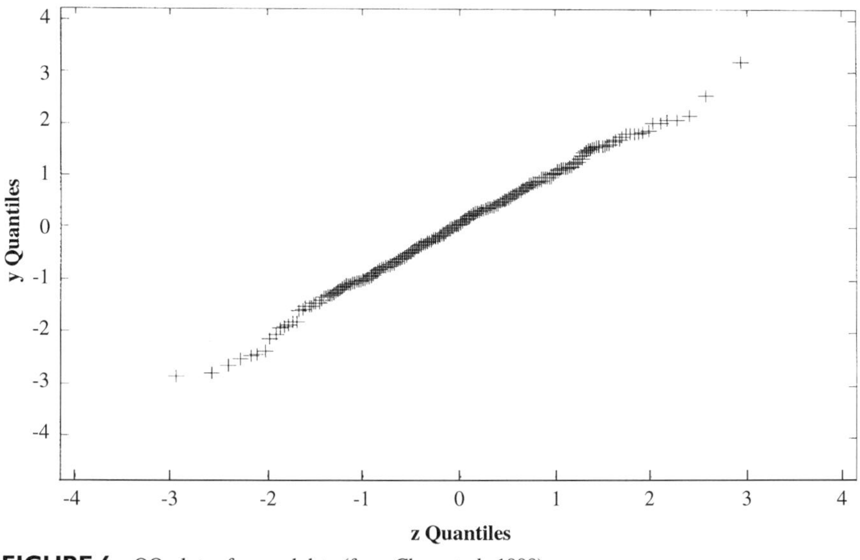

FIGURE 6 QQ-plots of normal data (from Chen *et al.*, 1998).

Frequently, the measurement error distributions arising in a practical data set deviate from the assumed Gaussian model, and they are often characterized by heavier tails (due to the presence of outliers). A typical heavy-tailed noise record is given in Fig. 7, while Fig. 8 shows the QQ-plots of this record, based on the hypothesized standard normal distribution.

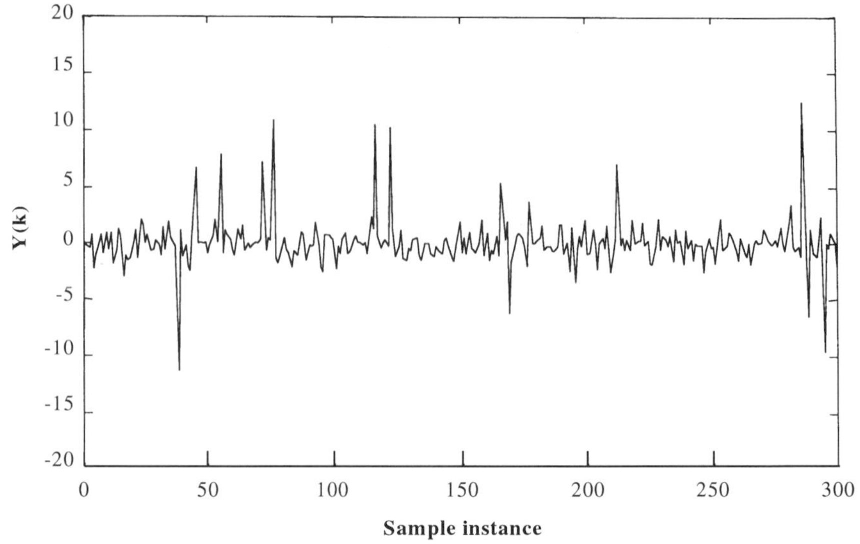

FIGURE 7 Gaussian-like data with heavier tail (from Chen *et al.*, 1998).

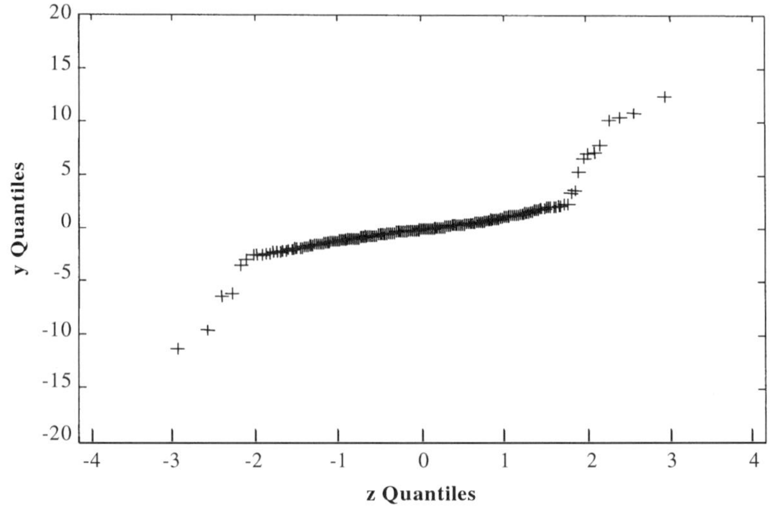

FIGURE 8 QQ-plots of Gaussian-like data with heavier tail (from Chen *et al.*, 1998).

As discussed before, the outliers generated by the heavy-tails of the underlying distribution have a considerable influence on the OLS problem arising in a conventional data reconciliation procedure. To solve this problem, a limiting transformation, which operates on the data set, is defined to eliminate or reduce the influence of outliers on the performance of a conventional rectification scheme.

For an ordered observation $\{y_i\}$, $i = 1, \ldots, M$, $\psi_i(y_i)$ is a suitably chosen limiting (nonlinear) transformation for cutting off the outlier's influence (Durovic and

Kovacevik, 1995; Chen *et al.*, 1998). In general, the form of this limiting transformation is

$$\psi_i(y_i) = \begin{cases} o_1(y_i, \alpha_i), & y_i \leq \alpha_i \\ o_2(y_i), & \alpha_i < y_i < \beta_i, \\ o_1(y_i, \beta_i), & y_i \geq \beta_i \end{cases} \tag{11.31}$$

where $o_k(^\circ)$, $k = 1, 2, 3$, are known deterministic functions, while α_i and β_i are adjustable parameters, depending on the rank i of the observation y in the ordered sequence $\{y_i\}$.

As pointed out by Durovic and Kovacevic (1995), $\psi_i(^\circ)$ should look like a linear function for small values of the argument, in order to preserve the Gaussian observations with high probability. Furthermore, $\psi_i(^\circ)$ should be a bounded and continuous function. Boundedness prevents a single observation from having an arbitrarily large influence, while continuity limits the effect of rounding, grouping, or quantization.

Once the type of limiting function has been selected, the parameters α_i and β_i, which quantitatively define the nonlinearity $\psi_i(^\circ)$ in Eq. (11.31), have to be chosen to give the desired efficiency at the nominal Gaussian model. An analytical procedure was developed by Chen *et al.* (1998) for choosing each set of tuning parameters, based on the conditional probability of the measurements.

EXAMPLE 11.3

Let us consider the same chemical reactor as in Example 11.1 (Chen *et al.*, 1998). Monte Carlo data for $\mathbf{y}$ were generated according to $\boldsymbol{\Psi}$ in order to simulate process sampling data. A window size of 25 was used here, and to demonstrate the performance of the robust approach two cases were considered, with and without outliers.

Without outliers. The results of data reconciliation from the conventional approach $\mathbf{x}_C$ and from the robust approach $\mathbf{x}_R$ are same, that is,

$$\mathbf{x}_C = [0.1731 \quad 5.0194 \quad 1.2116 \quad 3.9809]^T$$
$$\mathbf{x}_R = [0.1731 \quad 5.0194 \quad 1.2116 \quad 3.9809]^T.$$

With outliers. For comparison, we use the same data set (without outliers) and the same window size, 25, only we substitute one set of measurements with an outlier of $\mathbf{y} = [0.1858 \;\; 4.4735 \;\; 1.2295 \;\; 3.88]^T$. The outlier is assumed to be in the measurement of variable x_2. Figures 9 and 10 show the sampling data of variable x_2 and the corresponding QQ-plots. In Fig. 10, most of the measurements (24 of 25) approximate a straight line, which indicates that the main data has a Gaussian distribution. Only the 25th measurement is far away from this main structure and is obviously an outlier.

The results for the robust and the conventional approaches are, respectively,

$$\mathbf{x}_R = [0.1742 \quad 5.0508 \quad 1.2192 \quad 4.0058]^T$$
$$\mathbf{x}_C = [0.1593 \quad 4.6183 \quad 1.1148 \quad 3.6628]^T.$$

$\mathbf{x}_R$ from the robust approach, as expected, still gives the correct answer; however, the conventional approach fails to provide a good estimate of the process variables. Although the main part of the data distribution is Gaussian, the conventional approach fails in the task because of the presence of just one outlier. In a strict sense, the presence of this outlier results in the invalidation of the statistical basis of data reconciliation,

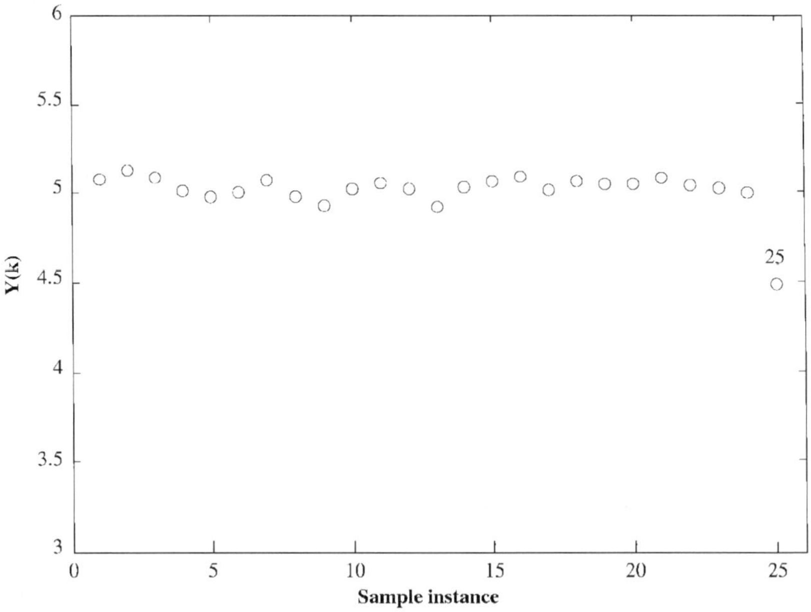

FIGURE 9 One set of sampling data of variable x_2 (from Chen *et al.*, 1998).

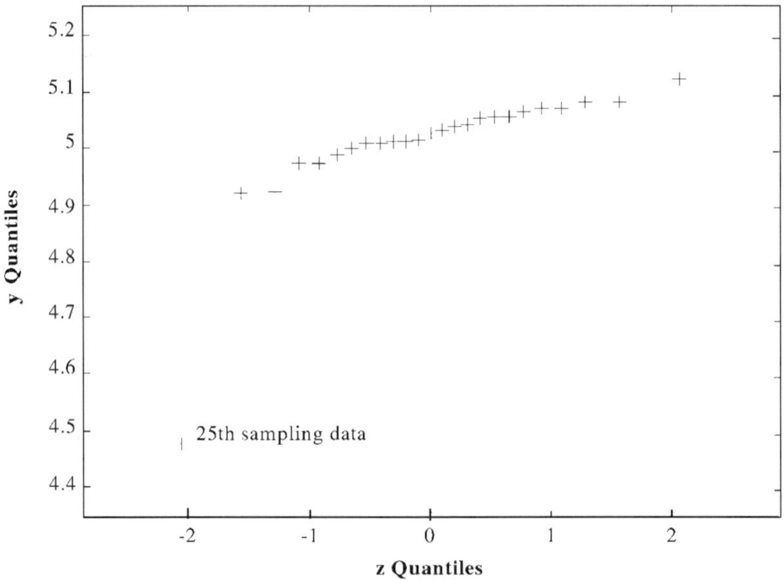

FIGURE 10 QQ-plots of one set sampling data of x_2 (from Chen *et al.*, 1998).

thus deteriorating its performance. Figures 11 and 12 show the sampling data of x_2 and the corresponding QQ-plots, after the application of the limiting transformation. From Fig. 12, one can easily conclude that the data now has a Gaussian distribution, thus providing some degree of confidence about the result of the data reconciliation.

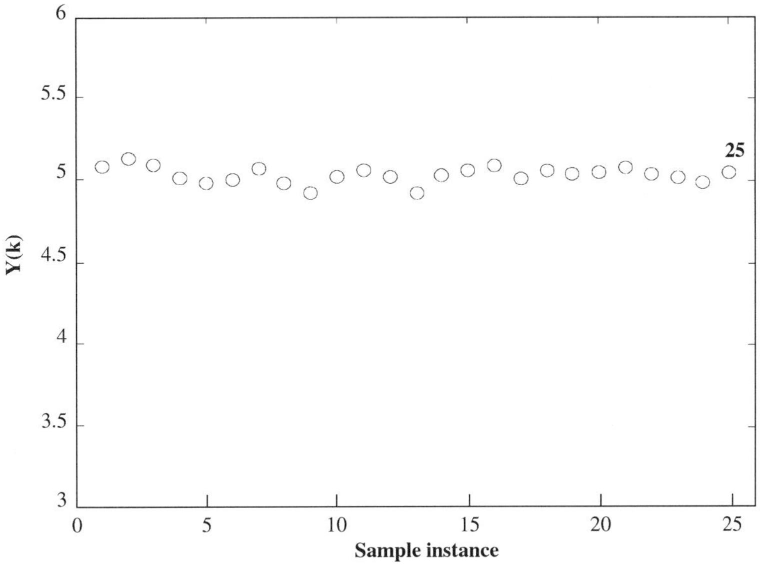

FIGURE 11 One set data of variable x_2 after limiting transform (from Chen *et al.*, 1998).

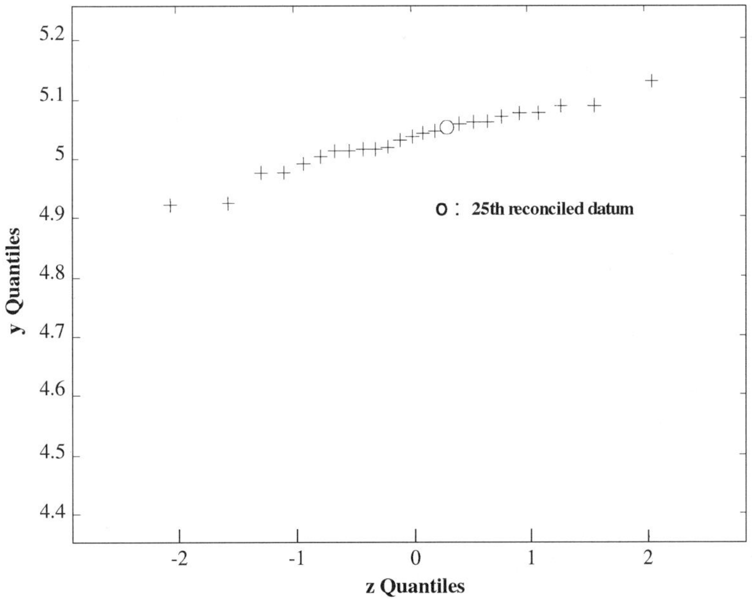

FIGURE 12 QQ-plots of one set data of variable x_2, after limiting transform (from Chen *et al.*, 1998).

11.3.3. The Trust Function Approach

This approach (Chen and Romagnoli, 1997) tries to assess directly the behavior of measurements from the temporal information (historical record). A trust degree on each measurement is calculated by referring to the main structure of the data. The trust function defined here is the measure of the confidence level of the measurement. Its value, between 0 and 1, stands for zero confidence and 100% trust in the measurement, respectively. Trust degree can also be viewed as the degree to which the individual data belongs to its own main structure. A moving window is used to capture the latest data structure. Generally, the more data that are employed, the more precise the information. However, this will increase the amount of computation.

The trust function should be selected in such way that it will decrease when measurements become less and less relevant to the main characteristic of the data set. Distance is used in this approach to describe the difference between, or the similarity of, data and can be expressed as

$$\mathbf{d}_i^2 = (\mathbf{y}_i - \mathbf{m})^{\mathrm{T}}\mathbf{V}^{-1}(\mathbf{y}_i - \mathbf{m}), \quad i = 1, \ldots, M. \tag{11.32}$$

It is suggested that the trust function, tf_1, should follow the distance-based function, that is,

$$\mathrm{tf}_1(\mathbf{d}_i) = \begin{cases} 1, & \mathbf{d}_i < k \\ k/\mathbf{d}_i & \text{otherwise} \end{cases}, \tag{11.33}$$

where $\mathbf{d}_i^2$ is the squared distance of the observation from the current estimate of location (mean) $\mathbf{m}$, $\mathbf{V}$ is the variance used to normalize the distance, and k is a parameter. k^2 is taken to be the 90% point of the χ^2 distribution. It worth noting that the trust function is a convex function of the residuals and belongs to the family of well-known Huber type functions (Huber, 1981).

In real practice, the location $\mathbf{m}$ and the variance have to be estimated from real data. An iterative algorithm, similar to the one used in Chapter 10 for the robust covariance estimation, is used to calculate the trust function. The main advantage of using this algorithm is that the convergence is warranted.

Once the trust function is calculated, the next step is to incorporate this information into the data reconciliation problem. The general idea of the approach is that the measurement's correction potential should be proportional to the trust function, that is, the more trust, the less correction needed. If the trust function equals or approaches zero, the measurement is deleted and estimated by using spatial redundancy. This kind of action brings the feature of robustness into data reconciliation, since the outliers will have little influence on the results.

EXAMPLE 11.4

The same problem discussed in Examples 11.1 and 11.3 is taken to illustrate the ideas described in this section (Chen and Romagnoli, 1997). To evaluate the performance of the proposed approach under different error distributions, Monte Carlo simulations have again been performed on the four previous distributions.

Only the results for the Cauchy and symmetric contaminated distributions are shown in Figs. 13 and 14, respectively. From these figures, it is clearly shown that the robust approach consistently and successfully performs the data reconciliation, regardless of the distributions of the data. This is a very desirable property in real applications, since in most cases the distribution is unknown or known only approximately.

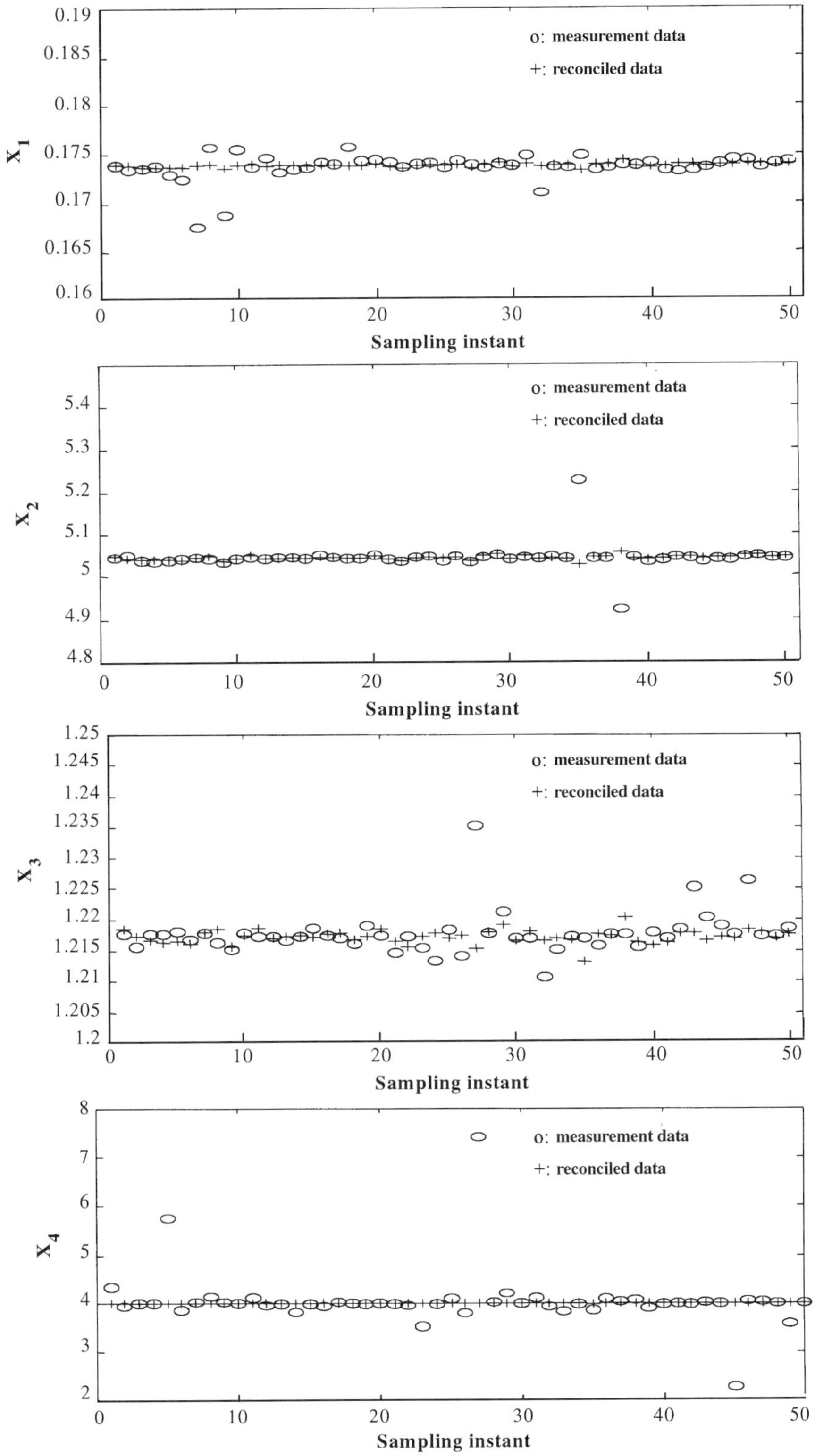

FIGURE 13 Cauchy distribution case (from Chen *et al.*, 1997).

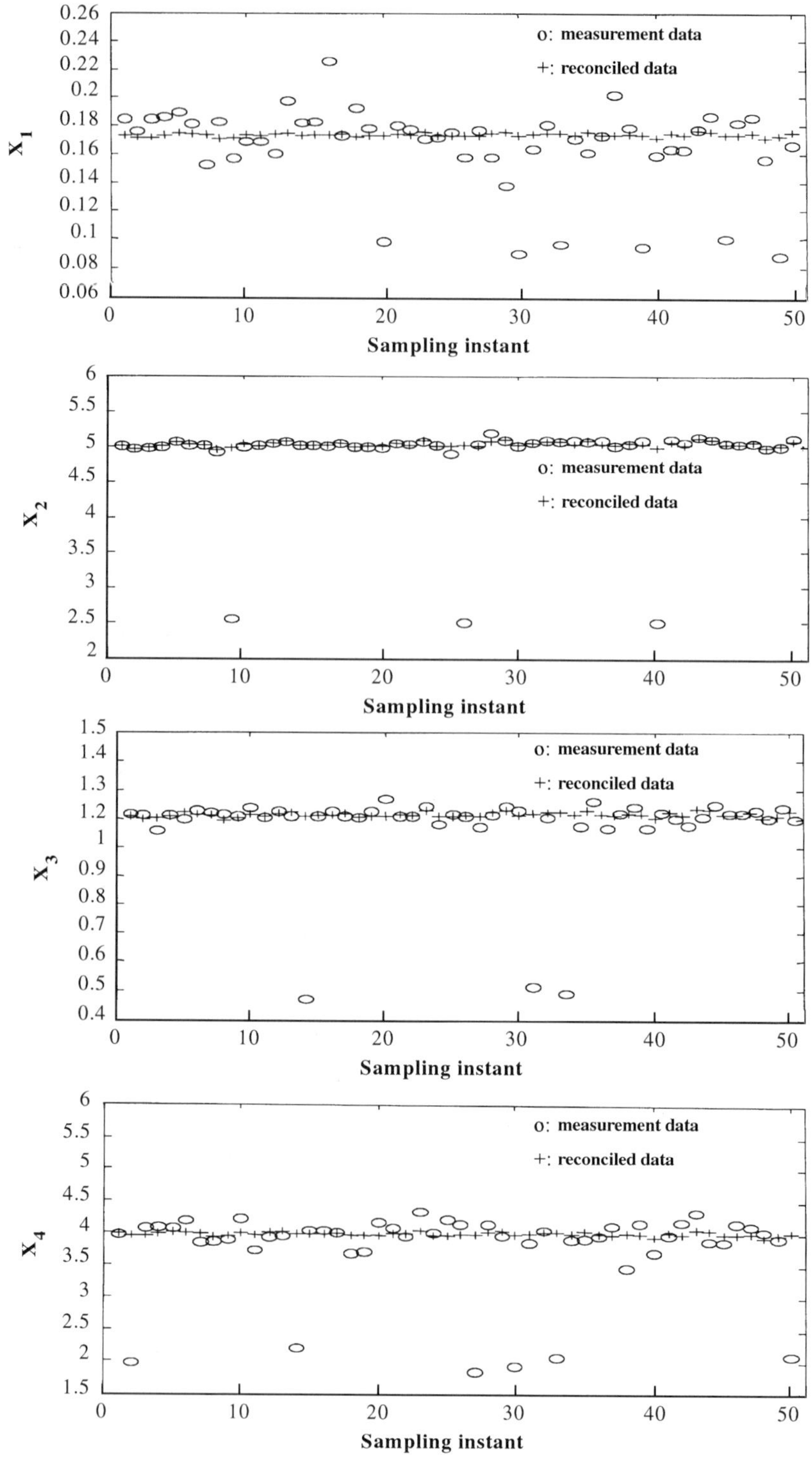

FIGURE 14 Asymmetric contaminated normal distribution case (from Chen *et al.*, 1997).

11.4. PRINCIPAL COMPONENT ANALYSIS IN DATA RECONCILIATION

Correlation is inherent in chemical processes even if one could assume that there is no correlation among the data. PCA is an effective tool in multivariate data analysis. It transforms a set of correlated variables into a new set of uncorrelated ones, known as principal components.

Principal components analysis (PCA) and project to latent structure (PLS) were suggested to absorb information from continued-process data (Kresta *et al.*, 1991; MacGregor and Kourti, 1995; Kourti and MacGregor, 1994). The key point of these approaches is to utilize PCA or PLS to compress the data and extract the information by projecting them into a low-dimension subspace that summarizes all the important information. Then, further monitoring work can be conducted in the reduced subspace. Two comprehensive reviews of these methods have been published by Kourti and Macgregor (1995) and Martin *et al.* (1996).

The key idea of this section is to combine PCA and the steady-state data reconciliation model to provide sharper and less confounding statistical tests for gross errors, through exploiting the correlation.

As shown in previous chapters, the residuals of the constraints for linear steady-state processes are defined as

$$\mathbf{r} = \mathbf{A}\mathbf{y}, \tag{11.34}$$

where $\mathbf{y}$ is the vector of measured variables and $\mathbf{A}$ is the balance matrix.

Note: We have considered that all variables are measured. If this is not the case, the unmeasured variables can be removed using some of the techniques described in Chapters 3 and 4. ♣

Assuming that the measurement errors follow a certain distribution with covariance $\mathbf{\Psi}$, then $\mathbf{r}$ will follow the same distribution with expectation and covariance given by

$$E(\mathbf{r}) = \mathbf{r}^* = \mathbf{0}, \quad \operatorname{cov}\{\mathbf{r}\} = \mathbf{\Phi} = \mathbf{A}\mathbf{\Psi}\mathbf{A}^{\mathrm{T}}. \tag{11.35}$$

Consider the eigenvalue decomposition of matrix $\mathbf{\Phi}$ as follows:

$$\mathbf{\Lambda} = \mathbf{U}^{\mathrm{T}}\,\mathbf{\Phi}\,\mathbf{U}, \tag{11.36}$$

where

$\mathbf{U}$ = matrix of orthonormalized eigenvectors of $\mathbf{\Phi}(\mathbf{U}\mathbf{U}^{\mathrm{T}} = \mathbf{I})$

$\mathbf{\Lambda}$ = diagonal matrix of the eigenvalues of $\mathbf{\Phi}$

Accordingly, the following linear combinations of $\mathbf{r}$ are proposed by Tong and Crowe (1995):

$$\mathbf{p} = \mathbf{W}^{\mathrm{T}}(\mathbf{r} - \mathbf{r}^*) = \mathbf{W}^{\mathrm{T}}\mathbf{r}, \tag{11.37}$$

where

$$\mathbf{W} = \mathbf{U}\mathbf{\Lambda}^{-1/2} \tag{11.38}$$

and $\mathbf{p}$ are the principal components of matrix $\mathbf{\Phi}$. Also,

$$\mathbf{r} \sim (\mathbf{0}, \mathbf{\Phi}) \Rightarrow \mathbf{p} \sim (\mathbf{0}, \mathbf{I}).$$

That is, a set of correlated variables $\mathbf{r}$ is transformed into a new set of uncorrelated variable $\mathbf{p}$.

Note: If the measurement errors are normally distributed, $\mathbf{y} \sim N(\mathbf{x}, \Psi) \Rightarrow \mathbf{p} \sim N(\mathbf{0}, \mathbf{I})$. ♣

Consequently, instead of looking at a statistical test for $\mathbf{r}$, we can perform the hypothesis test on $\mathbf{p}$. Tong and Crowe (1995) proposed the following test for a principal component:

$$p_i = (\mathbf{W}^{\mathrm{T}}\mathbf{r})_i \sim N(0, 1), \quad i = 1, \ldots, np, \tag{11.39}$$

which is tested against a threshold tabulated value. The constraints suspected to be in gross error can be identified by looking at the contribution from the jth residual in $\mathbf{r}$, r_j, to a suspect principal component, say p_i, which can be calculated by

$$g_j = (w_i)_j r_j, \quad j = 1, \ldots, m, \tag{11.40}$$

where w_i is the ith eigenvector in $\mathbf{W}$.

Principal component analysis can be further extended to study the chi-square statistic, since

$$\chi_m^2 = \mathbf{r}^{\mathrm{T}}\mathbf{\Phi}^{-1}\mathbf{r} = \mathbf{p}^{\mathrm{T}}\mathbf{p}, \tag{11.41}$$

where $\mathbf{p} \in R^m$. If $np < m$ principal components are retained, that is, $\mathbf{p} \in R^{np}$, we have

$$\chi_{np}^2 = \mathbf{p}^{\mathrm{T}}\mathbf{p}. \tag{11.42}$$

This is called the truncated chi-squares test (Tong and Crowe, 1995).

An interesting analysis was carried out by Tong and Crowe in comparing three different tests, the univariate test, the maximum power (MP) test, and the principal component (PC) test, for two cases: (1) when the $\mathbf{\Phi}$ matrix is diagonal and (2) when the $\mathbf{\Phi}$ matrix is not diagonal. This comparison is depicted in Figs. 15 and 16 for a two-dimensional problem, together with the chi-square statistic. As shown in Fig. 15, when $\mathbf{\Phi}$ is a diagonal matrix all the three statistics are identical. As $\mathbf{\Phi}$ shifts away

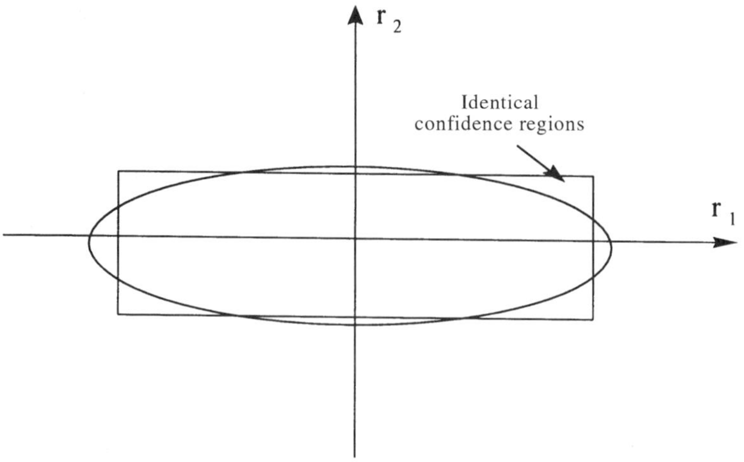

FIGURE 15 Confidence regions for diagonal $\mathbf{\Phi}$ (from Tong and Crowe, 1995).

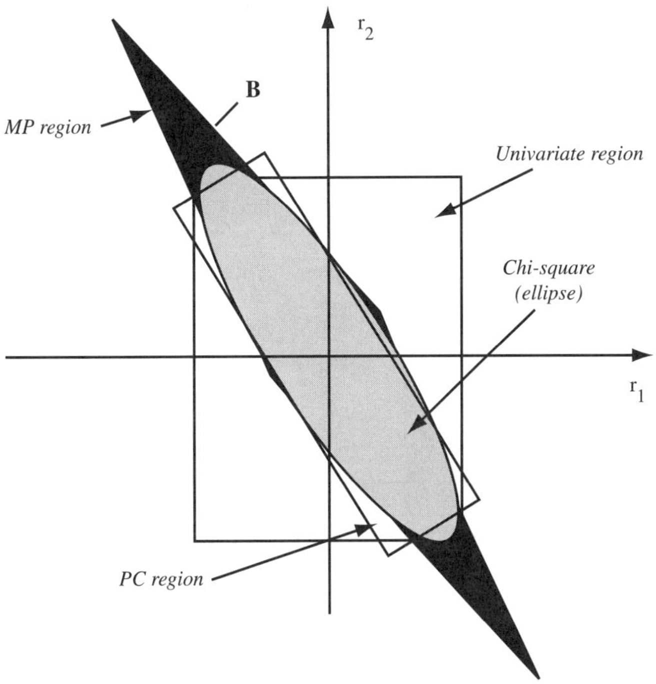

FIGURE 16 Confidence regions when Φ is not diagonal (from Tong and Crowe, 1995).

from diagonal form, it can be seen that the univariate and PC tests remain rectangular while the MP test changes gradually from a rectangle to a flattened parallelogram.

Tong and Crowe analyzed the behavior of the tests under the presence of different types of errors. They came to the following conclusions:

- Because of the flattened shape of the MP region, it has the potential to wrongly accept large gross errors, which would be rejected by the other tests, such as point B in Fig. 16.
- On the other hand, the relative positions of the PC region and the ellipse are fixed under any chosen type I error, regardless of the structure of Φ, thus leading to consistent performance.

A sequential procedure was further developed by Tong and Crowe (1996) by applying sequential analysis of the principal component test using the sequential probability ratio test (SPRT). Dunia *et al.* (1996) also used PCA for sensor fault identification via reconstruction. In that paper it was assumed that one sensor had failed and the remaining sensors are used for reconstruction. Furthermore, the transient behavior of a number of sensor faults in various types of residuals is analyzed, and a sensor validity index is suggested, determining the status of each sensor.

EXAMPLE 11.5

Let us consider the process flowsheet that was presented in Fig. 3 of Chapter 7, which consists of a recycle system with four units and seven streams. In this case, two measurement errors are simulated in order to show the application of principal component strategy in their identification. Fixed gross error magnitudes of 5 and 7 standard deviations are considered for streams 1 and 2, respectively.

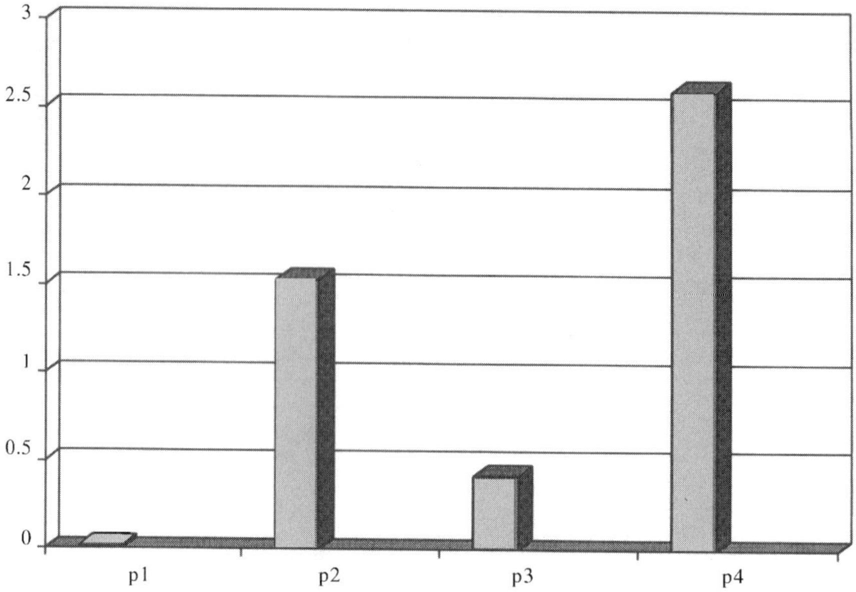

FIGURE 17 Principal components of the covariance matrix of the residuum.

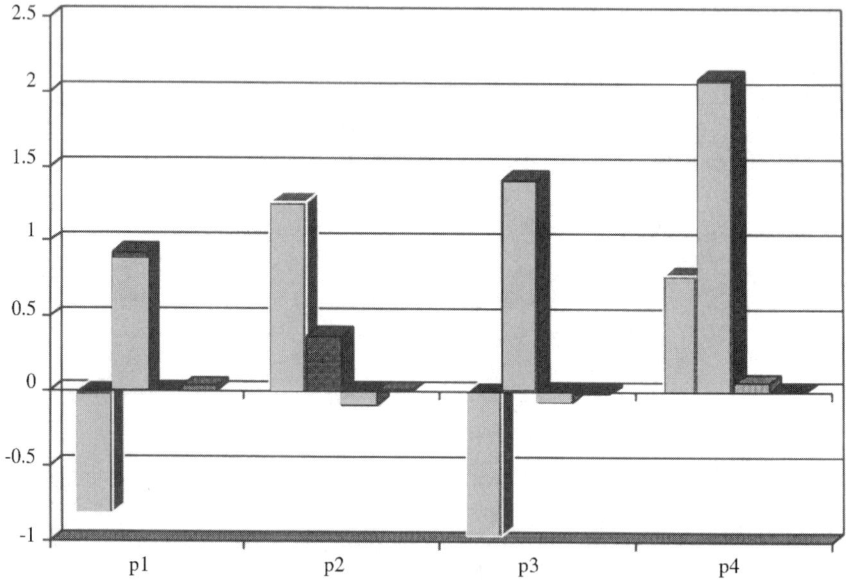

FIGURE 18 Contributions from the residuals of the constraints to **p**.

The measurement vector, the covariance matrix of the measurement errors, and the residuum vector are as follows:

$$\mathbf{y} = [5.5797 \quad 17.6334 \quad 14.9127 \quad 5.0336 \quad 10.0096 \quad 5.0652 \quad 4.9845]^{\mathrm{T}}$$
$$\boldsymbol{\Psi} = \mathrm{diag}(0.0016 \quad 0.0141 \quad 0.0141 \quad 0.0016 \quad 0.0062 \quad 0.0016 \quad 0.0016)$$
$$\mathbf{r} = [-1.9550 \quad 2.7207 \quad -0.1305 \quad -0.0401]^{\mathrm{T}}.$$

After performing the eigenvalue decomposition of matrix $\mathbf{\Phi}$, the vector $\mathbf{p}$ of principal components of $\mathbf{\Phi}$ is calculated through Eq. (11.37). The elements of $\mathbf{p}$ are represented in Fig. 17.

Each p_i statistic is tested against the threshold value of 1.96, which corresponds to a 95% confidence level. The last principal component results in suspect. The contributions from each residual of the constraints to the principal components are given in Fig. 18. From this figure we see that the residuals of units 1 and 2 are the main contributions to all the principal components and in particular to p_4. The flowrates involved in units 1 and 2 are f_1, f_2, f_3, f_4, f_6. Because f_3 and f_4 are related to unit 3, which is not suspect, and f_6 participates in the unsuspected unit 4, we can conclude that the only bias measurements are f_1, f_2, as was simulated.

11.5. CONCLUSIONS

In this chapter, some of the most recent approaches for dealing with different aspects of the data reconciliation problem were discussed. A more general formulation in terms of a probabilistic framework was first introduced and its application in dealing with gross error was discussed in particular. It was shown through this formulation that the conventional approach is just a special case of the problem when certain conditions are assumed.

Robust estimation approaches were then considered; in this case, the estimators are designed so they are insensitive to outliers. That is, they will give unbiased results in the presence of the ideal distribution, but will try to minimize the sensitivity to deviations from ideality.

Finally, a method for dealing with the inherent correlation existing in chemical processes was discussed. This method combines principal component analysis (PCA) and the steady-state data reconciliation model to provide sharper and less confounding statistical tests for gross errors.

NOTATION

$\mathbf{A}$	balance matrix for linear constraints
a	constant
b	constant
c	constant
$\mathbf{d}$	distance of the observation vector from the current estimate of the location
$\mathbf{e}$	noise vector
F	presumed distributions
$\mathbf{f}$	vector of modeling equations
f	probability density function
g	number of system imputs
G	gross error occurred
h	index
H	size of the window
k	constant
l	number of system outputs
$\mathbf{m}_h$	unit center
$\mathbf{m}$	current location vector
M	number of samples

n	number of parameters
np	dimension of $\mathbf{p}$
o	function defined by Eq. (11.31)
P	probability
p	order of the L_p estimator
$\mathbf{p}$	principal components of $\boldsymbol{\Phi}$
p_c	cumulative probability
q_h	mixing proportion
Q_h	unit covariance matrix
r	constant
$\mathbf{r}$	vector of residuum
R	random error occurred
rst	ratio of standard deviations
t	time
tf	trust function
$\mathbf{U}$	matrix of orthogonalized eigenvectors of $\boldsymbol{\Phi}$
$\mathbf{u}$	residual vector
v	function in Eq. (11.27)
$\mathbf{V}$	current variance
w	function in Eq. (11.27)
$\mathbf{W}$	matrix defined by Eq. (11.38)
$\mathbf{w}$	eigenvector in $\mathbf{W}$
$\mathbf{x}$	vector of state variables
$\mathbf{y}$	vector of measurements

Greek Symbols

α	parameter
β	parameter
$\boldsymbol{\Psi}$	covariance matrix of measurement errors
$\boldsymbol{\Phi}$	covariance matrix of residuum
φ	equality constraints
ω	inequality constraints
δ	measurement errors
η	probability of occurrence of gross errors
σ	standard deviation
ϕ_h	function defined by Eq. (11.14)
θ	parameters
ρ	convex function involved in Eq. (11.21)
ψ	influence function
ξ	scalar constant
$\boldsymbol{\Lambda}$	diagonal matrix of the eigenvalues of $\boldsymbol{\Phi}$

Subscripts

C	conventional estimation
R	robust estimation

REFERENCES

Albuquerque, J. S., and Biegler, L. T. (1996). Data reconciliation and gross error detection in dynamic systems. *AIChE J.* **42**, 2841–2856.

Butler, R., McDonald, J., Nelson, R., and White, S. (1990). Robust and partially adaptive estimation of regression models. *Rev. Econ. Stat.* **72**, 321.

Chen, J., and Romagnoli, J. A. (1997). Data reconciliation via temporal and spatial redundancies. *IFAC-ADCHEM*, pp. 647–653.

Chen, J., Bandoni, A., and Romagnoli, J. A. (1996). Robust PCA and normal region in multivariable statistical process monitoring. *AIChE J.* **42**, 3563–3566.

Chen, J., Bandoni, A., and Romagnoli, J. A. (1998). Outlier detection in process plant data. *Comput. Chem. Eng.* **22**, 641–646.

Dunia, R., Qin, S., Edgar, T., and McAvoy, T. (1996). Identification of faulty sensors using principal component analysis. *AIChE J.* **42**, 2797–2812.

Durovic, Z. M., and Kovacevic, B. D. (1995). QQ-plot approach to robust Kalman filtering. *Int. J. Control* **61**, 837–857.

Forsythe, A. B. (1972). Robust estimation of straight line regression coefficients by minimizing p-th power deviations. *Technometrics* **14**, 159–166.

Gnanadesikan, R. (1977). "Method for statistical data analysis of multivariate observations." Wiley, New York.

Hampel, F. R., Ronchetti, E. M., Rousseeuw, P. J., and Stahel, W. (1986). "Robust Statistics." Wiley, New York.

Huber, P. J. (1981). "Robust Statistics." Wiley, New York.

Johnston, L. P. M., and Kramer, M. A. (1994). Probability density estimation using elliptical basis function. *AIChE J.* **40**, 1639–1649.

Johnston, L., and Kramer, M. A. (1995). Maximum likelihood data rectification: Steady state systems. *AIChE J.* **41**, 2415–2426.

Johnston, L., and Kramer, M. A. (1998). Estimating state probability distributions from noisy and corrupted data. *AIChE J.* **44**, 591–602.

Kourti, T., and MacGregor, J. F. (1994). Multivariate SPC methods for monitoring and diagnosing of process performance. *Proc. PSE'94*, 739–746.

Kourti, T., and MacGregor, J. F. (1995). Process analysis, monitoring and diagnosis, using multivariate projection methods. *Chem Intell. Lab. Syst.* **28**, 3.

Kresta, J. V., MacGregor, J. F., and Marlin, T. E. (1991). Multivariate statistical monitoring of process operating performance. *Can. J. Chem. Eng.* **69**, 35.

MacGregor, J. F., and Kourti, T. (1995). Statistical process control of multivariate processes. *Control Eng. Pract.* **3**, 403.

Martin, E. B., Morris, A. J., and Zhang, J. (1996). Process performance monitoring using multivariate statistical process control. *IEE Proc. Control Theory* **143**, 132.

Moberg, T. F., Ramberg, J. S., and Randles, R. H. (1980). An adaptive multiple regression procedure based on M-estimators. *Technometrics* **22**, 2.

Papoulis, A. (1991). "Probability, Random Variables, and Stochastic Process." McGraw-Hill, New York.

Safavi, A., Chen, J., and Romagnoli, J. A. (1997). Wavelet-based density estimation and application to process monitoring. *AIChE J.* **43**, 1227–1241.

Scott, D. W. (1992). "Multivariate Density Estimation: Theory Practice and Visualisation." Wiley, New York.

Silverman, D. W. (1986). "Density Estimation for Statistics and Data Analysis." Chapman Hall, London.

Tjoa, I. B., and Biegler, L. T. (1991). Simultaneous strategies for data reconciliation and gross error detection of nonlinear system. *Comput. Chem. Eng.* **15**, 679.

Tong, H., and Crowe, C. M. (1995). Detection of gross errors in data reconciliation by principal component analysis. *AIChE J.* **41**, 1712–1722.

Tong, H., and Crowe, C. M. (1996). Detecting persistent gross errors by sequential analysis of principal components. *Comput. Chem. Eng.* **20**, S733–S738.

Wang, D., and Romagnoli, J. A. (1998). "Wavelet Based Robust Estimation," Internal Rep. PSE-I-No. 5. University of Sydney, Laboratory of Process Systems Engineering, Sydney, Australia.

Wilk, M. B., and Gnanadesikan, R. (1968). Probability plotting methods for the analysis of data. *Biometrika* **55**, 1–17.

Wu, X., and Cinar, A. (1996). An adaptive robust M-estimator for nonparametric nonlinear system identification. *J. Proc. Control* **6**, 233–239.

12 CASE STUDIES

This chapter discusses different applications of the theoretical developments of previous chapters. Emphasis is given to industrial applications as well as on-line exercises.

12.1. INTRODUCTION

Several case studies will be discussed in this chapter to show the use of and possible implementation methods for the ideas discussed in the previous chapters in a practical environment. The examples of application consist of two industrial cases for which real plant data were available, and an on-line application for the monitoring and control of a distillation column through a distributed control system.

The first case study consists of a section of an olefin plant located at the Orica Botany Site in Sydney, Australia. In this example, all the theoretical results discussed in Chapters 4, 5, 6, and 7 for linear systems are fully exploited for variable classification, system decomposition, and data reconciliation, as well as gross error detection and identification.

The second case study corresponds to an existing pyrolysis reactor also located at the Orica Botany Site in Sydney, Australia. This example demonstrates the usefulness of simplified mass and energy balances in data reconciliation. Both linear and nonlinear reconciliation techniques are used, as well as the strategy for joint parameter estimation and data reconciliation. Furthermore, the use of sequential processing of information for identifying inconsistencies in the operation of the furnace is discussed.

The third case study consists of a well-instrumented experimental distillation column that has been interfaced to an industrial distributed control system. In this example the use of the techniques described in previous chapters is shown in an actual on-line framework, using industrial hardware. Furthermore, the usefulness of data reconciliation, prior to process modeling and optimization, is clearly demonstrated.

12.2. DECOMPOSITION/RECONCILIATION IN A SECTION OF AN OLEFIN PLANT

12.2.1. Process Description

This section of the olefin plant includes ethylene refrigeration and compression to C_2-splitter sections (Sánchez and Romagnoli, 1996). A simplified node diagram of the process is given in Fig. 1. Cracked gases leaving the gas compressor enter the precooling and drying sections. The cooled cracked streams enter the deethanizer columns (nodes 3 and 4) where C_3 and higher hydrocarbons are separated as bottom product. The top product of this column, consisting of C_2 and lower hydrocarbons (C_2H_6, C_2H_4, C_2H_2, CH_4, H_2, etc.), enters the acetylene hydrogenation reactor, where acetylene is primarily hydrogenated to ethylene.

The hydrogenated gaseous stream enters the cold section, where it is passed through a number of heat exchangers and separators. A portion of the liquid stream from unit 10 is used as the recycle stream. Hydrogen is separated as a gaseous stream in unit 12.

The liquid streams from separators 7–11 enter the demethanizer column (unit 13). The top product of this column is methane, which is sent to the fuel gas stream via the cold section and drying/precooling section.

The bottom product of the demethanizer column enters the C_2 splitter column (unit 15) as a feed. The top product of this column is cooled and compressed and subsequently stored as ethylene product. The bottom product of the C_2 splitter column is

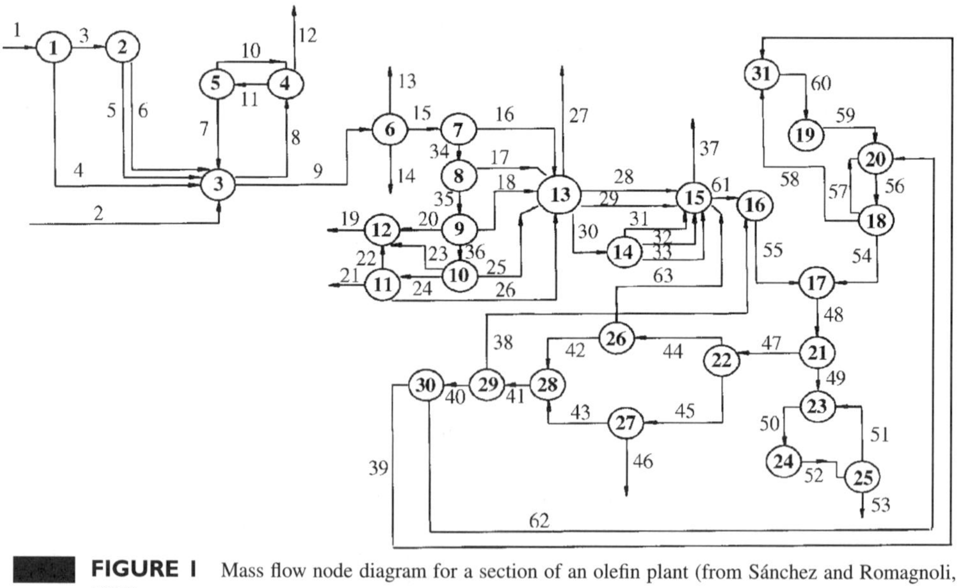

FIGURE 1 Mass flow node diagram for a section of an olefin plant (from Sánchez and Romagnoli, 1996).

ethane, which is sent back to the cracking furnace as feed stock through the precooling section.

12.2.2. Application of Decomposition and Data Reconciliation

Three different cases were considered in this example:

1. A small subsection of the whole system (consisting of units 1 to 6) to show stage-by-stage calculations
2. A larger section, consisting of 14 units and 35 process variables, that contains most of the redundancy in the system
3. The whole section of the plant (consisting of 31 units and 62 process variables, of which only 28 are measured).

The corresponding vectors for measured ($\mathbf{x}$) and unmeasured ($\mathbf{u}$) flowrates are

$$\mathbf{x} = [f_1 \; f_2 \; f_7 \; f_8 \; f_9 \; f_{12} \; f_{13} \; f_{14} \; f_{15} \; f_{16} \; f_{17} \; f_{18} \; f_{19} \; f_{20} \; f_{21} \; f_{22} \; f_{24} \; f_{25} \; f_{26} \; f_{27} \; f_{28} \; f_{29} \; f_{30} \; f_{31} \; f_{32} \; f_{36} \; f_{45} \; f_{52}]$$

$$\mathbf{u} = [f_3 \; f_4 \; f_5 \; f_6 \; f_{10} \; f_{11} \; f_{23} \; f_{33} \; f_{34} \; f_{35} \; f_{37} \; f_{38} \; f_{39} \; f_{40} \; f_{41} \; f_{42} \; f_{43} \; f_{44} \; f_{46} \; f_{47} \; f_{48} \; f_{49} \; f_{50} \; f_{51} \; f_{53} \; f_{54} \; f_{55} \; f_{56} \; f_{57} \; f_{58} \; f_{59} \; f_{60} \; f_{61} \; f_{62}],$$

and the values for measured variables and variances are included in Table 1.

TABLE 1 Data for Example 12.1 (from Sánchez and Romagnoli, 1996)

Stream	Variance	Measured value
1	10.870	70.490
2	0.2030	7.1030
7	0.3970	13.040
8	2.6240	35.380
9	5.7600	53.210
12	0.9220	23.900
13	0.0006	0.0765
14	5.7600	54.590
15	1.4400	12.780
16	0.7060	23.420
17	0.0170	0.2378
18	0.1300	8.6570
19	0.0900	5.0870
20	0.0140	1.7400
21	0.0002	0.0255
22	0.0180	3.1130
24	0.0900	5.4070
25	0.0140	2.8980
26	0.3600	15.830
27	0.5630	8.1970
28	0.0230	1.3640
29	1.1030	20.940
30	0.0080	1.0510
31	0.3970	12.580
32	0.1520	4.9990
36	0.0900	5.7300
45	0.1300	4.2500
52	1.2320	16.340

Case I

For the subsystem comprising units 1 to 6, we have eight measured and six unmeasured process variables. Matrices $\mathbf{A}_1$ and $\mathbf{A}_2$ are, for this case,

$$\mathbf{A}_1 = \begin{bmatrix} 1 & & & & & & & & \\ & & & & & & & & \\ & 1 & 1 & -1 & -1 & & & & \\ & & & 1 & & -1 & & & \\ & & -1 & & & & & & \\ & & & & 1 & & -1 & -1 & -1 \end{bmatrix},$$

$$\mathbf{A}_2 = \begin{bmatrix} -1 & -1 & & & & \\ 1 & & -1 & -1 & & \\ & 1 & 1 & 1 & & \\ & & & & 1 & -1 \\ & & & & -1 & 1 \\ & & & & & \end{bmatrix}.$$

Applying the orthogonal projection approach, the following $\mathbf{Q}_{u1}$, $\mathbf{Q}_{u2}$, $\mathbf{R}_{u1}$, and $\mathbf{R}_{u2}$ matrices are obtained:

$$\mathbf{Q}_{u1} = \begin{bmatrix} -0.7071 & 0 & 0.4082 \\ 0.7071 & 0 & 0.4082 \\ 0 & 0 & -0.8165 \\ 0 & -0.7071 & 0 \\ 0 & 0.7071 & 0 \\ 0 & 0 & 0 \end{bmatrix}, \quad \mathbf{Q}_{u2} = \begin{bmatrix} 0.5774 & 0 & 0 \\ 0.5774 & 0 & 0 \\ 0.5774 & 0 & 0 \\ 0 & 0.7071 & 0 \\ 0 & 0.7071 & 0 \\ 0 & 0 & 1.0 \end{bmatrix}$$

$$\mathbf{R}_{u1} = \begin{bmatrix} 1.4142 & 0 & -0.7071 \\ 0 & -1.4142 & 0 \\ 0 & 0 & -1.2247 \end{bmatrix}, \quad \mathbf{R}_{u2} = \begin{bmatrix} 0 & -0.7071 & 0.7071 \\ 1.4142 & 0 & 0 \\ 0 & -1.2247 & -1.2247 \end{bmatrix}.$$

Matrix $\mathbf{G}_x$ now becomes,

$$\mathbf{G}_x = \begin{bmatrix} 0.5774 & 0.5774 & 0.5774 & -0.5774 & -0.5774 & 0. & 0. & 0. \\ 0. & 0. & -0.7071 & 0.7071 & 0. & -0.7071 & 0. & 0. \\ 0. & 0. & 0. & 0. & 1. & 0. & -1. & -1. \end{bmatrix}.$$

Thus, we have identified the subset of redundant equations containing only measured (redundant) process variables. Applying the data reconciliation procedure to this reduced set of balances we obtain for the estimate of the measured variables

$$\hat{\mathbf{x}}^T = [70.4013 \quad 7.1013 \quad 12.8822 \quad 36.4232 \quad 53.9617 \quad 23.5410 \quad 0.0764 \quad 53.8853].$$

Furthermore, the rank of $\mathbf{R}_{u1}$ is equal to 3; therefore, only three unmeasured variables can be written in terms of the others and the reconciled measurements, following the procedure described in Chapter 4, Section 2:

$$\mathbf{u}_{r_u} = -\mathbf{R}_{u1}^{-1}\mathbf{Q}_{u1}^{T}\mathbf{A}_1\hat{\mathbf{x}} - \mathbf{R}_{u1}^{-1}\mathbf{R}_{u2}\mathbf{u}_{n-r_u},$$

where the components $\mathbf{u}_{n-r_u}$ are arbitrarily set. In this case, from the orthogonal transformation the subsets of **u** are defined as

$$\mathbf{u}_{r_u} = [f_3 \quad f_{10} \quad f_5], \quad \mathbf{u}_{n-r_u} = [f_4 \quad f_{11} \quad f_6].$$

Because the $\mathbf{R}_{IU}$ matrix does not contain zero rows,

$$\mathbf{R}_{IU} = \begin{bmatrix} 0. & 1. & 0. \\ -1. & 0. & 0. \\ 0. & 1. & 1. \end{bmatrix},$$

the vector $\mathbf{u}_{r_u}$ cannot be evaluated in terms of reconciled measurements.

Case 2

In this case a larger section of the plant was considered. It consists of 14 (1–14) units and 35 process streams, of which 25 are measured. The corresponding data is given in Table 1. Using a similar analysis to the previous case, the following characteristics were identified:

- There are 7 redundant equations containing all of the 25 measured variables.
- The rank of $\mathbf{A}_2$ is equal to 7; therefore, at least 3 unmeasured process variables are indeterminable.

From the orthogonal decomposition of the $\mathbf{A}_2$ matrix, the unmeasured process variables are divided in two subsets:

$$\mathbf{u}_{r_u} = [f_3 \quad f_{10} \quad f_{23} \quad f_{33} \quad f_4 \quad f_{34} \quad f_{35}], \quad \mathbf{u}_{n-r_u} = [f_{11} \quad f_6 \quad f_5].$$

The analysis of matrix $\mathbf{R}_{IU}$, in this case, shows that the rows associated with the unmeasured process streams f_{23}, f_{33}, f_{34}, and f_{35} are zero, and thus these unmeasured variables are determinable, since they do not depend on the assumed values of vector $\mathbf{u}_{n-r_u}$.

A data reconciliation procedure was applied to the subset of redundant equations; the results are displayed in Table 2. After the reconciliation, the estimation of the unmeasured variables was accomplished and the results are presented in Table 3.

Case 3

In this case the whole section of the plant was considered. It consists of 31 units and 62 process streams, of which 28 are measured. Using a similar analysis to the previous case the following characteristics were identified:

- There are 8 redundant equations containing all of the 28 measured variables.
- The rank of $\mathbf{A}_2$ is equal to 23; therefore, there are at least 11 unmeasured variables that are indeterminable.

The unmeasured process variables are divided into

$$\mathbf{u}_r = [f_3 \quad f_{10} \quad f_{23} \quad f_{33} \quad f_{37} \quad f_{38} \quad f_{41} \quad f_{44} \quad f_{47} \quad f_{49} \quad f_{55} \quad f_5 \quad f_{34} \quad f_{50} \quad f_{58} \quad f_{60} \quad f_{43} \quad f_{39} \quad f_{48} \quad f_{35} \quad f_{53} \quad f_{40} \quad f_{54}]$$

$$\mathbf{u}_{n-r} = [f_6 \quad f_4 \quad f_{61} \quad f_{56} \quad f_{11} \quad f_{57} \quad f_{59} \quad f_{51} \quad f_{42} \quad f_{46} \quad f_{62}].$$

TABLE 2 Results for Case 2 in Example 12.1

Streams	Measured values	Reconciled values
1	70.49	69.42
2	7.103	7.083
7	13.04	12.87
8	35.38	36.48
9	53.21	52.90
12	23.90	23.60
13	0.0765	0.076
14	54.59	52.82
15	12.78	11.44
16	23.42	22.76
17	0.238	0.222
18	8.657	8.448
19	5.087	5.273
20	1.740	1.746
21	0.025	0.0259
22	3.113	3.150
24	5.407	5.323
25	2.898	2.885
26	11.83	12.33
27	8.197	8.980
28	1.364	1.396
29	20.94	19.92
30	1.051	1.069
31	12.58	13.50
32	4.999	5.351

TABLE 3 Results for Case 2 in Example 12.1

Streams	Estimated values
f_{23}	4.657
f_{33}	41.386
f_{34}	18.625
f_{35}	13.130

Furthermore, from the analysis of the $\mathbf{R}_{IU}$ matrix, the unmeasured variables f_{23}, f_{33}, f_{34}, f_{35}, and f_{48} determinable.

A data reconciliation procedure was applied to the subset of redundant equations. The results are displayed in Table 4. A global test for gross error detection was also applied and the χ^2 value was found to be equal to 17.58, indicating the presence of a gross error in the data set. Using the serial elimination procedure described in Chapter 7, a gross error was identified in the measurement of stream 26. The procedure for estimating the amount of bias was then applied and the amount of bias was found

TABLE 4 Results for Case 3 in Example 12.1

Stream	Reconciled with bias	Reconciled without bias
1	68.639	70.898
2	7.068	7.110
7	12.867	12.886
8	36.522	36.395
9	52.052	54.500
12	23.655	23.509
13	0.0767	0.0763
14	51.975	54.424
15	10.881	12.485
16	22.489	23.276
17	0.215	0.234
18	8.470	8.408
19	5.290	5.238
20	1.752	1.737
21	0.026	0.0258
22	3.154	3.143
24	5.288	5.388
25	2.880	2.895
26	12.602	16.380
27	8.306	7.569
28	1.368	1.338
29	19.477	18.9920
30	1.0632	1.056
31	13.183	12.838
32	5.230	5.098
36	5.905	5.828
45	4.503	4.391
52	18.742	17.680

to be 4.56. By correcting the measured value by this amount, a new estimate of the measured process variables was calculated; this is also displayed in Table 4.

12.3. DATA RECONCILIATION OF A PYROLYSIS REACTOR

12.3.1. Process Description

The pyrolysis reactor is an important processing step in an olefin plant. It is used to crack heavier hydrocarbons such as naphtha and LPG to lower molecular weight hydrocarbons such as ethylene. The pyrolysis reactor, in this study, consists of two identical sides; each side contains four cracking coils in parallel (see Fig. 2).

The whole reactor is placed in a fuel-gas-fired furnace; the heat required for cracking the hydrocarbons is supplied from the hot flue gas by convection and radiation. Feed gases are divided and passed uniformly through each coil. Steam is also added to each coil in the required proportion as a cracking medium. Cracked gases from each pass (coil) in a side are mixed together and passed through a transfer line exchanger.

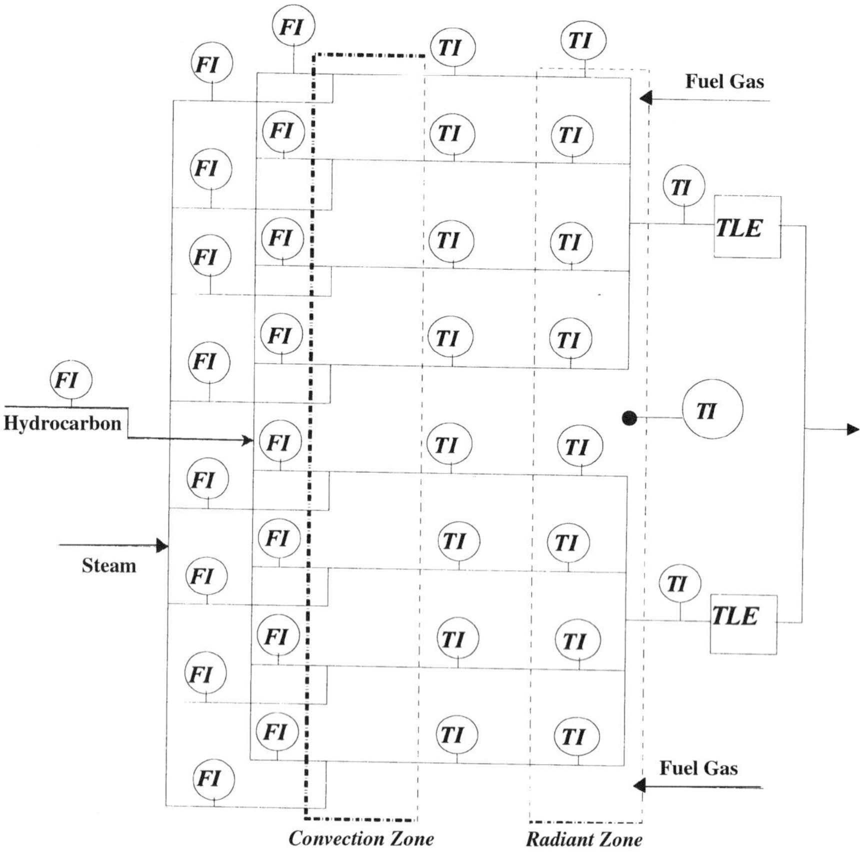

FIGURE 2 Simplified scheme for the industrial reactor (from Weiss *et al.*, 1996).

Then the cooled gases from the two sides are mixed together for further processing downstream. There are several pyrolysis reactors within the industrial complex.

Three major issues for efficient operation of the pyrolysis reactor are (1) maintaining equal cracking temperature in all the reactor coils, which is necessary for better product distribution; (2) controlling cracking severity to maximize yield; and (3) measuring the rate of coking of both the cracking coils and the transfer line exchangers; these rates are used as the key input into a reactor optimizer. The first and second issues can be dealt with on the basis of measured quantities, whereas the third issue can only be inferred from other measurements. To tackle these issues, several variables such as hydrocarbon/steam flow in each coil, total hydrocarbon flow, the inlet and outlet temperatures of each coil, and coil skin temperature are measured continuously.

An advanced control system has been implemented for efficient operation of the pyrolysis reactor. However, it faced problems due mainly to the difficulty of measuring the high coil and coil skin temperature reliably and consistently, because of regular drifting of the high-temperature sensors. Thus, there is a need for a data reconciliation package (DRP) to increase the level of confidence in key measured variables, to indicate the status of sensors, and to estimate the value of some unmeasured variables and parameters (Weiss *et al.*, 1996).

12.3.2. Model of the Reactor

Data reconciliation of the pyrolysis reactor was performed on the basis of the following mass and energy balance equations (Weiss *et al.*, 1996).

Hydrocarbon mass flow:

$$f_{ht} = \sum_{p=1}^{N} f_{hp}.$$

N = number of coils = 8.
Total energy:

$$UA\left\{T_{FB} - \frac{(T_0 + T_i)}{2}\right\} - F_{tm}(T_0 - T_i) = 0.$$

Energy pass balances:

$$UA_A\left\{T_{FB} - \frac{(COT_A + T_{iA})}{2}\right\} - (F_{tm})_A(COT_A - T_{iA}) = 0$$

$$UA_B\left\{T_{FB} - \frac{(COT_B + T_{iB})}{2}\right\} - (F_{tm})_B(COT_B - T_{iB}) = 0.$$

Energy coil balances:

$$UA_p\left\{T_{FB} - \frac{(T_{0p} + T_{ip})}{2}\right\} - (F_{tm})_p(T_{0p} - T_{ip}) = 0,$$

where $p = 1$ to 8.

In deriving the constraints, it was assumed that all passes are identical and that the two sides are identical. In addition, it was assumed that the heat transfer coefficients for the sides are related to the heat transfer coefficients for the passes. *Effects such as the heat of reaction were not considered separately, but were lumped into an "effective" specific heat for the hydrocarbons.*

There are 12 equations in all (overall material and energy balances; side A and B energy balances; coil 1 to 8 energy balances) and 36 variables. However, the heat transfer coefficients are not known with any great accuracy. Further, both the side and coil heat transfer coefficients depend on the fire-box temperature. It is therefore necessary to calculate values for the heat transfer coefficients from the data. This effectively reduces the number of independent equations to 11.

12.3.3. Application of Data Reconciliation

The various solution methods were applied to a set of measurements taken at different reactor conditions. Table 5 gives the data used for the reconciliation. The value corresponding to the specific heat is an "effective" specific heat, calculated from rigorous simulations using the Phenics package, and takes into account the heat of reaction.

Table 6 shows a set of process measurements along with the reconciled values. Some interesting results appeared. It is evident that there are large discrepancies between the measured values and the reconciled values, assuming no gross errors are present.

TABLE 5 Data for Reconciliation of the Pyrolysis Reactor

Specific heat of hydrocarbon ($C_{p,\mathrm{HC}}$)	0.93 kcal/kg°C
Specific heat of steam ($C_{p,\mathrm{ST}}$)	0.1 kcal/kg°C
Heat transfer areas	
per coil	12.2 m^2
per side	62.4 m^2
Standard deviations for measurements:	
Coil temperatures (in and out)	3°C
Coil flowrates (HC and ST)	20 kg/h
Total flowrate	100 kg/h
Fire-box temperature	15°C

TABLE 6 Reconciliation Results (from Islam *et al.*, 1994)

Measurement		Measured value	Reconciled value	Coil 2 corrected
Total hydrocarbon flow		16700.2	16607.0	16607.0
Coil hydrocarbon flow:	1	1922.9	1963.2	1951.0
	2	2177.2	2118.5	2191.9
	3	2004.7	2047.4	2035.8
	4	2136.2	2146.0	2134.5
	5	2131.7	2140.9	2131.8
	6	2106.9	2125.3	2116.2
	7	2018.4	2030.6	2020.6
	8	2004.1	2035.3	2025.6
Coil cracking temperatures:	1	845.6	848.2	847.6
	2	845.6	835.8	826.1
	3	845.8	849.0	848.3
	4	845.6	844.5	843.9
	5	841.7	841.2	840.2
	6	841.7	842.4	841.5
	7	841.6	841.5	840.6
	8	841.4	843.6	842.7

The coil heat transfer coefficient had a value of 0.1159 and a χ^2 statistic of 27.8. This value suggests that a gross error is present. Some clues can be found from examination of the residuals of the balances. They are presented in Table 7. There appears to be a problem with the balances for coils 2 and 3; their residuals are different from the others (especially coil 2). The reconciled value for the coil 2 cracking temperature is also significantly different from the measured value, thus suggesting an abnormal situation associated with both coils 2 and 3.

The serial elimination strategy was implemented to identify the source of the bias, and the results are displayed in Figs. 3 and 4. As can be seen, a significant improvement in the χ^2 value is obtained when one of the measurements associated with coil number 2 is deleted from the reconciliation procedure. Applying the gross error estimation, a value of $m_{\mathrm{b}} = -18.9$°C was obtained when the gross error was assumed to be in the cracking temperature of coil 2. On the other hand, correction factors of $m_{\mathrm{b}}(1) = -200$ kg/h and $m_{\mathrm{b}}(2) = +90$ kg/h were obtained when simultaneous gross errors

TABLE 7 Residuals of the Balances (from Islam et *al.*, 1994)

Balance	Residuals (scaled)
Hydrocarbon flow	0.1981
Side B enthalpy	0.021856
Side A enthalpy	0.009385
Coil enthalpy balances	0.016069
	−0.04789
	0.018285
	−0.00315
	−0.00169
	0.004556
	0.000618
	0.012461

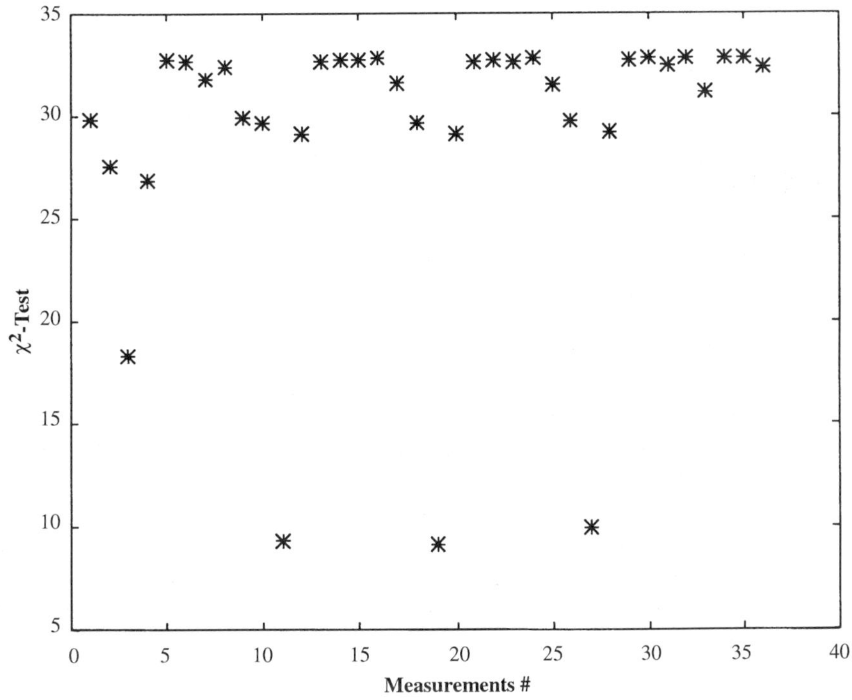

FIGURE 3 Serial elimination of the measurements.

were assumed in the flow rates of coils 2 and 3, respectively. Indeed, if the measured value for the coil 2 cracking temperature is reduced by 18.9°C and the new data set is reconciled, a χ^2 value of 6.9 results. The reconciled values for this case also presented in Table 6.

These results would be satisfactory, but for the arbitrary nature of the adjustment to the coil 2 cracking temperature. The gross error could also be eliminated by adjusting either the flow rates through the coil 2 and 3 or the inlet temperature to the radiant zone (crossover temperature) for coil 2. As it happens, the best reduction in

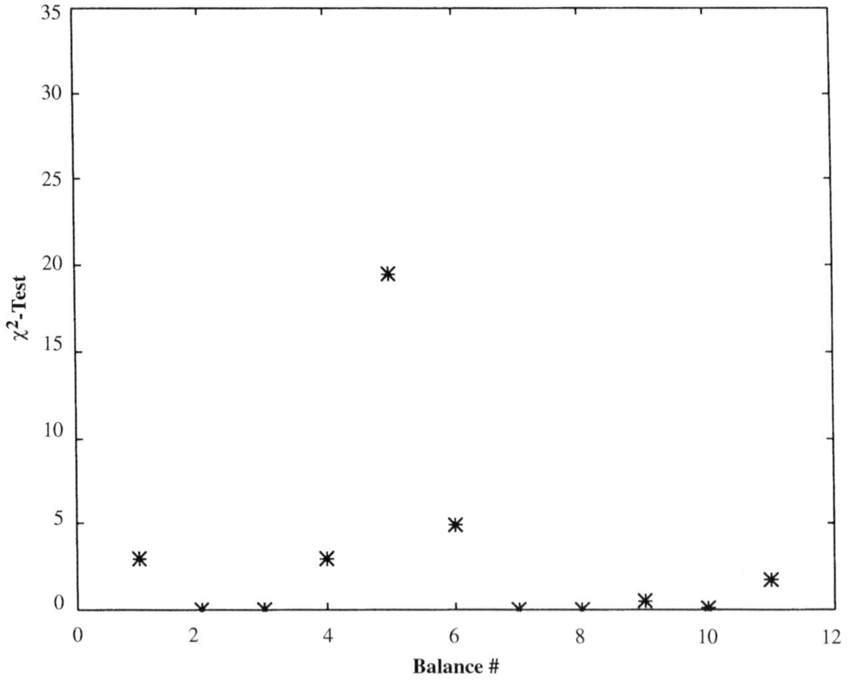

FIGURE 4 Serial elimination of the balances.

the χ^2 value comes from altering the coil cracking temperature; however, there is no real justification for doing this.

Analysis of Daily Averages

As described before, there is only one balance per coil and a residual in that equation may be due to an error in one of several variables. A further set of equations describing the coils is required, and they may be found in the long-term trends for that coil.

An indication of errors in the measurement of the coil flow rate can be found in the relative value of the crossover temperature for that coil. The absolute value of the crossover temperature is of less value because it depends on the firing rate for the furnace, as well as the coil flow rate. With this in mind, an analysis of the daily averages of the furnace data was undertaken.

Relationships of the form

$$COT_i - COT_a = a_i + b[(F_{tm})_i - (F_{tm})_a]$$

were sought, where F_i is the thermal mass flow in the convection section of the furnace, COT_i is the crossover temperature for the coil, COT_a and F_a are average values, a_i is an offset for coil i, and b is the regression coefficient for the relationship and is common to all coils.

Figures 5 and 6 present the results. Two features are apparent. First, a common pattern can be seen: The relationships for each coil do appear to have similar slopes. Second, the two sides behave in a different way. Side A is well behaved in that the average flow for each coil does not differ greatly from the average flow for all coils, and the

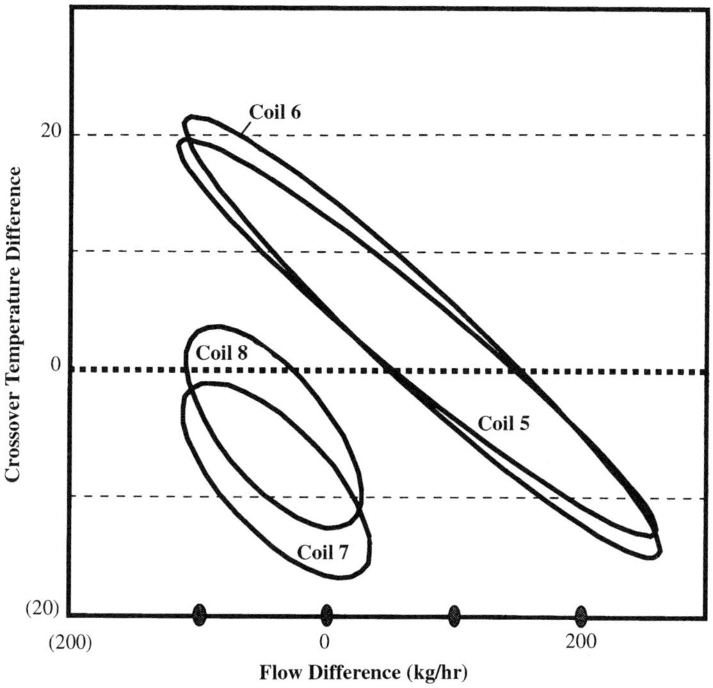

FIGURE 5 Crossover temperature data clusters—Side A (from Islam *et al.*, 1994).

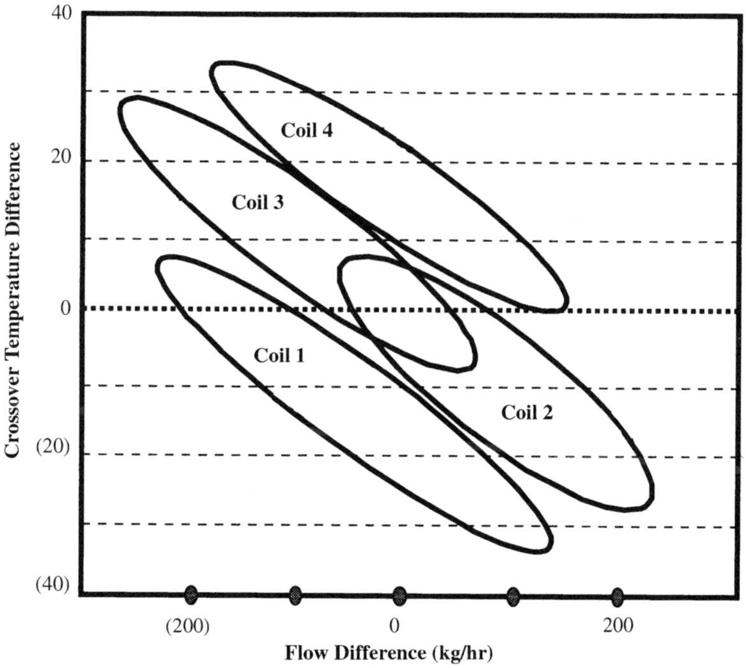

FIGURE 6 Crossover temperature data clusters—Side B (no flow adjustments) (from Islam *et al.*, 1994).

TABLE 8 Coil Energy Balances (from Islam *et al.*, 1994)

	Thermal mass flowrate	Coil heat transfer coefficient	Deviation from side average	Deviation student's *t* value
Coil 1	1823	0.111	−0.0002	−0.03
Coil 2	1982	0.122	0.010	2.7
Coil 3	1796	0.103	−0.0092	−2.5
Coil 4	1920	0.111	−0.0008	−0.2
Side B	1880	0.112		
Coil 5	1936	0.114	0.0017	0.6
Coil 6	1934	0.114	0.0016	0.7
Coil 7	1837	0.111	−0.0011	−0.4
Coil 8	1838	0.111	−0.0022	−0.9
Side A	1886	0.112		

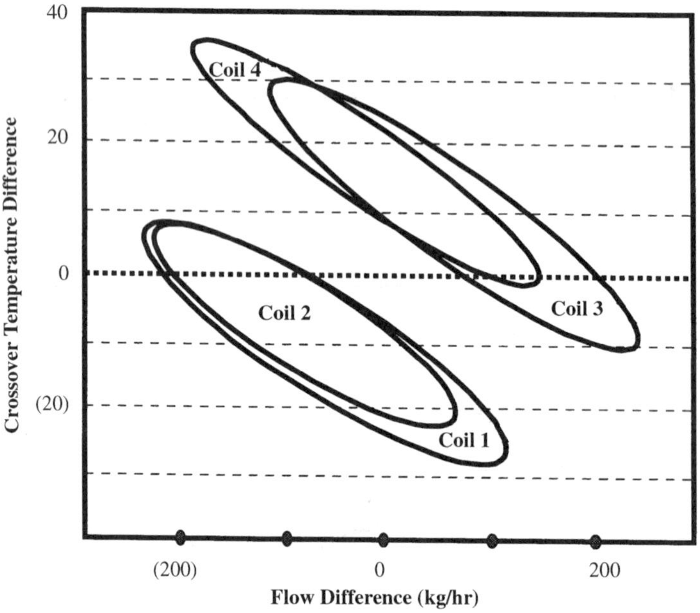

FIGURE 7 Crossover temperature data cluster—Side B (after flow adjustment) (from Islam *et al.*, 1994).

average coil crossover temperature differs from the average of all crossover temperatures in a manner that is consistent with the geometry of the furnace. This is not the case for side B. The average flows through coils 2 and 3 differ from the average of all flows by a larger amount. An indication of the likely cause of this problem can be found after an examination of the coil heat balances for the daily averages. Table 8 gives some results. There are no statistically significant deviations between the coil heat transfer coefficients for side A, but yet again, coils 2 and 3 give anomalous results for side B.

Interesting results are obtained if 160 kg/h is deducted from the measured hydrocarbon flow for coil 2, and 210 kg/h added to the hydrocarbon flow for coil 3. The anomaly between the coil heat transfer coefficients for coils 2 and 3 has disappeared. Figure 7 shows the effect that the adjustment to the coil flow has on the crossover

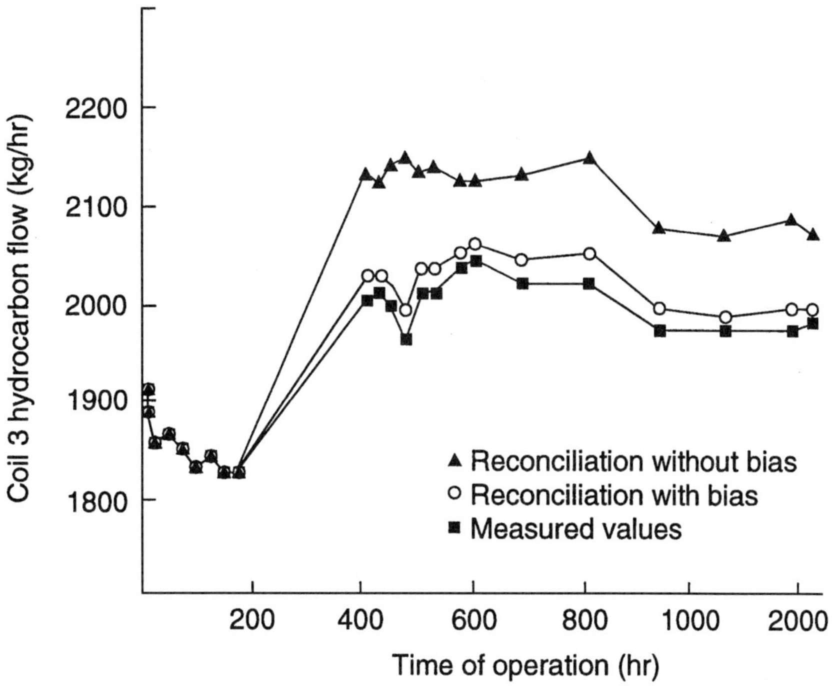

FIGURE 8 Time evolution of measured and reconciled values.

temperature correlations. Now the pattern seen for pass B is similar to that seen for pass A. This adds to the validity of the adjustment to the hydrocarbon flows, and suggests that these flow sensors show a consistent bias.

Analysis over the Whole Operating Cycle

For comparison purposes, plant data and reconciled data, both with and without bias treatment, over the whole operating cycle of the pyrolysis reactor for some variables are plotted in Fig. 8. This figure clearly shows that the measurements obtained from coil 2 are biased. There were not many differences between measured and reconciled values for the other variables.

The overall heat transfer coefficient calculated using the joint parameter estimation and data reconciliation approach is shown in Fig. 9. It is evident from this figure that the overall heat transfer coefficient remains fairly constant throughout the whole operating cycle of the pyrolysis reactor. Near the end of the cycle, the heat transfer coefficient drops to a comparably low value, signifying that the reactor needs to be regenerated.

12.4. DATA RECONCILIATION OF AN EXPERIMENTAL DISTILLATION COLUMN

The final case study consists of a well-instrumented experimental distillation column that has been interfaced to an industrial distributed control system (Nooraii

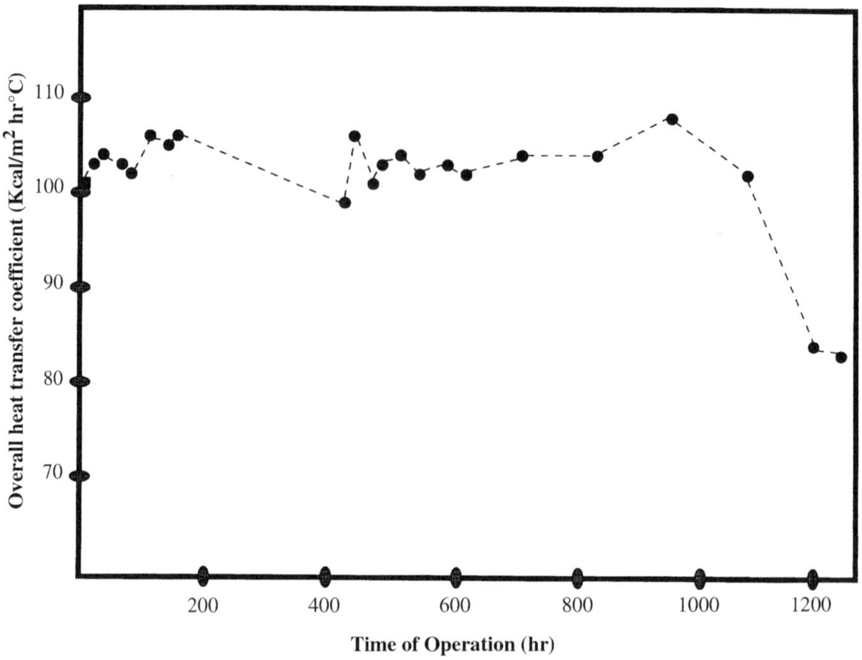

FIGURE 9 Time evolution of the estimated heat transfer coefficient (from Weiss *et al.*, 1996).

et al., 1994). This example illustrates the use of the techniques described in previous chapters in an actual on-line framework, using industrial hardware. Furthermore, the usefulness of data reconciliation prior to process modeling and optimization is clearly demonstrated.

12.4.1. Experimental Setup Description

The experimental facility is a pilot-scale distillation column connected to an industrial ABB MOD 300 distributed control system, which in turn is connected to a VAX cluster. The control system consists of a turbo node (configuration, history, console) remote I/O, and an Ethernet gateway, which allows communication with the VAX-station cluster through the network. This connection allows time-consuming and complex calculations to be performed in the VAX environment. Figure 10 shows the complete setup.

The 23-cm-diameter distillation column under study is used to separate ethanol and water. It contains 12 sieve trays with a 30-cm spacing (Fig. 11) as well as three possible feed locations, an external reboiler, and two condensers, which are used at the bottom and the top of the column. The second condenser is also used as a reflux drum; a pump sends the reflux back to the column (tray 1) and the product to the product tank.

The following variables are monitored through the DCS:

- Temperatures at trays 12, 11, 9, 7, 5, 3, 1
- Temperatures of feed, distillate, bottoms, and water in and out of the condenser

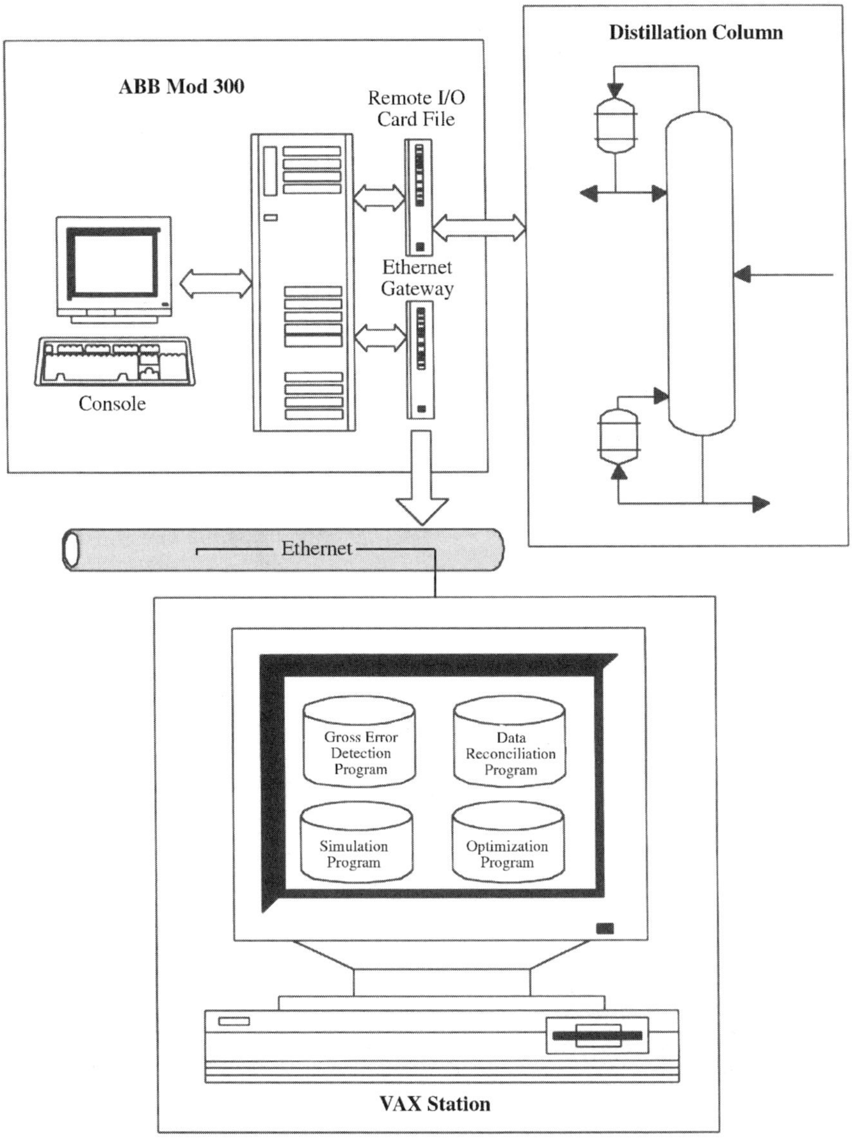

FIGURE 10 The setup of the column, DCS, and interface with VAX-station and supervisory control.

- Flowrates of steam to the reboiler, water to the condenser, feed, distillate, bottoms, and reflux
- Pressure at the bottom of the column
- Liquid levels in the condenser and the bottom of the column

To predict the composition, a density meter is installed beside the column. Samples from the feed, distillate, bottoms, and trays 2, 4, 6, 8, 10, and 12 are sent to this density meter through a sampling pump. The samples can be selected from different points by the use of several valves available on the sampling lines. In order to make this sampling continuous, a Taylor Control Language program has been written in

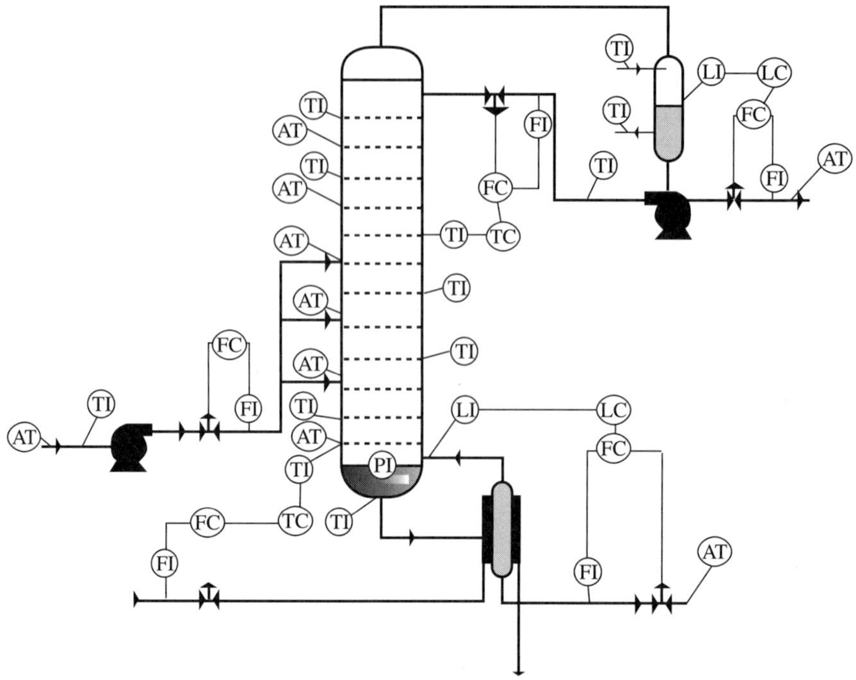

FIGURE 11 The pilot scale distillation column under study (from Nooraii, 1996).

the DCS. This program opens the valve for each sampling point for 2 min while the other valves are closed. The results from the density meter are sent back to the DCS, converted to weight percent composition, and presented on the screen. Consequently, when only samplings from the feed, distillate, and bottoms are performed, it will take around 6 min for each concentration to be updated.

12.4.2. Model for Data Reconciliation

For the case of the distillation column under study, the following constraints were used as equality constraints in the reconciliation procedure (Nooraii, 1996):

1. Total mass balance:

$$F = D + B.$$

2. Component mass balance:

$$Fx_f = Dx_d + Bx_b.$$

3. Total energy balance:

$$FH_f + Q_b = DH_d + BH_b + Q_c,$$

where

$$Q_b = S\lambda_s$$
$$Q_c = wC_p(T_{\text{out}} - T_{\text{in}}).$$

4. Energy balance around the column and reboiler:

$$FH_f + Q_b + RH_r = VH^V_{\text{tray}} + BH_b$$
$$V = D + R.$$

5. Bubble point calculation at the bottom of the column:

$$T_b = f(x_b, P).$$

The following equations were used to calculate the bubble point in our reconciliation procedure:

$$P = P^*_{\text{water}} \upsilon_{\text{water}}(1 - x_b) + P^*_{\text{EtOH}} \upsilon_{\text{EtOH}} x_b$$

and

$$P^*_{\text{water}} = \exp\left(a^{\text{o}}_w + \frac{b^{\text{o}}_w}{T} + c^{\text{o}}_w \ln(T) + d^{\text{o}}_w T^2\right).$$

A linearized version of this model was also obtained, represented by

$$\mathbf{Ay} + \mathbf{e} = \mathbf{0}.$$

Matrices **A** and **e** are

$$\mathbf{A} = \begin{bmatrix} 0 & 0 & 0 & 0 & 0 & 0 & -1 & 1 & 1 & 0 & 0 & 0 & 0 & 0 \\ 0 & 0 & 0 & \tilde{F} & \tilde{D} & \tilde{B} & \tilde{x}_f & \tilde{x}_d & \tilde{x}_b & 0 & 0 & 0 & 0 & 0 \\ -E1 & E2 & E3 & -E4 & E5 & E6 & -\tilde{H}_f & \tilde{H}_d & \tilde{H}_b & 1 & 1 & 0 & 0 & 0 \\ E1 & E11 & -E3 & E4 & E7 & -E6 & \tilde{H}_f & -\tilde{H}_d & -\tilde{H}_b & 0 & 1 & -E8 & 0 & \tilde{H}_d - \tilde{H}_{\text{tray}} \\ 0 & -E9 & 0 & 0 & -E10 & 0 & 0 & 0 & 0 & 0 & 0 & 1 & 0 & 0 \end{bmatrix}$$

$$\mathbf{e} = \begin{bmatrix} 0 \\ \tilde{F}\tilde{x}_f - \tilde{D}x_d - \tilde{B}\tilde{x}_b \\ E1\tilde{T}_f - E2\tilde{T}_d - E3\tilde{T}_b + E4\tilde{x}_f - E5\tilde{x}_d - E6\tilde{x}_b \\ -E1\tilde{T}_f - E11\tilde{T}_d + E3\tilde{T}_b - E4\tilde{x}_f - E7\tilde{x}_d + E6\tilde{x}_b + E8\tilde{T}_{\text{tray}} \\ E9\tilde{T}_b + E10\tilde{x}_b + \tilde{P}^*_{\text{water}}\tilde{\upsilon}_{\text{water}}(1 - \tilde{x}_b) + \tilde{P}^*_{\text{EtOH}}\tilde{\upsilon}_{\text{EtOH}}\tilde{x}_b \end{bmatrix},$$

where $\sim$ denotes the values of the measurements at the linearization point and $E_1, \ldots, E_9$ are given in Nooraii (1996). The vector, **y**, of measurements is

$$\mathbf{y}^{\text{T}} = [T_f \quad T_d \quad T_b \quad x_f \quad x_d \quad x_b \quad F \quad D \quad B \quad Q_c \quad Q_b \quad T_{\text{tray}} \quad P_b \quad R].$$

To carry out data reconciliation, as discussed in Chapter 10, the variance–covariance matrix of the measurements is essential. In reconciling process data, it is important that the estimated variance incorporate the process variability as well as the measurement errors. Otherwise, the adjustment of an accurate measurement in a highly variable process could be seem to be excessively large and wrongly be taken as a gross error. The approach would be to characterize the nonrandom variation by any of a number of techniques, such as time series analysis or principal component. Use of the variance–covariance matrix calculated from the data themselves is a practical and easy, if not always fully justified, alternative (Crowe, 1996). Thus, the most appropriate weighting matrix, **W**, was determined here by trial and error, using available information on the accuracy of the various measurement devices used. Obviously, this matrix may change during the operating life of the column, especially if some of the measurement devices change, or if there is any kind of maintenance on the measurement or related devices. The matrix used in this work is

$$\mathbf{W} = \begin{bmatrix} 0.08 \\ 0.077 \\ 0.06 \\ 8 \\ 238 \\ 300 \\ 0.05 \\ 0.07 \\ 0.6 \\ 2 \times 10^{-5} \\ 1.5 \times 10^{-5} \\ 0.07 \\ 9 \\ 0.5 \end{bmatrix}.$$

The optimization package GAMS was used to solve the data reconciliation problem, using the MINOS subroutine that solves the optimization problem using a reduced-gradient algorithm (Wolfe, 1962), combined with the quasi-Newton algorithm (Davison, 1959) that generally leads to superlinear convergence. The linearized version of the model was used to implement the sequential processing of the information for gross error detection and identification.

12.4.3. Application of the Data Reconciliation and Gross Error Detection Procedure within a Supervisory Control Scheme for the Column

In supervisory control, process and economic models of the plant are used to optimize the plant operation by maximizing daily profit, yields, or production rates. The computer program reviews operating conditions periodically, computes the new conditions that optimize a chosen objective function, and adjusts plant controller set points, thus implementing the new improved conditions. This scheme will obviously require a model of the plant, current information about operating conditions from the plant's control system, and finally, sophisticated optimization software.

The accurate and efficient calculation of optimal operating conditions is critically dependent on the following basic components:

- The availability and accuracy of plant model(s)
- The availability and accuracy of measured plant data
- Robust, flexible, and fast optimization software
- The effective integration of all these

In the case of the distillation column under study, several options are available within the environment of the VAX cluster, for both simulation and optimization studies. PROCESS and SPEEDUP simulation programs were used throughout these studies.

The critical part of such an optimization scheme is the analysis and reconciliation of the measurements, to ensure accurate and consistent data and the detection of instrument errors and faults. The overall scheme for the on-line monitoring and optimization of the column is also shown in Fig. 10. Data from the plant are first

TABLE 9 Comparison among Raw, Reconciled, and Bias Estimation Plus Reconciliation Data and the Results from Simulation (from Nooraii, 1996)

	Plant raw data	PROCESS with raw data	Reconciled data without bias deletion	Linear reconciliation with bias deletion	Nonlinear reconciliation with bias deletion	PROCESS with nonlinear reconciliation
Feed rate (kg/h)[a]	67.94	67.94	69.7	68.226	68.225	68.228
Feed temperature[a]	27.4	27.4	18.7	27.4	27.4	27.4
%(w) EtOH in feed[a]	36	36	33.3	34.4	34.4	34.4
Distillate rate (kg/h)[a]	27.24	27.24	27.18	27.134	27.133	27.133
Distillate temperature[a]	33.6	33.6	36	33.6	33.6	33.6
%(w) EtOH in distillate	85.47	86.2	84.97	85.49	85.5	85.45
Reflux rate (kg/h)[a]	43.58	43.58	43.59	43.58	43.58	43.58
Bottom rate (kg/h)	41.11	41.11	42.53	41.09	41.09	41.09
Bottom temperature	102	99.6	109.7	102	102	102
%(w) EtOH in bottom	0.67	2.3	0.34	0.67	0.7	0.7123
Bottom pressure[a]	1.09	1.09	1.42	1.1168	1.109	1.109
Condenser duty (mm kJ/h)	127	85.2	127	85	85	85.9
Reboiler duty (mm kJ/h)	144.5	97.6	144	98.3	98.3	99

[a] Inputs to PROCESS.

subjected to analysis for faults and gross measurement errors. If the former are detected, the operator is alerted and can either proceed with the reconciliation, prior to bias identification and estimation, or stop the calculations and repair the instruments. After a consistent set of data is obtained, the operator can continue with the simulation and optimization studies.

Table 9 presents a summary of the variables in a typical real-time run. The raw measurements are initially used to run the simulation with PROCESS (therefore, only the simulation switch is activated). The first column of the table shows the raw measurements, and the second indicates the results from PROCESS. It is clear that the results from the simulation are not in agreement with the measurements.[1] It can be seen from Table 9 that the measurements of the condenser and reboiler duties are quite different from the simulation results. This suggests that there are gross errors in those measurements. The gross error detection and data reconciliation modules are then activated. The third, fourth, and fifth columns show the rectified and reconciled data.

By direct application of the statistical criterion introduced in Chapter 7 to the overall system,

$$\tau = \mathbf{r}^{\mathrm{T}}\mathbf{\Phi}^{-1}\mathbf{r} = 12.03.$$

If an error probability of 0.1 is considered, then $12.03 > 9.24$ (from statistical tables) and one may say that inconsistency is important at this error probability level. After sequential processing of the measurements, as is shown in Table 10, feed temperature and reboiler and condenser duties are suspected to contain gross errors. Since feed temperature and reboiler duty do not appear in separate equations, it is difficult to isolate the gross error when it happens in one or both of duties. In this case, we need

[1] Note that only some measurements are used as inputs for simulation.

TABLE 10 Least Squares Objective with Each Deletion of Measurements (from Nooraii, 1996)

Deleted measurements	Least squares objective	Deleted measurements	Least squares objective
T_f	0.26	X_{d}	12.
T_d	0.41	X_{b}	4.2
T_b	0.7	Q_{c}	0.3
F	12.	Q_{r}	0.26
D	11.6	P_{B}	11.7
B	4.9	T_{tray}	12.
X_f	11.8	R	11.8

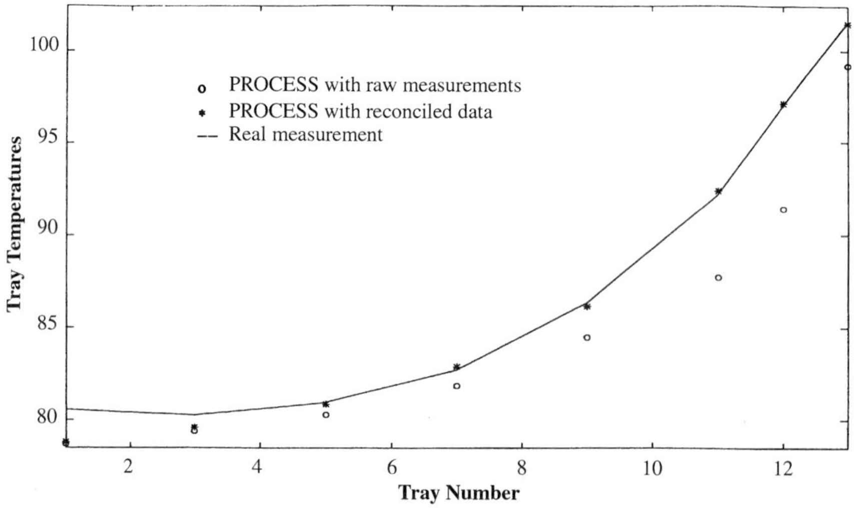

FIGURE 12 Comparison of tray temperatures from simulation (reconciled and raw data) with real-time temperatures (from Nooraii, 1996).

either additional information (additional relationships) or input from the operator's experience to decide which measurements to treat for bias. In this case, reboiler and condenser duties are treated as biased measurements and the corresponding biases are estimated.

After the estimation of the bias, we have two options for data reconciliation: Either we can use the results from the linear reconciliation problem, or we can implement a nonlinear reconciliation procedure after subtracting the estimated bias from the measurements. Although both options are carried out in this case, the nonlinear reconciliation problem is quite fast and is therefore used in the on-line program. Table 9 compares the results using the two options. It can also be seen in Table 9 that, by using the refined data as input to the simulation module, a better agreement between measurement and simulation has been achieved. The prediction of the tray temperature distribution along the column, obtained from the simulation using the reconciled data, is presented in Fig. 12. These are in better agreement with the measurements coming from the column than those obtained from the simulation using

TABLE 11 On-line Optimization Results after Reconciliation (from Nooraii, 1996)

	Plant raw data	PROCESS with nonlinear reconciliation	Optimum point
Feed rate (kg/hr)[a]	67.94	68.228	68.228
Feed temperature[a]	27.4	27.4	27.4
%(w) EtOH in feed[a]	36	34.4	34.4
Distillate rate (kg/hr)[a]	27.24	27.133	23.24
Distillate temperature[a]	33.6	33.6	33.6
%(w) EtOH in distillate	85.47	85.45	86.83
Reflux rate (kg/hr)[a]	43.58	43.58	38.23
Bottom rate (kg/hr)	41.11	41.09	44.988
%(w) EtOH in bottom	0.67	0.7123	7.3
Bottom pressure[a]	1.09	1.109	1.109
Condenser duty (mm kJ/h)	127	85.9	73.8
Reboiler duty (mm kJ/h)	144.5	99	86.6
Objective function	—	—	61.057

[a] Inputs to PROCESS.

the raw data. As not all of these tray temperatures have been used in the data reconciliation procedure, this agreement confirms the claim that data reconciliation allows us to obtain a better understanding of the actual state of the process. Finally, Table 11 gives the on-line optimization results from the previous run, using the reconciled data as current inputs to the simulation program. These results can be used by the operator as a valuable guide toward ideal operation.

12.5. CONCLUSIONS

The case studies discussed in this chapter show clearly the importance of the techniques developed throughout this book, when used in a practical environment. In particular, they provide additional insight into different ways of implementing these ideas.

The first case study, using simple material balances within a section of an olefin plant, fully exploited the theoretical results discussed for linear systems, namely, variable classification, system decomposition, data reconciliation, and gross error detection and identification.

In the second example, that of an industrial pyrolysis reactor, simplified material and energy balances were used to analyze the performance of the process. In this example, linear and nonlinear reconciliation techniques were used. A strategy for joint parameter estimation and data reconciliation was implemented for the evaluation of the overall heat transfer coefficient. The usefulness of sequential processing of the information for identifying inconsistencies in the operation of the furnace was further demonstrated.

Finally, a well-instrumented experimental distillation column that has been interfaced to an industrial distributed control system was used to show the implementation of the techniques described in previous chapters in an actual on-line framework, using industrial hardware. In this case, the usefulness of data reconciliation, prior to process modeling and optimization, was clearly demonstrated.

NOTATION

a	regression coefficient
A	area
$\mathbf{A}_1$	matrix for measured variables $(m \times g)$
$\mathbf{A}_2$	matrix for unmeasured variables $(m \times n)$
b	regression coefficient
B	bottom product
COT	cross over temperature
C_p	specific heat
D	distillate
$\mathbf{f}$	vector of mass flowrate
f_{ht}	total hydrocarbon flowrate
f_{hp}	coil flowrate
F_{tm}	thermal mass flow
F	feed of the distillation column
$\mathbf{G}_\mathrm{x}$	reduced matrix for measurement reconciliation $[(m - r_\mathrm{u}) \times g]$
H	specific enthalpy
m_b	bias magnitude
N	number of coils
P	pressure
Q_c	heat transferred in the condenser
Q_b	heat transferred in the boiler
$[\mathbf{Q}_\mathrm{u}, \mathbf{R}_\mathrm{u}, \mathbf{\Pi}_\mathrm{u}]$	$\mathrm{QR}(\mathbf{A}_2)$
r_u	$\mathrm{rank}(\mathbf{A}_2)$
$\mathbf{R}_\mathrm{IU}$	matrix defined by Eq. (4.20) $[r_\mathrm{u} \times (n - r_\mathrm{u})]$
$\mathbf{r}$	residuum
R	reflux
S	steam flowrate
T	temperature
T_out	outlet temperature of cooling fluid
T_in	inlet temperature of cooling fluid
U	overall heat transfer coefficient
$\mathbf{u}$	vector of unmeasured variables $(n \times 1)$
$\mathbf{u}_{r_\mathrm{u}}, \mathbf{u}_{n-r_\mathrm{u}}$	partitions of $\mathbf{u}$
V	vapor
w	flowrate of cooling fluid
$\mathbf{W}$	weighting matrix
$\mathbf{x}$	vector of measured variables $(g \times 1)$
x	molar fraction
$\mathbf{y}$	vector of measurements

Greek

λ	heat of vaporization
$\mathbf{\Phi}$	covariance matrix of the residuum
τ	global test statistic

Superscripts

$\sim$	with measured values
$\frown$	with reconciled values
$\mathbf{v}$	vapor
$*$	pure component

Subscripts

a	average value
A	side A
B	side B
b	bottom product
d	distillate product
f	feed of the distillation column
FB	fired box
HC	hydrocarbon
i	inlet
o	outlet
p	coil index
r	reflux
ST	steam
tray	first tray
w	water

REFERENCES

Crowe, C. M. (1996). Data reconciliation—Progress and challenges. *J. Proc. Control* **6**, 89–98.

Davison, W. C. (1959). "Variable Metric Methods for Minimization," A.E.C. Res. Dev. Rep. ANL-5990. Argonne Natl. Lab., Argonne, IL.

Islam, K., Weiss, G., and Romagnoli, J. A. (1994). Nonlinear data reconciliation for an industrial pyrolysis reactor. *Comput. Chem. Eng.* **18**, S 217–221.

Nooraii, A. (1996). Implementation of advanced operational techniques for an experimental distillation column within a DCS environment. Ph.D. Thesis, University of Sydney, Department of Chemical Engineering, Sydney, Australia.

Nooraii, A., Barton, G., and Romagnoli, J. A. (1994). On-line data reconciliation, simulation and optimising control of a pilot-scale distillation column. *Process Syst. Eng. PSE'94* **2**, 1265–1268.

Sánchez, M., and Romagnoli, J. A. (1996). Use of orthogonal transformations in data classification—Reconciliation. *Comput. Chem. Eng.* **20**, 483–493.

Weiss, G., Islam, K., and Romagnoli, J. A. (1996). Data reconciliation—An industrial case study. *Comput. Chem. Eng.* **20**, 1441–1449.

Wolfe, P. (1962). "The Reduced–Gradient Method." RAND Corporation (unpublished).

APPENDIX
STATISTICAL CONCEPTS

Here, several general statistical concepts are briefly discussed as a complement to the material covered in this book. The books of Davis and Goldsmith (1972) and Mikhail (1976) are excellent sources for such information. Most of the concepts and definitions presented in this Appendix were extracted and summarized from these references, and for more detailed information, the reader is referred to these publications.

I. FREQUENCY DISTRIBUTIONS

The collection of data often results in a somewhat randomly organized list of observations distributed in some way around a central value.

In analyzing such a set of data, it is helpful to arrange the data in a manner that aids the determination of the important features of the distribution, such as the central tendency and the spread. One of the simplest ways to do this is to produce a *frequency table*. This is achieved by dividing the range of observations into a number of subranges, tallying the number of observations which fall into each subrange, and tabulating the results. The distribution of the results is called a *frequency distribution*.

Consider as an illustration measurements of the carbon content of a mixed powder fed to a plant over a period of a month (Davis and Goldsmith, 1972). We have a total of 178 measurements, covering a range of 4.1% to 5.2%. This range was divided into groups such as 4.10–4.19 and so on. The results are displayed in Table 1.

The features of the distribution can be seen clearly if the results are plotted as shown in Fig. 1. This plot of the distribution is called a histogram.

TABLE 1 Percentage of Carbon in a Mixed Powder (from Davis and Goldsmith, 1972)

Ranges of values of % carbon	Tally	Number of results	Proportion of results
4.10–4.19	*L*	1	0.006
4.20–4.29	*Ll*	2	0.011
4.30–4.39	*lllll ll*	7	0.039
4.40–4.49	*lllll lllll lllll lllll*	20	0.112
4.50–4.59	*lllll lllll lllll lllll llll*	24	0.135
4.60–4.69	*lllll lllll lllll lllll lllll lllll l*	31	0.174
4.70–4.79	*lllll lllll lllll lllll lllll lllll lllll lll*	38	0.214
4.80–4.89	*lllll lllll lllll lllll llll*	24	0.135
4.90–4.99	*lllll lllll lllll lllll l*	21	0.118
5.00–5.09	*lllll ll*	7	0.29
5.10–5.19	*Lll*	3	0.017
Total		178	1.00

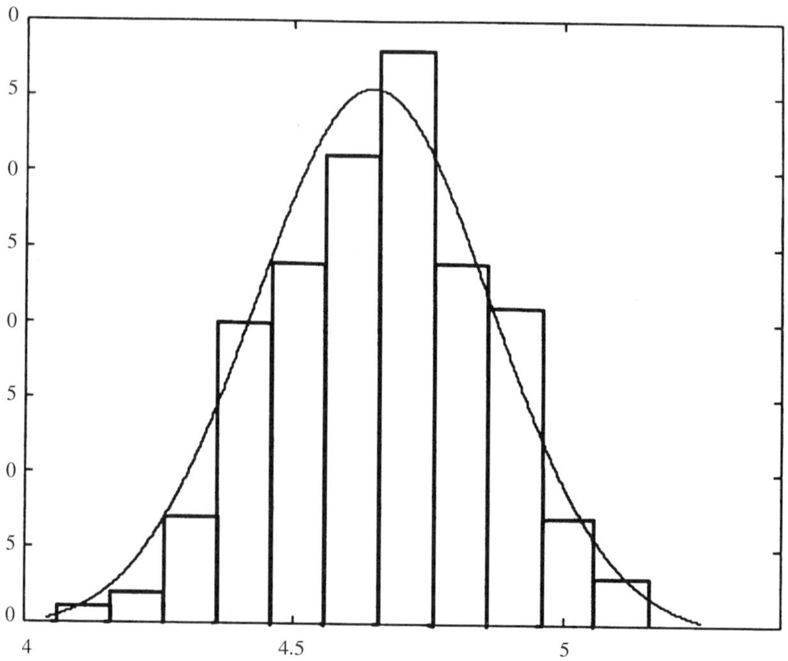

FIGURE 1 Histogram and frequency curve of the data in Table 1 (from Davis and Goldsmith, 1972).

Now if the number of observations is very large and the groups are made much narrower, the previous irregular step form of the histogram can be represented as a smoothed curve known as the *frequency curve*, also shown in Fig. 1.

Some important considerations regarding the frequency distribution are as follows:

1. The area under the curve, lying between the limits of each of the subranges, is approximately equal to the frequency of occurrence of that subrange.

2. The ability of the frequency curve to accurately represent the underlying distribution increases with the number of observations. With a small number of results only an approximation is possible, and the divergence may be relatively large.

Sometimes it is preferable to express the frequencies in terms of proportional frequencies rather than actual frequencies. For each given observation, the probability of occurrence is defined as the proportional frequency with which it occurs in a large number of observations. The resulting curve is referred to as the *probability distribution*.

2. MEASURES OF CENTRAL TENDENCY AND SPREAD

The frequency curve and the histogram described previously have very explicit attributes which can be used as the basis of criteria to characterize the distribution:

1. There is a certain value (for example the arithmetic mean) that represents the *center* of the distribution and serves to locate it.
2. The values are *spread* around this central value, extending over a range.
3. The spread which may or may not be uniform, reflecting the *shape* of the curve. Usually, most values lie close to the central value.

The distribution can then often be defined sufficiently in terms of the central value, a quantity expressing the degree of spread, and the general form of the distribution (the shape of the curve). The equation of the curve may be then written as (Davis and Goldsmith, 1972)

$$y = p(x; \theta_1, \theta_2), \tag{A.1}$$

where θ_1 and θ_2 are constant parameters, measuring the central value and the degree of spread, respectively, and the function p defines the shape.

2.1. The Central Value

The arithmetic mean ("mean" or "average") is the commonest measure of location or central value and is given by the sum of all observations divided by their number. That is

$$\text{Arithmetic mean} = \bar{x} = \frac{\sum_{i=1}^{N} x_i}{N} \tag{A.2}$$

As an example, the arithmetic mean for the carbon content of the mixture displayed in Table 1 is

$$\bar{x} = \frac{\sum_{i=1}^{178} x_i}{178} = 4.698$$

The mean of a set of observed values is an estimate only of the mean of the underlying probability distribution or population. The sample mean, $\bar{x}$, becomes a better estimate of the population mean or true value, μ, as N increases, approaching μ as N approaches infinity. That is

$$\bar{x} \to \mu \quad \text{as} \quad N \to \infty. \tag{A.3}$$

2.2. Measures of Dispersion

The spread is most usefully defined in terms of the standard deviation, which for a sample of N observations $x_1, x_2, \ldots, x_N$ is given by

$$\text{Standard deviation} = s = \sqrt{\frac{\sum_{i=1}^{N}(x_i - \bar{x})^2}{N-1}}, \tag{A.4}$$

where $\bar{x}$ is the mean of the sample. For the data presented in Table 1, the standard deviation is calculated as follows

$$s = \sqrt{\frac{\sum_{i=1}^{178}(x_i - \bar{x})^2}{178-1}} = 1.94$$

Most of the observations are likely to be within the range $\bar{x} \pm 2s$, and practically all within the range $\bar{x} \pm 3s$. In common with the sample mean, the value of the standard deviation calculated from a set of observations is only an estimate of the true or population value of the standard deviation: s becomes a better estimate of σ(population value) as N increases and

$$s \to \sigma \quad \text{as} \quad N \to \infty. \tag{A.5}$$

The variance of a population is another useful measure of dispersion and reflects the extent of the differences between the data. Denoted by σ^2, it is equal to the mean squared deviation of the individual values from the population mean. Usually, the symbols V and s^2 are used for the variance deduced from sample data. Thus, for a sample of N data drawn from a population with mean μ, the estimated variance is

$$V = \frac{\sum_{i=1}^{N}(x_i - \mu)^2}{N}. \tag{A.6}$$

Since, in general μ is not known, the estimated mean, $\bar{x}$, based on the sample must be used. If this substitution is made, the variance is underestimated as it can be shown that

$$\sum(x-a)^2 \quad \text{is a minimum when } a = \bar{x}. \tag{A.7}$$

and thus, $\sum(x-\bar{x})^2$ is less than $\sum(x-\mu)^2$. It can also be shown that this bias is removed when $(N-1)$ is used as the divisor in place of N. The best estimate of the population variance from the available data is therefore given by

$$V = \frac{\sum_{i=1}^{N}(x_i - \bar{x})^2}{N - 1}. \tag{A.8}$$

The *degrees of freedom* of the estimate of variance is given by the divisor $(N - 1)$. This is the number of independent comparisons that can be made between N observations since $\bar{x}$ is calculated from the observations. If $\bar{x}$ and $(N - 1)$ of the values of x are given, the other can be determined.

The definition of the sample covariance is also important. Given a set of N pairs of values $((x_1, y_1), (x_2, y_2), \ldots, (x_n, y_n)$, of the random vectors $(\mathbf{x}, \mathbf{y})$, we can compute, in addition to the variances deduced from sample data V_x, V_y, the sample covariance as

$$V_{xy} = \frac{\sum_{i=1}^{N}(x_i - \bar{x})(y_i - \bar{y})}{N - 1}. \tag{A.9}$$

2.3. The Form of the Distribution

While the form of the distribution is more difficult to determine than either the mean or the standard deviation, the data may give some idea of the probable shape. Assumptions about the actual shape of the distribution usually require additional information such as that obtained from past experience. These assumptions should concur with the observations.

A number of types of distributions have been fully studied, because they, or at least close approximations to them, frequently arise in practice. In connection with the theory of measurement errors and least squares adjustments, the normal and chi-square distributions are often used, so they are briefly discussed in the following paragraphs.

Normal or Gaussian Distribution

Many distributions obtained in experimental and observational work are found to have a more or less bell-shaped probability curve. These distributions are described by the *normal* or *gaussian distribution* shown in Fig. 2. This theoretical distribution is extremely important in statistics, and its use is not limited to data which are exactly, or very nearly normal.

For normally distributed data, the relative frequency or probability, dP, for the range $(x_1 - \frac{1}{2}\,dx)$ to $(x_1 + \frac{1}{2}\,dx)$ is given by

$$dP = \frac{1}{\sigma\sqrt{2\pi}} \exp\left\{ -(x_1 - \mu)^2 \Big/ 2\sigma^2 \right\} dx, \tag{A.10}$$

where, μ is the population mean, σ is the population standard deviation, and the integral over all values of x is equal to unity. The value of $p(x) = dP/dx$ is a maximum when $x = \mu$ and falls off symmetrically on both sides.

The use of this equation to calculate frequencies requires the determination of two parameters, μ and σ. It can be simplified by using the following substitution

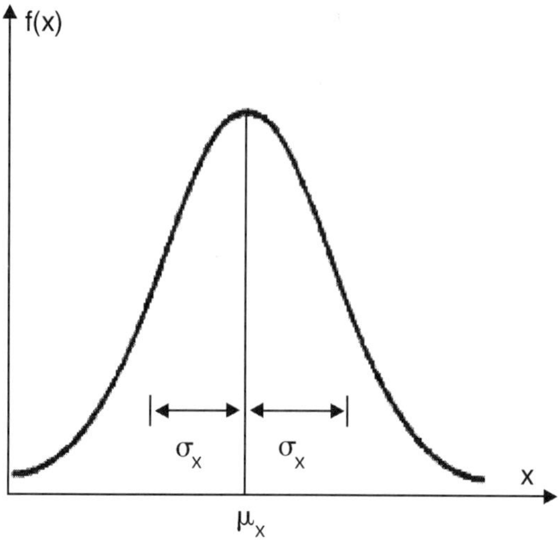

FIGURE 2 Normal distribution (from Mikhail, 1976).

based on the deviation of x from μ

$$u = {}^{(x-\mu)}\big/_{\sigma}. \tag{A.11}$$

Then, the form known as the *standardized form of the normal distribution*, is obtained

$$dP = \frac{1}{\sqrt{2\pi}} \exp\left\{{}^{-u^2}\big/_{2}\right\} du, \tag{A.12}$$

where the variable u is said to be in *standard measure*. This equation contains no adjustable parameters, and if the value of u is known, $dP/du = p(u)$ is determined uniquely.

Equation A.12 gives the probability of occurrence of observations whose standardized value lies between $(u - \frac{1}{2}\,du)$ and $(u + \frac{1}{2}\,du)$. Integration yields the proportion of observations expected to fall between any two values of u and hence between the corresponding two values of x. While this cannot be performed explicitly, the definite integral has been tabulated for a wide range of values of u.

Since the normal curve is symmetrical about $u = 0$, the area under the curve to the right of a given value u_1 (i.e., the probability of occurrence, α, of a value greater than u_1) is equal to the area to the left of $-u_1$. It is therefore sufficient to tabulate only the values corresponding to positive values of u_1. For example, there is a probability of 0.05 than u will exceed 1.64, and hence that x will exceed $\mu + 1.64\sigma$.

For a normal distribution, the probability that values of x will deviate from the mean by more than u_1 multiples of the standard deviation is 2α, corresponding to the probability that $|u| > u_1$. The complement $(1 - 2\alpha)$, is the probability that $|u|$ does not exceed u_1 (i.e., that observations will not deviate far from the mean).

For normally distributed variates, values of $|u|$ greater than 3 seldom occur, and values greater than 4 occur very rarely (once in about 10,000 observations). It is usual to consider that if an isolated value greater than 2 is obtained there is some doubt as to whether it represents an observation from a normal population with the given mean and standard deviation. The odds against finding such a value in a single trial are about 19 to 1.

A random p-dimensional vector $\mathbf{x}$ follows a p-variate normal distribution, $\mathbf{x} \sim N_p(\boldsymbol{\mu}, \boldsymbol{\Sigma})$ if its probability density function is given by

$$f(\mathbf{x}) = \left[\frac{1}{(2\Pi)^{p/2}\sqrt{|\Sigma|}}\right] \exp\left\{-\frac{1}{2}(\mathbf{x}-\boldsymbol{\mu})^{\mathrm{T}} \sum^{-1} (\mathbf{x}-\boldsymbol{\mu})\right\}, \tag{A.13}$$

where $\boldsymbol{\mu}$ and $\boldsymbol{\Sigma}$ are the mean vector and the covariance matrix, respectively.

Chi-Square Distribution

Let $X_1, X_2, \ldots, X_\nu$ be ν independent and normally distributed random variables. The square of the standardized variable U_i is defined as

$$U_i^2 = \left(\frac{X_i - \mu_i}{\sigma_i}\right)^2. \tag{A.14}$$

The sum of the U_i^2 yields a new random variable for which the symbol χ^2 is customarily used:

$$\chi^2 = U_1^2 + U_2^2 + \cdots + U_\nu^2. \tag{A.15}$$

This random variable has a χ^2 distribution with ν degrees of freedom. Its density function is given by

$$f(\chi^2) = \frac{1}{(2)^{\nu/2}\Gamma(\nu/2)}(\chi^2)^{\nu/2-1}e^{-x^2/2} \tag{A.16}$$

The χ^2 distribution has simple central moments: mean $= \nu$ and variance $= 2\nu$.

3. ESTIMATION

Let X be a random variable with some probability distribution depending on an unknown parameter θ. Let $X_1, X_2, \ldots, X_N$ be a sample of X and let $x_1, x_2, \ldots, x_N$ be the corresponding sample values. If $g(X_1, X_2, \ldots, X_N)$ is a function of the sample to be used for estimating θ, then g is considered an estimator of θ. The value $g(x_1, x_2, \ldots, x_N)$ is designed as an estimate of θ and it is usually written as $\theta = g(x_1, x_2, \ldots, x_N)$.

Because estimates can be obtained in different ways, certain criteria are needed to judge the quality of the estimation. We consider here the following three criteria extracted from Mikhail (1976):

1. *Consistence*: An estimator is called consistent if for $n \to \infty$, the probability that the estimator $\hat{\theta}$ approaches the parameter θ converges toward 1. Thus, for any small $\varepsilon > 0$,

$$\lim_{n\to\infty} \Pr\{|\hat{\theta} - \theta| < \varepsilon\} = 1 \tag{A.17}$$

2. *Unbiased estimation*: An estimator $\hat{\theta}$ is called an unbiased estimator for θ if it satisfies $E(\hat{\theta}) = \theta$. If this property holds only for $n \to \infty$, the estimator is said to be asymptotically unbiased.

3. *Minimum variance*: We say that $\hat{\theta}$ is an unbiased minimum variance estimate of θ if for all estimates θ^* such that $E(\theta^*) = \theta$, we have that $\sigma^2(\hat{\theta}) \leq \sigma^2(\theta^*)$ for all θ. That is, of all unbiased estimates of θ, $\hat{\theta}$ has the smallest variance.

EXAMPLE A.1

Let $x_1, x_2, \ldots, x_N$ be a random sample of N observations from an unknown distribution with mean μ and variance σ^2. It can be demonstrated that the sample variance V, given by equation A.8, is an unbiased estimator of the population variance σ^2.

By the definition of sample variance, we have

$$V = \frac{1}{N-1}\left[\sum_{i=1}^{N} x_i^2 - N(\bar{x})^2\right].$$

Furthermore, $E(x^2) = \sigma^2 + \mu^2$ for a random variable x. Since each x value, x_1, $x_2, \ldots, x_N$, was randomly selected from a population with mean μ and variance σ^2, it follows that

$$E\left(x_i^2\right) = \sigma^2 + \mu^2 \quad (i = 1, 2, \ldots, N)$$

$$E\left(\bar{x}^2\right) = \sigma_{\bar{x}}^2 + (\mu_{\bar{x}})^2 = \sigma^2/N + \mu^2.$$

Taking the expected value of V and substituting it into these expressions we obtain

$$E(V) = E\left\{\frac{1}{N-1}\left[\sum_{i=1}^{N} x_i^2 - N(\bar{x})^2\right]\right\} = \frac{1}{N-1}\left\{\sum_{i=1}^{N} E\left[x_i^2\right] - NE[(\bar{x})^2]\right\}$$

$$= \frac{1}{N-1}\left\{\sum_{i=1}^{N}(\sigma^2 + \mu^2) - N\left(\frac{\sigma^2}{N} + \mu^2\right)\right\} = \left(\frac{N-1}{N-1}\right)\sigma^2 = \sigma^2.$$

This shows that V is an unbiased estimator of σ^2, regardless of the nature of the sample population.

3.1. Methods of Estimation

There are a number of different methods for finding point estimators of parameters. Following Mikhail (1976) we mention here:

Moment method: Takes the kth sample moment:

$$m_k = \frac{1}{N}\sum_{i=1}^{N} x_i^k \qquad \text{(A.18)}$$

as an estimate for the kth moment of the probability distribution.

Maximum likelihood method: The estimate of a parameter θ, based on a random sample $X_1, X_2, \ldots, X_N$, is that value of θ which maximizes the likelihood function $L(X_1, X_2, \ldots, X_N, \theta)$ which is defined as

$$L(X_1, X_2, \ldots, X_N, \theta) = f(X_1, \theta)f(X_2, \theta)\ldots f(X_N, \theta). \qquad \text{(A.19)}$$

If X is discrete, $L(X_1, X_2, \ldots, X_N, \theta)$ represents $P[X_1 = x_1, \ldots, X_N = x_N]$, while if X is continuous, it represents the joint probability density function of $(X_1, X_2, \ldots, X_N)$.

EXAMPLE A.2

Let $x_1, x_2, \ldots, x_N$ be a random sample of N observations on a random variable x with exponential density function

$$f(x) = \begin{cases} \dfrac{e^{-x/\beta}}{\beta} & \text{if } 0 \leq x \leq \infty \\ 0 & \text{elsewhere} \end{cases}$$

In order to determine the maximum likelihood estimator of β, we have

$$L = f(x_1)f(x_2)\cdots f(x_N) = \left(\frac{e^{-x_1/\beta}}{\beta}\right)\left(\frac{e^{-x_2/\beta}}{\beta}\right)\cdots\left(\frac{e^{-x_N/\beta}}{\beta}\right) = \frac{e^{-\sum_{i=1}^{N} x_i/\beta}}{\beta^N},$$

because $x_1, x_2, \ldots, x_N$ are independent random variables.

Setting the derivative of the natural logarithm of L with respect to β equal to 0 and solving for $\hat{\beta}$ yields

$$\hat{\beta} = \frac{\sum_{i=1}^{N} x_i}{N} = \bar{x}.$$

That is, the maximum likelihood of β is the sample mean.

Least squares method: Finds the estimate of θ that minimizes the mean square error

$$\text{MSE} = E(\hat{\theta} - \theta)^2 = \int_{-\infty}^{\infty} (\hat{\theta} - \theta)^2 f(\hat{\theta})\, d\hat{\theta}, \tag{A.20}$$

where $f(\hat{\theta})$ is the density function of $\hat{\theta}$. If the random variables to which the observations refer are normally distributed, the least square method gives results identical to those of the maximum likelihood method.

4. CONFIDENCE INTERVALS

The estimation of means, variances, and covariances of random variables from the sample data is called *point estimation*, because one value for each parameter is obtained. By contrast, *interval estimation* establishes confidence intervals from sampling.

After a point estimation is performed, the question is how much the deviation of the estimate is likely to be from the still unknown parameter. As it was pointed out by Mikhail (1976), it is only possible to estimate the probability that the true value of the parameter is likely to be within a certain interval around the estimate if the cumulative distribution function $F(x)$ of the random variable is given.

TABLE 2 Data for Example A3

2.15	2.06	2.33	1.96	1.96	1.89	2.20	2.24	2.16	2.48
2.32	1.95	2.30	1.85	2.24	2.13	2.14	2.15	2.31	2.49
2.12	2.16	2.25	2.05	2.20	1.90	2.00	2.08	2.06	2.24

The probability statement for a confidence interval of a parameter θ is:

$$\Pr\{\theta_1 < \hat{\theta} < \theta_2\} = 1 - \alpha \tag{A.21}$$

where $(1 - \alpha)$ is called the *confidence level*. The values θ_1 and θ_2 are the lower and upper confidence limits for the parameter $\hat{\theta}$. Eq. (A.21) defines the confidence interval for θ as the interval around the estimate $\hat{\theta}$, such that the probability that this interval includes the value of the parameter is $(1 - \alpha)$.

In constructing confidence intervals, it is essential to use suitable random variables whose values are determined by the sample data as well as by the parameters, but whose distributions do not involve the parameters in question.

EXAMPLE A.3

The following data (Table 2) represent the weight percent measurements of an undesirable product in a process stream. Find a 90% confidence interval for its true mean. Assume normality and $\sigma = 0.25$.

The $100(1-\alpha)\%$ confidence interval for μ is given by $(\bar{x} \pm K_{\alpha/2}\sigma/\sqrt{N})$, where $K_{\alpha/2}$ is the $100\alpha/2$ percentage point of the normal distribution. This result is obtained by considering the distribution of $[(\bar{x}-\mu)\sqrt{N}/\sigma]$, which is normally distributed with mean 0 and standard deviation 1. Hence,

$$\bar{x} \pm K_{\alpha/2}\frac{\sigma}{\sqrt{N}} = 2.1457 \pm 1.645\frac{0.25}{\sqrt{30}} = 2.1457 \pm 0.075$$

5. TESTING OF STATISTICAL HYPOTHESES

Another way of dealing with the problem of making a statement about an unknown parameter associated with a probability distribution, based on a random sample, is the testing of statistical hypotheses. First, a value for the parameter is hypothesized; then, the information from the sample is used to confirm or discard the hypothesized value.

Two hypotheses are considered. Based on a random sample, the validity of the null hypothesis (H_0) is tested against the alternate hypothesis (H_1) in order to either reject or accept the first one.

The procedure for testing hypotheses includes the following steps:

- Postulation of the null and alternative hypotheses.
- Collection of a sample of size N from the random variable X.
- Selection of the test statistic τ, with a known distribution under the assumption that the null hypothesis holds.
- Selection of the significance level of the test (α) and determination of the critical statistic value τ_c from the corresponding distribution table. τ_c is such that

$$F(\tau_c) = P(\tau \leq \tau_c) = (1 - \alpha).$$

- Calculation of τ as function of the measurements.
- If $\tau > \tau_c$, H_0 is rejected. In the opposite case H_0 is accepted.

A test will not always lead to the right decision. It is possible that H_0 is true and is rejected, or that H_0 is false and is accepted.

Wrongly rejecting a true hypothesis is referred to as committing a type I error. Its probability is designed by α, the significance level of the test.

The wrong acceptance of a false hypothesis leads to committing a type II error. The probability of a type II error is designated β and $(1 - \beta)$ is called the power of the test. For a certain H_1, it is not possible to make both α and β arbitrarily small. Decreasing the probability of one type of error increases the probability of the other and vice versa. The balance between both types of errors depends on the purpose of the test.

EXAMPLE A.4

Let us suppose that the maximum allowable concentration of a contaminant in a workplace is 1 ppm. Twenty air samples are examined in order to establish if the mean contaminant concentration exceeds the allowable value. The mean and standard deviation of the sample are $\bar{x} = 1.9$ ppm and $s = 1.6$ ppm.

For this example, the postulated null and alternative hypotheses are the following:

$$H_0\colon \mu = 1 \text{ ppm}$$

$$H_a\colon \mu > 1 \text{ ppm}$$

The selected test statistic is

$$\tau = \frac{\bar{x} - \mu}{s/\sqrt{N}},$$

which has a t distribution with $(N - 1)$ degrees of freedom. It is assumed that the relative frequency distribution of the population from which the sample is selected is approximately normal.

For a significance level of the test $\alpha = 0.05$, the critical value of the test statistic is 1.729. Because the calculated value of τ is $2.51 > 1.729$, we can conclude that $\mu > 1$ ppm, so the concentration of contaminant in this workplace is unacceptable.

REFERENCES

Davis, O. L., and Goldsmith, P. L. (1972). "Statistical Methods in Research and Production." Published by Oliver & Boyd Tweedale Court, Edinburgh for Imperial Chemical Industries Ltd. (ICI).

Mikhail, E. (1976). "Observations and Least Squares." IEP Ser. Harper & Row, New York.

AUTHOR INDEX

SUBJECT INDEX